Telecom Protocol Finder

RAD COM
TEST-OF-THE-ART

A comprehensive review of data- and telecommunications protocols and technologies for network managers and field service engineers

McGraw-Hill

New York Chicago San Francisco Lisbon London
Madrid Mexico City Milan New Delhi San Juan Seoul
Singapore Sydney Toronto

Library of Congress Cataloging-in-Publication Data

Telecommunications protocol finder/Radcom Ltd.
p. cm.
ISBN 0-07-138044-2
1. Telecommunication–Standards–Encyclopedias. 2. Computer network protocols–Encyclopedias. I. Radcom Ltd.

TK5102.T45 2001
621.382′12–dc21

2001030795

1 2 3 4 5 6 7 8 9 0 DOC/DOC 0 7 6 5 4 3 2 1

ISBN 0-07-138044-2

The sponsoring editor for this book was Marjorie Spencer and the production supervisor was Pamela A. Pelton.

Printed and bound by R.R. Donnelley & Sons Company.

Table of Contents

PROTOCOLS

Protocol Index

Preface

The data communications industry has experienced unprecedented growth in the last decade. This growth has not been solely in numbers, but more importantly in complexity. Networks once consisted of one mainframe computer and many terminals. The introduction of personal computers, although providing power and versatility which had been until then unattainable, created problems of communication. For purposes of file management, printing, etc., it was imperative that these many stations speak to each other.

Data communications in the 80's

In the 80's, computerization in the workplace and data communications in general required the combining of workstations into local area networks. However, communication beyond this closed world of the LAN was nonexistent. Printing and local communications were possible, but there was no need to communicate with systems outside an individual LAN. Ethernet and Token Ring technologies flourished. Many applications were developed to run on these technologies.

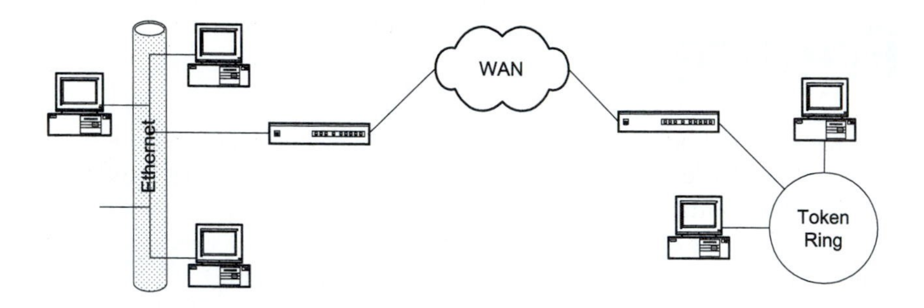

Data communications in the early 90s

The 90's brought the advancement of wide area networks to connect isolated LANs. As businesses, institutes and universities expanded geographically, it became imperative to connect these networks. X.25 and then Frame Relay became popular protocols for interconnecting LANs.

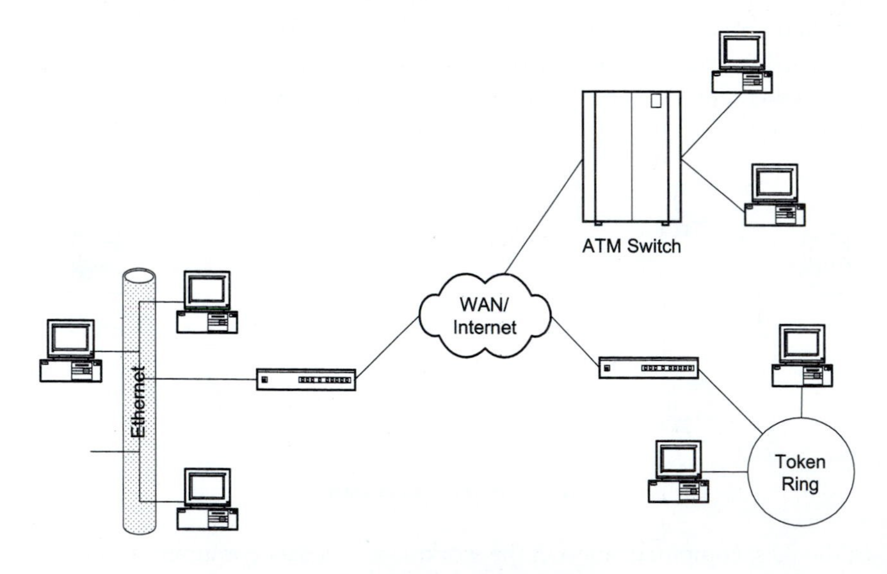

Data communications today

Now, as we find ourselves entering the 21st century, newer, faster and converging technologies have entered the data communications picture.

With the advent of the Internet and the need to send information ever faster, ATM, ISDN, Fast Ethernet, Gigabit Ethernet, GPRS, UMTS and WAP are just some of the most talked about technologies. New protocol standards are being developed which allow users to take advantage of the faster high-bandwidth technologies, with minimum investment in infrastructure.

The pace is mind-boggling. New technologies and protocols are introduced even before we've learned the previous ones. Keeping up with the industry is ever more important for both equipment manufacturers and service providers. No one can afford to be left behind; however, the investment required in knowledge and infrastructure is often significant.

The collection of both new and old technologies and protocols presented in this book, although limited in scope, is meant to assist data communications professionals make sense of this changing industry.

OSI Model

Data communications protocols are generally grouped according to their function within the 7-layer OSI (Open System Interconnection) model.

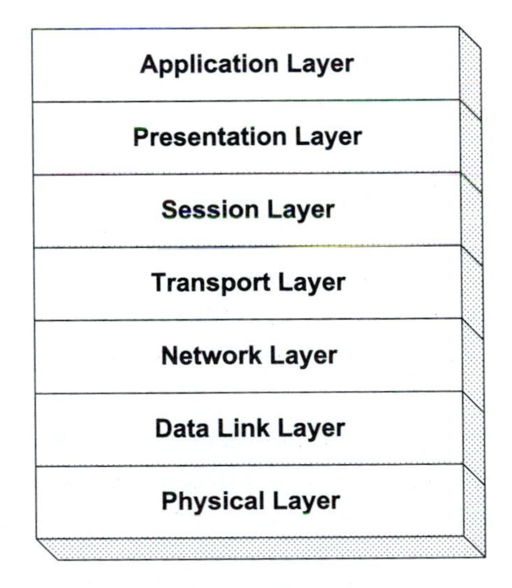

7-layer OSI model

The OSI model consists of the following layers from bottom to top:

- Physical Layer - Responsible for the connection to the physical media (copper, fiber, etc.).
- Data Link Layer - Provides error-free transmission and establishes logical connection between stations.
- Network Layer - Provides addressing and control functions (e.g., routing) necessary to move data through the network.
- Transport Layer - Defines message protocols and provides end-to-end flow control services to optimize flow through the network.
- Session Layer - Provides support for connections between sessions, administrative tasks and security.
- Presentation Layer - Responsible for meaningful exchange of data.
- Application Layer - Responsible for interaction with the operating system and providing the user interface to the system (e.g., FTP, TELNET, SMTP).

The physical, data link and transport layers are all that are needed for reliable connections through a network. Together, they form the Logical Link Control Function.

About this Book

The first section of this book lists the most common data communications protocols in use today and indicates their function in respect to the OSI model. In particular, it provides information concerning the structure of the protocol (header, PDU, etc.), various errors and parameters. Protocols are listed according to protocol suite. Protocol suites are listed in alphabetical order.

The second section of this book describes numerous physical layer technologies which are used today in the data communications industry.

Telecom Protocol Finder

PROTOCOLS

1

AppleTalk Protocols

Apple Computer developed the AppleTalk protocol suite to implement file transfer, printer sharing, and mail service among Apple systems using the LocalTalk interface built into Apple hardware. AppleTalk ports to other network media such as Ethernet by the use of LocalTalk to Ethernet bridges or by Ethernet add-in boards for Apple machines.

AppleTalk is a multi-layered protocol providing internetwork routing, transaction and data stream service, naming service, and comprehensive file and print sharing. In addition, many third-party applications exist for AppleTalk protocols.

To extend the addressing capability of AppleTalk networks and provide compliance with the IEEE 802 standard, Apple Computers introduced AppleTalk Phase 2 in 1989. AppleTalk Phase 2 differs primarily in the range of available network layer addresses and the use of the IEEE 802.2 Logical Link Control (LLC) protocol at the Data Link Layer.

The AppleTalk protocol suite includes the following protocols:

- AARP: AppleTalk Address Resolution Protocol.
- DDP: Datagram Delivery Protocol.
- RTMP: Routing Table Maintenance Protocol.
- AEP: AppleTalk Echo Protocol.
- ZIP: Zone Information Protocol.
- ATP: AppleTalk Transaction Protocol.
- ADSP: AppleTalk Data Stream Protocol.
- NBP: Name-Binding Protocol.
- ASP: AppleTalk Session Protocol.
- PAP: Printer Access Protocol.
- AFP: AppleTalk Filing Protocol.

The following diagram illustrates the AppleTalk protocol suite in relation to the OSI model:

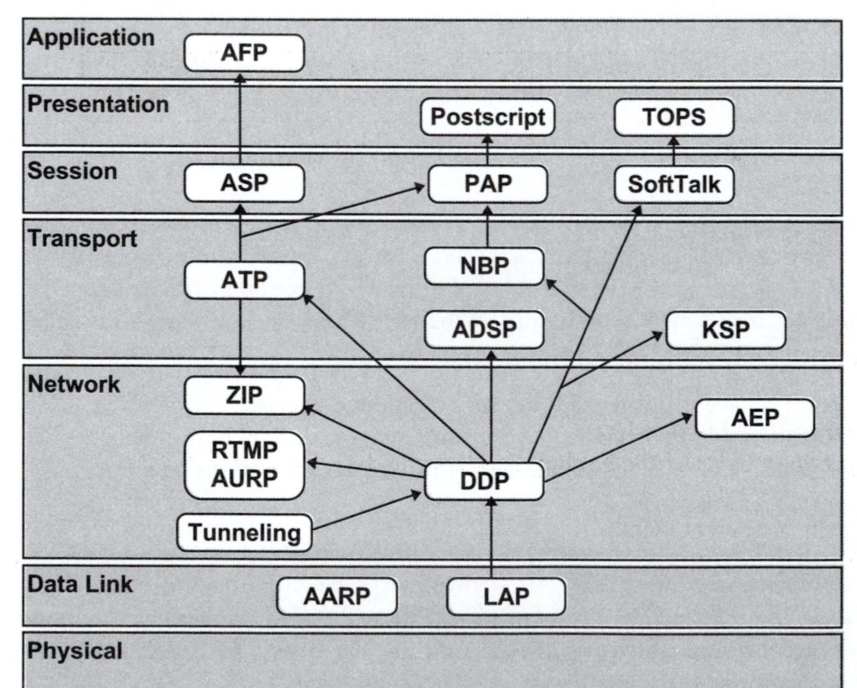

AppleTalk protocol suite in relation to OSI model

AARP

AARP (AppleTalk Address Resolution Protocol) maps between any two sets of addresses at any level of one or more protocol stacks. Specifically, it is used to map between AppleTalk node addresses used by the Datagram Delivery Protocol (DDP), as well as higher-level AppleTalk protocols, and the addresses of the underlying data link that is providing AppleTalk connectivity. AARP makes it possible for AppleTalk systems to run on any data link.

The AARP packet structure is shown below:

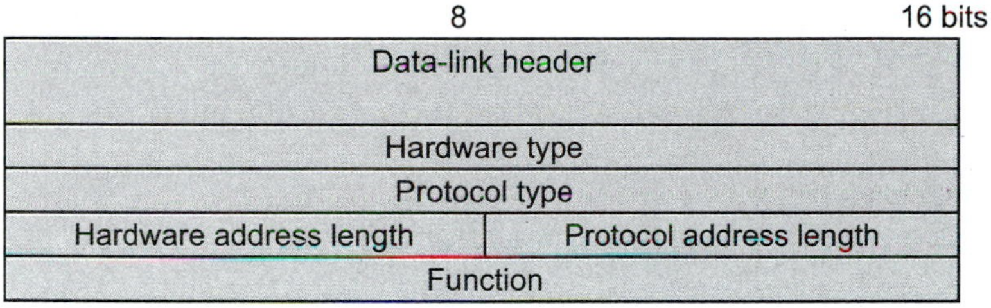

8	16 bits
Data-link header	
Hardware type	
Protocol type	
Hardware address length	Protocol address length
Function	

AARP packet structure

Hardware type
Identifier for the data-link type.

Protocol type
Identifier for the protocol family.

Hardware address length
Length in bytes of the hardware address field.

Protocol address length
Length in bytes of the protocol address field.

Function
Indicates the packet function (1=AARP request, 2=AARP response and 3-AARP probe).

Following the header are the hardware and protocol addresses according to the values of the function field.

DDP

The Datagram Delivery Protocol (DDP) provides a datagram delivery and routing service to higher layer protocols. DDP frame headers can use the long or short format. Short format DDP headers carry only the source and destination service socket numbers, while long format DDP headers also carry the source and destination network and node addresses needed for routing capability. Because AppleTalk Phase 2 does not use LLAP to identify source and destination nodes, it supports only the long format DDP header for Phase 2.

Short Format Frames

The following fields are present for short format DDP frames:

Destination socket
Destination service socket address used by the frame.

Source socket
Source service socket address used by the frame.

Length
Total length of the datagram.

DDP type
Code used to indicate the higher layer protocol used.

Long Format Frames

The following additional fields are present for long format DDP frames:

Destination
Destination network/node/socket. Destination network number, node address, and socket address used by the frame, displayed in the following format: NNNN.nn (ss), where NNNN is the network number, nn is the node address and ss is the socket address.

Source

Source network/node/socket. Source network number, node address, and socket address used by the frame, in the same format as D.

Checksum

Checksum of the entire datagram beginning after the checksum field. A checksum of zero implies that the checksum is not in use.

Hop count

Number of routers encountered by the frame. After 16 hops, the protocol discards the frame.

RTMP

The Routing Table Maintenance Protocol (RTMP) manages routing information for AppleTalk networks. RTMP communicates known network numbers and data concerning accessibility between networks. AppleTalk Phase 2 allows for split horizon routing where the protocol transfers only routing data about directly connected networks in an effort to reduce the traffic overhead imposed by routing updates.

Frames

RTMP frames may be one of the following types:

[request]	Requests network number and local router ID.
[reply]	Supplies network number and local router ID.
[data]	Carries the current routing table data.
[RDR split]	Requests immediate routing data using split horizon (only Phase 2).
[RDR full]	Requests full table of routing data (only Phase 2).

Frame Parameters

The following parameters are present in Apple RTMP frames:

Source network/node ID

Network number and address of the system sending the RTMP [reply] or [data] frame. The network/node ID is displayed in the following format: NNNN.nn, where NNNN is the network number and nn is the node address.

Routing table

List of known network nodes and accessibility number representing the relative routing cost to the network. The routing table displays items in the following format: NNNN(cc), where NNNN is the network number and cc is the routing cost. AppleTalk Phase 2 RTMP frames can specify a range of network numbers and a protocol version number for the first term, as follows:
NNNN-NNNN(cc) [V=x], where NNNN is the network number, cc is the routing cost in hops and x is the protocol version (2 for Phase 2).

AEP

The AppleTalk Echo Protocol (AEP) provides an echo service to AppleTalk hosts. It can specify up to 585 bytes of data for an echo transaction.

AEP frames may be one of the following types:

[echo reqst]	Request to echo the specified data.
[echo reply]	Echo response containing echo data.

ATP

The AppleTalk Transaction Protocol (ATP) provides reliable delivery service for transaction-oriented operations. ATP uses a bitmap token to handle acknowledgement and flow control and a sequence of reserved bytes for use by higher level protocols.

Frames

ATP frames may be of the following types:

[request]	Requests data specified by the bitmap.
[reply]	Returns requested data.
[release]	Indicates end of transaction.

Frame Parameters

ATP frames contain the following parameters:

Transaction ID

Reference code used to match ATP requests with ATP replies.

Transaction bitmap

Bitmap requests a specific data frame and provides acknowledgment for received data. 1 in the bitmap indicates an outstanding request for a data segment; 0 indicates that the system satisfied the request. The bitmap position corresponds to the data segment position. The bit on the far right represents the first data segment with successive segments indicated to the left. The bitmap is 8 bits wide, permitting ATP to send up to 8 data segments per transaction request.

Sequence number

Sequence number corresponding to the data segment of the current response frame.

User bytes

Four bytes reserved for use by higher level protocols.

Control flags

The following control flags display in upper-case when set and in lower-case when inactive:

x, X When set, exactly-once mode is set, ensuring that the current transaction is performed only once.

e, E When set, the frame is the end of a data response.

s, S When set, the bitmap status requests reuse of buffers already acknowledged.

NBP

The AppleTalk Name Binding Protocol (NBP) manages the use of names on AppleTalk networks. NBP maintains a names directory that includes names registered by hosts and bound to socket addresses. After a name is registered, the AppleTalk host can perform a name lookup to find the socket address associated with the name. When the host issues a name lookup on the Internet, NBP sends a broadcast lookup to a router that generates name lookup requests to each network within the zone specified in the name.

Frames

NBP frames may be one of the following types:

[brdcast lookup]	Broadcast search for the specified name.
[name lookup]	Local search for the specified name.
[lookup reply]	Reply to a name lookup.

Frame Parameters

NBP frames have the following parameters:

Number of names

Number of socket/name pairs contained in the message.

Transaction ID

Reference code used to match NBP replies with NBP requests.

ZIP

The AppleTalk Zone Information Protocol (ZIP) manages the relationship between network numbers and zone names. AppleTalk networks primarily implement ZIP in routers that gather network number information by monitoring RTMP frames.

Frames

ZIP frames may be one of the following types:

[zonename query]	Requests zone name for a network number.
[zonename reply]	Supplies zone name for network number.
[zonelist query]	Requests the complete list of known zones.
[zonelist reply]	Supplies the complete zone list.
[get zone reqst]	Requests the local zone ID.
[get zone reply]	Supplies the local zone ID.
[takedown zone]	Removes a zone from the zone list.
[bring up zone]	Adds a zone to the zone list.
[local zone req]	Requests local zones on extended networks.
[ext name reply]	Zone name replies too long for one frame.
[change notify]	Alerts nodes of a zone name change.
[net info reqst]	Requests network information for a zone name.
[net info reply]	Supplies network range and multicast address for zones on extended nets.

Frame Parameters

Apple ZIP frames contain the following parameters:

Number
Number of networks for the request or zone information reply.

Start index
The starting zone for the zone list request.

Zone name
The name associated with the specified zone.

Multicast
Multicast address assigned to the specified zone.

Default zone
The local zone name.

Old zone name
The previously used name for the specified zone.

New zone name
New zone name for the specified zone.

Network range
The range of network numbers associated with the specified zone display in the format: SSSS-EEEE where SSSS is the starting network number and EEEE is the ending network number.

Network/zone list
List of networks and zone names represented as follows:
NNNN = zonename, where NNNN is the network number and zonename is the zone name.

Messages

Apple ZIP [net info reply] and [change notify] frames can contain the following messages:

{invalid zone}	Specified zone name does not exist.
{one zone}	Specified zone is the only zone.
{use broadcast}	Local network does not support multicasting, use broadcasting.

ASP

The AppleTalk Session Protocol (ASP) manages sessions for higher layer protocols such as AFP. ASP issues a unique session identifier for each logical connection and continuously monitors the status of each connection. It maintains idle sessions by periodically exchanging keep alive frames in order to verify the session status.

Frames

ASP frames can be one of the following types:

[open session reqst]	Requests to open an ASP session.
[close session reqst]	Requests to close an ASP session.
[command call reqst]	Calls to higher level protocol.
[status request]	Requests server status.
[session keep alive]	Maintains idle connections.
[session write reqst]	Requests to perform a write operation.
[write continue req]	Begins the transfer of write data.
[attention request]	Send urgent data.
[close session reply]	Acknowledges session close.
[command call reply]	Reply from higher level protocol.
[server status reply]	Reply containing server information.
[open session reply]	Reply to open session request.
[session write reply]	Reply to session write request.
[write continue rply]	Session write data.
[attention reply]	Acknowledges receipt of attention request.

Frame Parameters

Apple ASP frames can contain the following parameters:

Session ID

A reference code used to identify the session.

Sequence number

Used by command, write, and write continue frames to maintain data order.

Server session socket

The socket number in use by the server end of the connection.

Workstation session socket

Workstation session socket. The socket number in use by the workstation end of the connection.

Version number

ASP version number currently in use.

Buffer size

Buffer size available for receiving command blocks.

Messages

Apple ASP reply frames can contain the following messages:

{OK}	Command completed successfully.
{xxxx bytes written}	Number of bytes written for [write continue rply] frames.
{bad version number}	ASP version not supported.
{buffer too small}	Request buffer too small for command block.
{no more sessions}	Server cannot open any more sessions.
{no servers}	Server not responding.
{parameter error}	ASP parameter values invalid.
{server is busy}	Server too busy to open another session.
{session closed}	Referenced session has been closed.
{size error}	Command block larger than maximum.
{too many clients}	Client number limit exceeded.
{no acknowledgement}	Workstation did not acknowledge.
{unknown error}	Unknown error condition.

PAP

The Printer Access Protocol (PAP) manages the virtual connection to printers and other servers. PAP is used to convey connection status and coordinate data transfer.

Frames

PAP frames can be one of the following types:

[open connection rqst]	Request to open a PAP connection.
[open connection rply]	Reply to open connection request.
[send data request]	Request to send PAP data.
[PAP data segment]	Segment of PAP data transfer.
[session keep alive]	Verify connection status.
[close connection req]	Request to close a PAP connection.
[close connection rep]	Reply to close connection request.
[send server status]	Request server status.
[server status reply]	Reply to server status request.

Frame Parameters

PAP frames can contain the following parameters:

Connection ID
Reference code used to identify the PAP connection.

ATP responding socket
ATP socket number used for PAP status and data transfers.

Maximum buffer size
Maximum amount of data in bytes that the protocol can send in response to each [send data request] (also known as the Flow Quantum).

Wait time
Length of time that a workstation waits for a connection.

Sequence number
Used in send data request frames to maintain data order.

EOF

End-of-file indicator. Used to indicate the end of a data transfer.

Result

Result code indicating the outcome of an [open connection rqst]:
0000 Connect OK.
FFFF Printer busy.

Status

Status message returned by status and open connection reply frames.

ADSP

The AppleTalk Data Stream Protocol (ADSP) provides a data channel for the hosts. It is a connection-oriented protocol that guarantees in-sequence data delivery with flow control.

Frames

ADSP frames can be one of the following types:

[acknowledge/probe]	Acknowledges data or requests acknowledge.
[open connect reqst]	Requests an ADSP connection.
[open connect ackn]	Acknowledges ADSP connection.
[open request & ackn]	Acknowledges inbound connection and requests an outbound connection.
[open connect denial]	Refuses an inbound connection.
[close connection]	Requests to close an ADSP connection.
[forward reset]	Requests to ignore specific data.
[forward reset ackn]	Acknowledges forward reset of data stream.
[retransmit advise]	Requests to retransmit data.

Frame Parameters

ADSP frames can contain the following parameters:

Source connection ID

Reference code used to identify the sending side of a connection.

Destination connection ID

Reference code identifying the receiving side of a connection.

Send sequence number

Sequence number used for the outbound data stream.

Receive sequence number

Sequence number used for the inbound data stream.

Receive window size

Amount of unacknowledged data that the other side of a connection can send.

Version
ADSP version in use.

Attention sequence number
Lowest byte sequence number for which the protocol can send an attention frame.

Code
Attention code supplied by attention frames.

Control flag
When set (value=1), frame is a control frame with no data.

Ack request flag
When set (value=1), sender requests an acknowledgment.

End of message flag
When set (value=1), the current frame is the end of a data message.

Attention flag
When set (value=1), the frame is an attention frame.

AFP

The AppleTalk Filing Protocol (AFP) is the file sharing protocol of the AppleTalk architecture. It provides a native mode interface to Apple file system resources.

Apple files are comprised of two data structures called forks. An Apple file may be accessed by its data fork or its resource fork. The data fork holds raw file data while the resource fork contains information used by the operating system to manage icons and drivers.

Frames

AFP frames can be one of the following commands:

[lock/unlock bytes]	Locks or unlocks a specified byte range.
[close volume]	Closes the specified volume resource.
[close directory]	Closes the specified directory.
[close fork]	Closes the specified fork (file).
[copy file]	Copies the specified file.
[create directory]	Creates the specified directory.
[create file]	Creates the specified file.
[delete file]	Deletes the specified file or directory.
[list directory]	Lists the specified directory.
[flush to disk]	Writes data held in RAM to disk.
[flush fork]	Writes data to disk for the specified fork.
[get fork params]	Retrieves parameters for the specified fork.
[get server info]	Retrieves server information.
[get server params]	Retrieves server parameters.
[get volume params]	Retrieves volume parameters.
[consumer login]	Begins workstation log-in.
[login continue]	Continues workstation log-in.
[logout]	Workstation log-out.
[map user/group ID]	Gets ID associated with user/group name.
[map user/grp name]	Gets name associated with user/group ID.
[move and rename]	Moves and renames a file.
[open volume]	Opens the specified volume.
[open directory]	Opens the specified directory.
[open fork]	Opens the specified fork (file).
[read from fork]	Reads from the specified fork (file).
[rename file/dir]	Renames a file or directory.

[set dir params]	Sets directory parameters.
[set file params]	Sets file parameters.
[set fork params]	Sets fork parameters.
[set volume params]	Sets volume parameters.
[write to fork]	Writes to the specified fork (file).
[get file/dir pars]	Gets file or directory parameters.
[set file/dir pars]	Sets file or directory parameters.
[change password]	Changes user password.
[get user info]	Retrieves user information.
[open database]	Opens the desktop database.
[close database]	Closes the desktop database.
[get icon]	Retrieves an icon from the desktop database.
[get icon info]	Retrieves icon information.
[add APPL mapping]	Adds application information.
[remove APPL]	Removes application information.
[get APPL mapping]	Retrieves application information.
[add comment]	Adds a comment to a file or directory.
[remove comment]	Removes a comment from a file or directory.
[get comment]	Retrieves comment text from a file/directory.
[add icon]	Adds an icon for an application.

Frame Parameters

Apple AFP frames can contain the following parameters:

APPL index

Index, beginning with 1, of the first application mapping contained in the frame.

APPL tag

Tag information associated with the application mapping contained in the frame.

Attributes

Attributes of a file or directory are as follows:

Directory attributes:

Inv	Invisible to workstation user.
Sys	System directory.
Bk	Backup is needed (dir modified).
RI	Rename inhibit mode set.
DI	Delete inhibit mode set.

File attributes:

Inv	Invisible to workstation user.
MU	Multi-user application.
RAO	File resource fork already open.
DAO	File data fork already open.
RO	Read-only mode set for both forks.
WI	Cannot write to either fork.
Sys	File is system file.
Bk	Backup is needed (file modified).
RI	Rename inhibit mode set.
DI	Delete inhibit mode set.
CP	Copy protect mode set.

Backup date
Date of the last time the system backed-up the volume or directory.

Bitmap
Field containing bits used to indicate the parameters present in request or reply.

Request count
Maximum number of files to return for list directory requests.

Creation date
Date that the system created the file or directory.

File creator
ID string of the application or device that created a file.

Destination directory ID
Destination directory ID for a file copy or move.

Data fork length
Data fork length. Length of the file.

Destination volume ID
Destination volume ID for a file copy or move.

Directory bitmap
Field with bits that indicate which directory parameters are present in AFP frames.

Directory ID
Identifier associated with the specified directory.

Desktop database reference number
Reference number used to access the desktop database.

File bitmap
Bits that indicate which file parameters are present in AFP frames.

Free bytes
Number of bytes free on the volume.

Open fork reference number
Reference code used to access the open fork.

Group ID
Group ID used for authentication.

Group name
Group name used for authentication.

Icon tag
Tag information associated with the specified icon.

Icon size
Size of the specified icon, in bytes.

Icon type
Type code identifying the specified icon.

Long name
Long file name (maximum 31 characters).

Machine type
Type of AFP server in use.

Maximum reply size
Maximum number of bytes this protocol returns for list directory requests.

Access mode

Open mode attributes for a fork, represented as follows:

R	Allows everyone read access.
W	Allows everyone write access.
Deny-R	Denies read access if the file is open.
Deny-W	Denies write access if the file is open.

Modification date

Date the system last modified the file or directory.

New line character

Character used to indicate a new line (CR, LF) for read data.

New line mask

Value used to mask data for comparison to the new line character.

Offset

Starting file offset for write commands.

Offspring count

Number of files returned for list directory requests.

Owner ID

ID of the file or directory.

Volume password

Password required for access to the volume.

Parent directory ID

ID of the parent directory.

ProDOS information

ProDOS file type and Aux type for use by ProDOS workstations.

Resource fork length

Length of the file resource fork, in bytes.

Source directory ID

Source directory ID for a file copy or move.

Short name
Short file name (maximum 12 characters).

Signature
Identifies the volume type, as follows:
1 Flat, no support for directories.
2 Fixed directory ID.
3 Variable directory ID.

Source volume ID
Source volume ID for a file copy.

Start index
Start index, beginning with 1, of the requested file list for list directory commands and replies.

Total bytes
Total number of bytes on the volume.

User authentication method
Type of user authentication in effect.

User ID
User ID number used for authentication.

User name
User name used for authentication.

Version
Version number of AFP in use.

Volume bitmap
Field with bits that indicate which volume parameters are present in AFP frames.

Volume ID
Identifier associated with the specified volume.

Volumes
Number of volumes contained on the server.

Messages

AFP [get server params] replies contain a listing in the format: VolName(P,II), where VolName is a list of the volume names, P indicates password-protection and II indicates Apple II configuration information present.

The following status and error messages may be displayed for AFP replies:

Status	Error Message
{OK}	Command completed successfully.
{Object locked}	Specified object locked.
{Volume locked}	Specified volume locked.
{Icon type error}	Icon size mismatch.
{Directory not found}	Specified directory does not exist.
{Can't rename}	Cannot rename volume or root directory.
{Server going down}	The server is no longer active on the network.
{Too many open files}	Open file limit exceeded.
{Object type error}	Specified object invalid for operation.
{Call not supported}	AFP call unsupported by this version.
{User not authorized}	User has insufficient access rights.
{Session closed}	Specified session ID has been closed.
{Byte range overlap}	Lock conflicts with existing lock.
{Range not locked}	Attempt to unlock an unlocked byte range.
{Parameter error}	Specified parameters invalid for operation.
{Object not found}	Specified object does not exist.
{Object exists}	Specified object already exists.
{No server}	AFP server is not responding.
{No more locks}	Number of server locks exceeded.
{Miscellaneous error}	General command error.
{Lock error}	Byte range already locked by another user.
{Item not found}	Specified item not found.
{Flat volume}	Volume does not support directories.
{File busy}	Specified file is currently open.
{EOF error}	End of fork reached unexpectedly.
{Disk full}	Volume is out of disk space.
{Directory not empty}	Attempt to delete a non-empty directory.
{Deny conflict}	Specified deny rights conflict.
{Cannot move}	Cannot move directory to a descendent directory.
{Bitmap error}	Invalid bitmap specified for object.
{Bad version number}	Specified version number is invalid.
{Bad User Authentic}	User authentication failed.

{Continue Authentic}	Authentication not completed.
{Access denied}	User does not have permission for operation.

For further information on AppleTalk, refer to "Inside AppleTalk Second Edition" by Gursharan S. Sidhu, Richard F. Andrews and Alan B. Oppenheimer, Apple Computer, Inc.

2

ATM

ATM relies on cell-switching technology. ATM cells have a fixed length of 53 bytes which allows for very fast switching. ATM creates pathways between end nodes called virtual circuits which are identified by the VPI/VCI values.

This chapter describes the ATM UNI and NNI cell header structures and the PDU structures for the various ATM/SAR formats including: AAL0, AAL1, AAL2, AAL3/4 and AAL5.

UNI/NNI Cells

The UNI or NNI cell header comprises the first 5 bytes of the ATM cell. The remaining 48 bytes comprise the payload of the cell whose format depends on the AAL type of the cell. The structure of the UNI and NNI cell headers are given here:

	4	8 bits
GFC	VPI	
VPI	VCI	
VCI		
VCI	PTI (3 bits)	CLP
HEC		

UNI cell header

	4	8 bits
VPI		
VPI	VCI	
VCI		
VCI	PTI (3 bits)	CLP
HEC		

NNI cell header

GFC
Generic flow control (000=uncontrolled access).

VPI
Virtual path identifier.

VCI
Virtual channel identifier.
Together, the VPI and VCI comprise the VPCI. These fields represent the routing information within the ATM cell.

PTI
Payload type indication.

CLP
Cell loss priority.

HEC
Header error control.

AAL1 PDU

The structure of the AAL1 PDU is given in the following illustration:

←SN→		←SNP→		
CSI	SC	CRC	EPC	SAR PDU Payload
1 bit	3 bits	3 bits	1 bit	47 bytes

AAL1 PDU

SN
Sequence number. Numbers the stream of SAR PDUs of a CPCS PDU (modulo 16). The sequence number is comprised of the CSI and the SN.

CSI
Convergence sublayer indicator. Used for residual time stamp for clocking.

SC
Sequence count. The sequence number for the entire CS PDU, which is generated by the Convergence Sublayer.

SNP
Sequence number protection. Comprised of the CRC and the EPC.

CRC
Cyclic redundancy check calculated over the SAR header.

EPC
Even parity check calculated over the CRC.

SAR PDU payload
47-byte user information field.

AAL2

ITU-T I.366.2

AAL2 provides bandwidth-efficient transmission of low-rate, short and variable packets in delay sensitive applications. It supports VBR and CBR. AAL2 also provides for variable payload within cells and across cells. AAL type 2 is subdivided into the Common Part Sublayer (CPS) and the Service Specific Convergence Sublayer (SSCS).

AAL2 CPS Packet

The CPS packet consists of a 3 octet header followed by a payload. The structure of the AAL2 CPS packet is shown in the following illustration.

CID	LI	UUI	HEC	Information payload
8 bits	6 bits	5 bits	5 bits	1-45/64 bytes

AAL2 CPS packet

CID

Channel identification. Values may be as follows:

0 Not used
1 Reserved for layer management peer-to-peer procedures
2-7 Reserved
8-255 Identifies AAL2 user (248 total channels)

LI

Length indicator. This is the length of the packet payload associated with each individual user. Value is one less than the packet payload and has a default value of 45 bytes (may be set to 64 bytes).

UUI

User-to-user indication. Provides a link between the CPS and an appropriate SSCS that satisfies the higher layer application. Values may be:
1-15 Encoding format for audio, circuit mode data and demodulated fascimile image data using SSCS type 1 packets.
16-22 Reserved.
23 Reserved for SSCS type 2 packets.
24 SSCS type 3 packets except alarm packets.

25 Non-standard extension.
26 Framed mode data, final packet.
27 Framed mode data, more to come.
28-30 Reserved.
31 Alarm packets.

HEC

Header error control.

Information payload

Contains the CPS/SSCS PDU as described below.

AAL2 CPS PDU

The structure of the AAL2 SAR PDU is given in the following illustration.

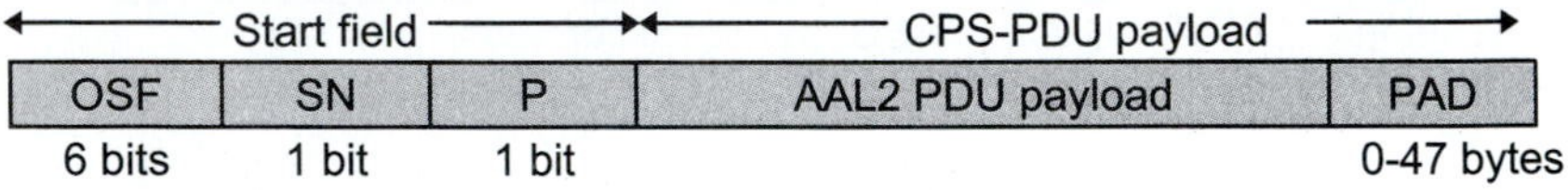

Start field			CPS-PDU payload	
OSF	SN	P	AAL2 PDU payload	PAD
6 bits	1 bit	1 bit		0-47 bytes

AAL2 CPS PDU

OSF

Offset field. Identifies the location of the start of the next CPS packet within the CPS-PDU.

SN

Sequence number. Protects data integrity.

P

Parity. Protects the start field from errors.

SAR PDU payload

Information field of the SAR PDU.

PAD

Padding.

AAL2 SSCS Packet

The SSCS conveys narrowband calls consisting of voice, voiceband data or circuit mode data. SSCS packets are transported as CPS packets over AAL2 connections. The CPS packet contains a SSCS payload. There are 3 SSCS packet types.

Type 1 Unprotected; this is used by default.

Type 2 Partially protected.

Type 3 Fully protected: the entire payload is protected by a 10-bit CRC which is computed as for OAM cells. The remaining 2 bits of the 2-octet trailer consist of the message type field.

AAL2 SSCS Type 3 Packets:
The type 3 packets are used for the following:

- Dialled digits
- Channel associated signalling bits
- Facsimile demodulated control data
- Alarms
- User state control operations.

The following illustration gives the general sturcture of AAL2 SSCS Type 3 PDUs. The format varies and each message has its own format according to the actual message type.

Redundancy	Time stamp	Message dependant information	Message type	CRC-10
2	14	16	6	10 bits

AAL2 SSCS Type 3 PDU

Redundancy
Packets are sent 3 times to ensure error correction. The value in this field signifies the transmission number.

Time stamp
Counters packet delay variation and allows a receiver to accurately reproduce the relative timing of successive events separated by a short interval.

Message dependant information
Packet content that varies, depending on the message type.

Message type

The message type code.

The following message type codes exist:

Information stream	Message type code	Packet format
Dialled digits	000010	Dialled digits
Channel associated signalling	000011	CAS bits
Facsimile demodulation control	100000	T.30 Preamble
	100001	EPT
	100010	Training
	100011	Fax Idle
	100100	T.30 Data
Alarms	000000	Alarm
User state control	000001	User state control

CRC-10

The 10-bit CRC.

AAL3/4

AAL3/4 consists of message and streaming modes. It provides for point-to-point and point-to-multipoint (ATM layer) connections. The Convergence Sublayer (CS) of the ATM Adaptation Layer (AAL) is divided into two parts: service specific (SSCS) and common part (CPCS). This is illustrated in the following diagram:

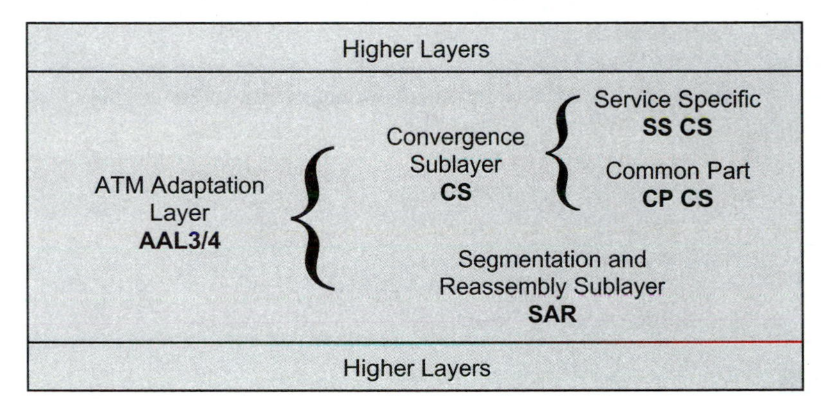

AAL3/4 packet

AAL3/4 packets are used to carry computer data, mainly SMDS traffic.

AAL3/4 CPCS PDU

The functions of the AAL3/4 CPCS include connectionless network layer (Class D), meaning no need for an SSCS; and frame relaying telecommunication service in Class C. The CPCS PDU is composed of the following fields:

Header	←	Info	→	←	Trailer		→
CPI	Btag	Basize	CPCS SDU	Pad	0	Etag	Length
1	1	2	0-65535	0-3	1	1	2 bytes

AAL3/4 CPCS PDU

CPI

Message type. Set to zero when the BAsize and Length fields are encoded in bytes.

Btag
Beginning tag. This is an identifier for the packet. It is repeated as the Etag.

BAsize
Buffer allocation size. Size (in bytes) that the receiver has to allocate to capture all the data.

CPCS SDU
Variable information field up to 65535 bytes.

PAD
Padding field which is used to achieve 32-bit alignment of the length of the packet.

0
All-zero.

Etag
End tag. Must be the same as Btag.

Length
Must be the same as BASize.

AAL3/4 SAR PDU

The structure of the AAL3/4 SAR PDU is illustrated below:

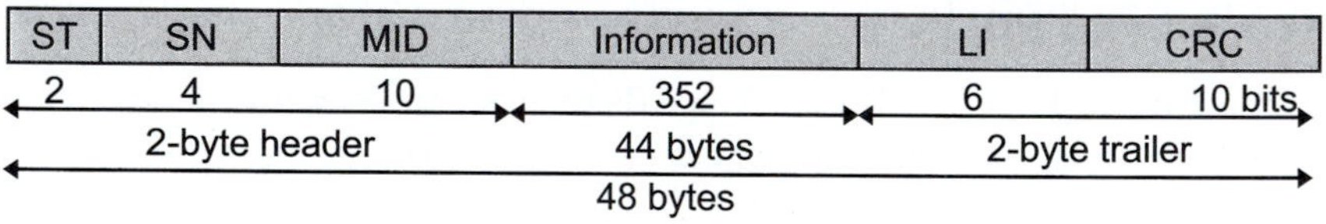

ST	SN	MID	Information	LI	CRC
2	4	10	352	6	10 bits
2-byte header			44 bytes	2-byte trailer	
48 bytes					

AAL3/4 SAR PDU

ST
Segment type. Values may be as follows:

Segment type	Value	Meaning
BOM	10	Beginning of message
COM	00	Continuation of message
EOM	01	End of message
SSM	11	Single segment message

SN

Sequence number. Numbers the stream of SAR PDUs of a CPCS PDU (modulo 16).

MID

Multiplexing identification. This is used for multiplexing several AAL3/4 connections over one ATM link.

Information

This field has a fixed length of 44 bytes and contains parts of CPCS PDU.

LI

Length indication. Contains the length of the SAR SDU in bytes, as follows:

Segment type	*LI*
BOM, COM	44
EOM	4, ..., 44
EOM (Abort)	63
SSM	9, ..., 44

CRC

Cyclic redundancy check.

Functions of AAL3/4 SAR include identification of SAR SDUs; error indication and handling; SAR SDU sequence continuity; multiplexing and demultiplexing.

AAL5

The type 5 adaptation layer is a simplified version of AAL3/4. It also consists of message and streaming modes, with the CS divided into the service specific and common part. AAL5 provides point-to-point and point-to-multipoint (ATM layer) connections.

AAL5 is used to carry computer data such as TCP/IP. It is the most popular AAL and is sometimes referred to as SEAL (simple and easy adaptation layer).

AAL5 CPCS PDU

The AAL5 CPCS PDU is composed of the following fields:

Info	Trailer					
CPCS payload	Pad	UU	CPI	Length	CRC	
0-65535	0-47	1	1	2	4	bytes

AAL5 CPCS PDU

CPCS payload
The actual information that is sent by the user. Note that the information comes before any length indication (as opposed to AAL3/4 where the amount of memory required is known in advance).

Pad
Padding bytes to make the entire packet (including control and CRC) fit into a 48-byte boundary.

UU
CPCS user-to-user indication to transfer one byte of user information.

CPI
Common part indicator is a filling byte (of value 0). This field is to be used in the future for layer management message indication.

Length
Length of the user information without the Pad.

CRC

CRC-32. Used to allow identification of corrupted transmission.

AAL5 SAR PDU

The structure of the AAL5 CS PDU is as follows:

Information	PAD	UU	CPI	Length	CRC-32
1-48	0-47	1	1	2	4 bytes

8-byte trailer

AAL5 SAR PDU

The fields are as described for the AAL5 CPCS PDU.

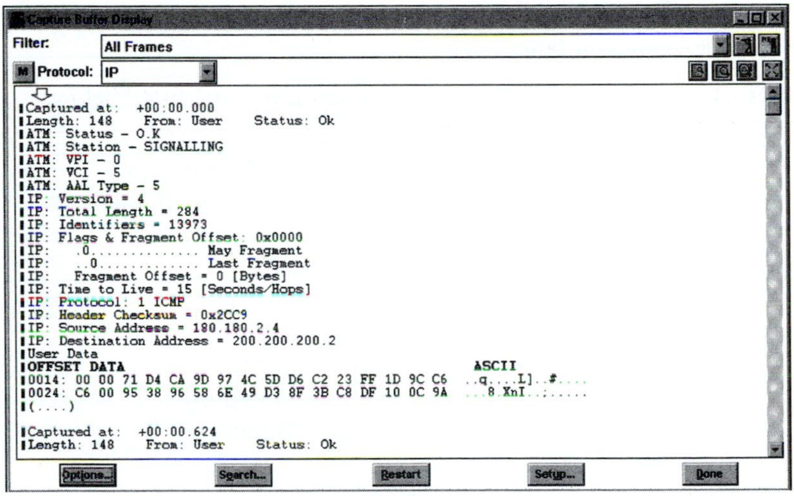

```
Capture Buffer Display
Filter: All Frames
Protocol: IP
Captured at:  +00:00.000
Length: 148    From: User    Status: Ok
ATM: Status - O.K
ATM: Station - SIGNALLING
ATM: VPI - 0
ATM: VCI - 5
ATM: AAL Type - 5
IP: Version = 4
IP: Total Length = 284
IP: Identifiers = 13973
IP: Flags & Fragment Offset: 0x0000
IP:    .0.............. May Fragment
IP:    ..0............. Last Fragment
IP:   Fragment Offset = 0 [Bytes]
IP: Time to Live = 15 [Seconds/Hops]
IP: Protocol: 1 ICMP
IP: Header Checksum = 0x2CC9
IP: Source Address = 180.180.2.4
IP: Destination Address = 200.200.200.2
User Data
OFFSET DATA                                             ASCII
0014: 00 00 71 D4 CA 9D 97 4C 5D D6 C2 23 FF 1D 9C C6  ..q....L]..#....
0024: C6 00 95 38 96 58 6E 49 D3 8F 3B C8 DF 10 0C 9A  ...8.XnI..;.....
(....)

Captured at:  +00:00.624
Length: 148    From: User    Status: Ok
Options...   Search...   Restart   Setup...   Done
```

IP frames encapsulated over ATM

F4/F5 OAM

The structure of the F4 and F5 OAM cell payload is given in the following illustration.

OAM type	Function type	Function specific	Reserved	CRC-10
4	4	360	6	10 bits

48 bytes

F4/F5 OAM PDU

CRC-10

Cyclic redundancy check: $G(x) = x^{10}+x^9+x^5+x^4+x+1$

OAM type / Function type

The possible values for OAM type and function type are listed below:

OAM type	*Value*	*Function type*	*Value*
Fault Management	0001	Alarm Indication Signal (AIS)	0000
		Far End Receive Failure (FERF)	0001
		OAM Cell Loopback	1000
		Continuity Check	0100
Performance Management	0010	Forward Monitoring	0000
		Backward Reporting	0001
		Monitoring and Reporting	0010
Activation/ Deactivation	1000	Performance Monitoring	0000
		Continuity Check	0001

OAM F4 cells operate at the VP level. They use the same VPI as the user cells, however, they use two different reserved VCIs, as follows:

VCI=3 Segment OAM F4 cells.
VCI=4 End-end OAM F4 cells.

OAM F5 cells operate at the VC level. They use the same VPI and VCI as the user cells. To distinguish between data and OAM cells, the PTI field is used as follows:

PTI=100 (4) Segment OAM F5 cells processed by the next segment.
PTI=101 (5) End-to-end OAM F5 cells which are only processed by end stations terminating an ATM link.

RM Cells

There are two types of Rate Management (RM) cells: RM-VPC, which manages the VP level and RM-VCC, which manages the VC level.

The format of RM-VPC cells is shown in the following illustration:

ATM Header: VCI=6 and PTI=110 (5 bytes)
RM protocol identifier (1 byte)
Message type (1 byte)
ER (2 bytes)
CCR (2 bytes)
MCR (2 bytes)
QL (4 bytes)
SN (4 bytes)
Reserved (30 bytes)
Reserved (6 bits) + CRC-10 (10 bits)

RM-VPC cell format

RM protocol identifier
Always 1 for ABR services.

Message type
This field is comprised of several bit fields:

Bit	*Name*	*Description*
8	DIR	Direction of the RM cells: 0=forward, 1=backward.
7	BN	BECN: 0=source is generated; 1=network is generated.
6	CI	Congestion Indication: 0=no congestion, 1=congestion.
5	NI	No increase: 1=do not increase the ACR.
4	RA	Not used.

ER
Explicit rate.

CCR
Current cell rate.

MCR
Minimum cell rate.

QL
Not used.

SN
Not used.

RM-VCC cells are exactly the same as RM-VPC cells, except that the VCI is not specified. The cell is identified solely by the PTI bits.

Reserved VPI/VCI Values

A number of VPI/VCI values are reserved for various protocols or functions, e.g., 0,5 is used for signalling messages. The following table contains a list of all reserved VPI/VCI values and their designated meanings:

VPI	VCI	Description
0	0	Idle cells. Must also have GFC set to zero. Idle cells are added by the transmitter to generate information for non-used cells. They are removed by the receiver together with bad cells.
0	1	Meta signalling (default). Meta-signalling is used to define the subchannel for signalling (default value: 0,5).
Non-zero	1	Meta signalling.
0	2	General broadcast signalling (default). Can be used to broadcast signalling information which is independent of a specific service. Not used in practice.
Non-zero	2	General broadcast signalling.
0	5	Point-to-point signalling (default). Generally used to set-up and release switched virtual circuits (SVCs).
Non-zero	5	Point-to-point signalling.
	3	Segment OAM F4 flow cell. OAM cells are used for continuity checks as well as to notify and acknowledge failures.
	4	End-to-end OAM F4 flow cell.
	6	RM-VPC cells for rate management.

VPI	VCI	Description
0	15	SPANS. The Simple Protocol for ATM Network Signalling is a simple signalling protocol, developed by FORE systems and used by FORE and other manufacturers working in cooperation with FORE, for use in ATM networks. Refer to Chapter 3 for more information.
0	16	ILMI. The Interim Local Management Interface is used to manage and compare databases across an ATM link. This is used for signalling address registration, RMON applications, SNMP, etc. Refer to *ILMI* in this book for more information.
0	18	PNNI signalling.

3

ATM Signalling and Routing Protocols

Signalling is the process by which ATM users and the network exchange the control of information, request the use of network resources or negotiate for the use of circuit parameters. The VPI/VCI pair and requested bandwidth are allocated as a result of a successful signalling exchange.

The protocols illustrated below support connection control signalling. These messages are sent over the Signalling ATM Adaptation Layer (SAAL), which ensures their reliable delivery. The SAAL is divided into a Service Specific Part and a Common Part. The Service Specific Part is further divided into a Service Specific Coordination Function (SSCF), which interfaces with the SSCF user; and a Service Specific Connection-Oriented Protocol (SSCOP), which assures reliable delivery.

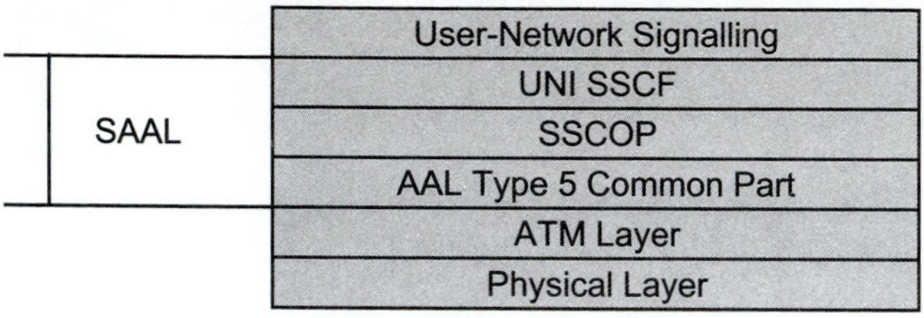

	User-Network Signalling
SAAL	UNI SSCF
SAAL	SSCOP
SAAL	AAL Type 5 Common Part
	ATM Layer
	Physical Layer

ATM signalling protocol stack

The UNI signalling protocols within the SAAL are responsible for ATM call and connection control, including call establishment, call clearing, status enquiry, and point-to-multipoint control.

UNI 3.0, UNI 3.1, Q.2931, Q.2971, UNI 4.0, IISP, PNNI, Q.SAAL and SPANS are described in this chapter. Refer to Chapter 17 for information on ILMI, which is used for address registration.

UNI 3.x Signalling

ATM Forum UNI 3.0 1993-10, UNI 3.1 1993-10

A signalling message uses the Q.931 message format. It is made up of a message header and a variable number of Information Elements (IEs). This is shown in the following figure:

Message header
IE
IE
...
IE

ATM signalling message structure

The message header is shown in the following diagram:

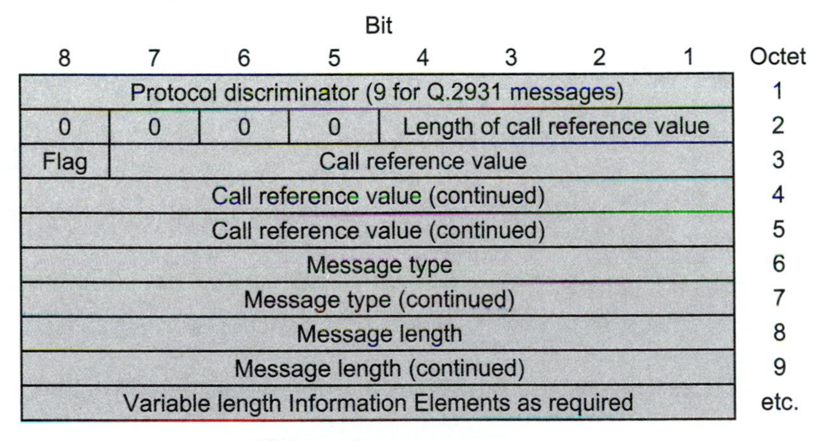

Bit 8	7	6	5	4	3	2	1	Octet
Protocol discriminator (9 for Q.2931 messages)								1
0	0	0	0	Length of call reference value				2
Flag	Call reference value							3
Call reference value (continued)								4
Call reference value (continued)								5
Message type								6
Message type (continued)								7
Message length								8
Message length (continued)								9
Variable length Information Elements as required								etc.

ATM signalling message structure

Protocol discriminator

Distinguishes messages for user-network call control from other messages.

Call reference

Unique number for every ATM connection which serves to link all signalling messages relating to the same connection. It identifies the call at the local user network interface to which the particular message applies. The call reference is comprised of the call reference value and the call reference flag. The call reference flag indicates who allocated the call reference value.

Message type

The message may be of the following types:

Call establishment messages:

CALL PROCEEDING, sent by the called user to the network or by the network to the calling user to indicate initiation of the requested call.
CONNECT, sent by the called user to the network or by the network to the calling user to indicate that the called user accepted the call.
CONNECT ACKNOWLEDGE, sent by the network to the called user to indicate that the call was awarded or by the calling user to the network.
SETUP, sent by the calling user to the network or by the network to the calling user to initiate a call.

Call clearing messages:

RELEASE, sent by the user to request that the network clear the connection or sent by the network to indicate that the connection has cleared.

RELEASE COMPLETE, sent by either the user or the network to indicate that the originator has released the call reference and virtual channel.
RESTART, sent by the user or the network to restart the indicated virtual channel.
RESTART ACKNOWLEDGE, sent to acknowledge the receipt of the RESTART message.

Miscellaneous messages:

STATUS ENQUIRY, sent by the user or the network to solicit a STATUS message.
STATUS, sent by the user or network in response to a STATUS ENQUIRY message.

Point-to-Multipoint messages:

ADD PARTY, adds a party to an existing connection.
ADD PARTY ACKNOWLEDGE, acknowledges a successful ADD PARTY.
ADD PARTY REJECT, indicates an unsuccessful ADD PARTY.
DROP PARTY, drops a party from an existing point-to-multipoint connection.
DROP PARTY ACKNOWLEDGE, acknowledges a successful DROP PARTY.

Message length

The length of the contents of a message.

Information elements

Described below.

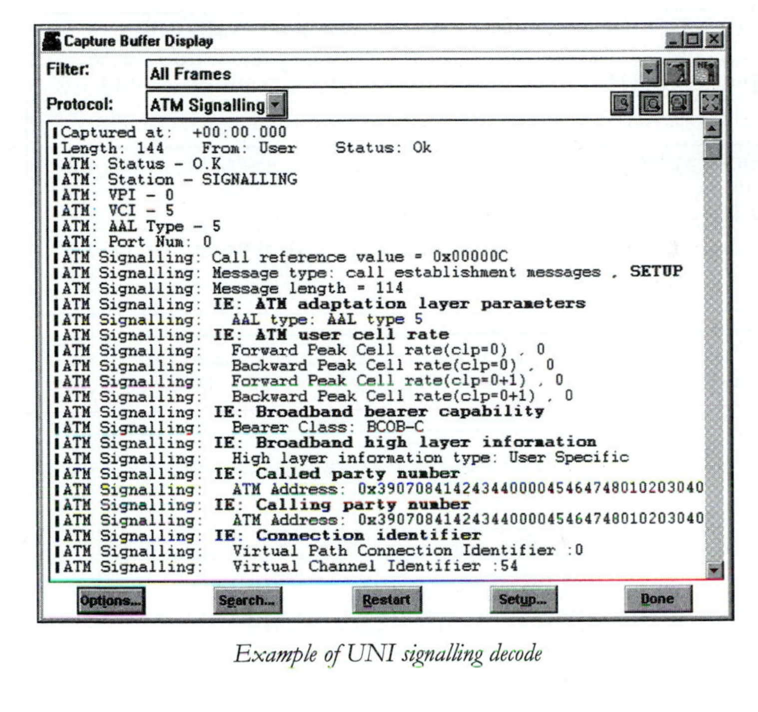

Example of UNI signalling decode

Types of Information Elements

There are several types of information elements. Some may appear only once in the message; others may appear more than once. Depending on the message type, some information elements are mandatory and some are optional. The order of the information elements does not matter to the signalling protocol. The information elements in UNI 3.0 are listed in the following table:

IE	Description	Max. No.
Cause	Gives the reason for certain messages. For example, the Cause IE is part of the release message, indicating why the call was released.	2
Call state	Indicates the current state of the call.	1
Endpoint reference	Identifies individual endpoints in a point-to-multipoint call.	1
Endpoint state	Indicates the state of an endpoint in a point-to-multipoint call.	1
AAL parameters	Includes requested AAL type and other AAL parameters.	1
ATM user cell rate	Specifies traffic parameters.	1
Connection identifier	Identifies the ATM connection and gives the VPI and VCI values.	1
Quality of Service parameter	Indicates the required Quality of Service class for the connection.	1
Broadband high-layer information	Gives information about the high-layer protocols for compatibility purposes.	1
Broadband bearer capacity	Requests a service from the network (such as CBR or VBR link, point-to-point and point-to-multipoint link).	1
Broadband low-layer information	Checks compatibility with layer 2 and 3 protocols.	3
Broadband locking shift	Indicates a new active codeset.	-
Broadband non-locking shift	Indicates a temporary codeset shift.	-
Broadband sending complete	Indicates the competition of sending the called party number.	1
Broadband repeat indicator	Indicates how IEs which are repeated in the message should be handled.	1

IE	Description	Max. No.
Calling party number	Origin of the call.	1
Calling party subaddress	Subaddress of calling party.	1
Called party number	Destination of the call.	1
Called party subaddress	Subaddress of the called party.	1
Transit network selection	Identifies one requested transit network.	1
Restart indicator	Identifies which facilities should be restarted (e.g., one VC, all VCs).	1

Types of UNI IEs

For further reference about the exact structure and parameters of the IEs, refer to the ATM Forum, ATM User-Network Interface Specifications 3.0 and 3.1.

ITU Q.2931 Signalling

ITU Q.2931 1995-02

This is the ITU version of signalling. The Q.2931 signalling protocol specifies the procedures for the establishment, maintenance and clearing of network connections at the B-ISDN user network interface. The procedures are defined in terms of messages exchanged. The basic capabilities supported by Q.2931 Signalling are as follows:

- Demand (switched virtual) channel connections.
- Point-to-point switched channel connections.
- Connections with symmetric or asymmetric bandwidth requirements.
- Single-connection (point-to-point) calls.
- Basic signalling functions via protocol messages, information elements and procedures.
- Class X, Class A and Class C ATM transport services.
- Request and indication of signalling parameters.
- VCI negotiation.
- Out-of-band signalling for all signalling messages.
- Error recovery.
- Public UNI addressing formats for unique identification of ATM endpoints.
- End-to-end compatibility parameter identification.
- Signalling interworking with N-ISDN and provision of N-ISDN services.
- Forward compatibility.

The message types for Q.2931 are the same as in UNI 3.0/3.1, with the exception of the point-to-multipoint messages which are not supported. The following are additional signalling messages specific to Q.2931:
ALERTING, sent by the called user to the network and by the network to the calling user indicating that the called user alerting has been initiated.
PROGRESS, sent by the user or the network to indicate the progress of a call in the event of interworking.
SETUP ACKNOWLEDGE, sent by the network to the calling user or by the called user to the network to indicate that call establishment has been initiated.
INFORMATION, sent by the user or the network to provide additional information.

NOTIFY, sent by the user or by the network to indicate information pertaining to a call connection.

The information elements in Q.2931 are as follows:

- Called party number.
- Called party sub-address.
- Transit network selection.
- Restart indicator.
- Narrow-band low layer compatibility.
- Narrow-band high layer compatibility.
- Broadband locking shift.
- Broadband non-locking shift.
- Broadband sending complete.
- Broadband repeat indicator.
- Calling party number.
- Calling party sub-address.
- ATM adaptation layer parameters.
- ATM traffic descriptor.
- Connection identifier.
- OAM traffic descriptor.
- Quality of Service parameter.
- Broadband bearer capability.
- Broadband Low Layer Information (B-LLI).
- Broadband High Layer Information (B-HLI).
- End-to-end transit delay.
- Notification indicator.
- Call state.
- Progress indicator.
- Narrow-band bearer capability.
- Cause.

ITU Q.2971 Signalling

ITU Q.2971 10/95

This is the ITU version of signalling. The Q.2971 signalling protocol specifies the procedures for the establishment, maintenance and clearing of point-to-multipoint virtual channel calls/connections by means of Digital subscriber signalling system 2 (DSS2) at the B-ISDN user network interface. The procedures are defined in terms of messages exchanged. Q.2971 uses the same message capabilities as Q.2931. However, in addition, it also supports point-to-multipoint unidirectional switched channel connections. A point-to-multipoint virtual channel connection is a collection of associated ATM virtual channel links connecting 2 or more endpoints. Q.2971 supports only unidirectional transport from the root to the leaves.

The following are the additional messages (not used in Q.2931) used with ATM point-to-mulitpoint call and connection control:
ADD PARTY, adds a party to an existing connection.
ADD PARTY ACKNOWLEDGE, acknowledges a successful ADD PARTY.
PARTY ALERTING
ADD PARTY REJECT, indicates an unsuccessful ADD PARTY.
DROP PARTY, drops a party from an existing point-to-multipoint connection.
DROP PARTY ACKNOWLEDGE, acknowledges a successful DROP PARTY.

UNI 4.0 Signalling

ATM Forum UNI 4.0 1996-07

UNI 4.0 provides the signalling procedures for dynamically establishing, maintaining and clearing ATM connections at the ATM User-Network Interface. UNI 4.0 applies both to Public UNI (the interface between endpoint equipment and a public network) and private UNI (the interface between endpoint equipment and a private network).

The following features are available within the UNI 4.0 signalling protocol:

- Leaf initiated join.
- Enhanced ATM traffic descriptor.
- Available bit rate capability.
- Individual QoS parameters.
- Narrow ISDN over ATM.
- AnyCast capability.
- New information elements.
- New VPI/VCI options.
- Proxy signalling capability.
- Virtual UNIs.
- Supplementary services such as direct dialing in, multiple subscriber number, calling line identification presentation, calling line identification restriction, connected line identification presentation, connected line identification rest, user-to-user signalling.
- Error handling for instruction indicators.
- Using setup for adding parties.
- Both NSAP and ASTM end system addresses.
- Network can support leaves that do not support P-PM.

The message types for UNI 4.0 are the same as in Q.2931, with the exception of the SETUP ACKNOWLEDGE and INFORMATION messages which are not supported. The following are new signalling messages specific to UNI 4.0: LEAF SETUP REQUEST and LEAF SETUP FAILURE.

The following are the information elements contained in UNI 4.0:

- Narrowband bearer capability.
- Cause.

- Call state.
- Progress indicator.
- Notification indicator.
- End-to end transit delay.
- Connected number.
- Connected subaddress.
- Endpoint reference.
- Endpoint state.
- ATM adaptation layer parameters.
- ATM traffic descriptor.
- Connection identifier.
- Quality of service parameter.
- Broadband high layer information.
- Broadband bearer capability.
- Broadband low layer information.
- Broadband locking shift.
- Broadband non-locking shift.
- Broadband repeat indicator.
- Calling party number.
- Calling party subaddress.
- Called party number.
- Called party subaddress.
- Transit network selection.
- Restart indicator.
- Narrowband low layer compatibility.
- Narrowband high layer compatibility.
- Generic identifier transport.
- Minimum acceptable traffic descriptor.
- Alternative ATM traffic descriptor.
- ABR setup parameters.
- Leaf initiated join call identifier.
- Leaf initiated join parameters.
- Leaf sequence number.
- Connection scope selection.
- ABR additional parameters.
- Extended QoS parameters.

Q.SAAL

The structure for each Q.SAAL message type is shown below.

BGN PDU (Begin)
The BGN PDU is used to initially establish an SSCOP connection or reestablish an existing SSCOP connection between two peer entities. The BGN requests the clearing of the peer's transmitter and receiver buffers, and the initialization of the peer's transmitter and receiver state variables.

Bytes

	1			2	3	4
1	N(UU)					
2	Rsvd	S	PDU Type	N(MR)		
	8 7 6	5	4 3 2 1			

Begin PDU (BGN PDU)

BGAK PDU (Begin Acknowledge)
The BGAK PDU is used to acknowledge the acceptance of a connection request by the peer entity.

Bytes

	1		2	3	4
1	N(UU)				
2	Rsvd	PDU Type	N(MR)		
	8 7 6 5	4 3 2 1			

Begin Acknowledged PDU (BGAK PDU)

BGREJ PDU (Begin Reject)
The BGREJ PDU is used to reject the connection establishment of the peer SSCOP entity.

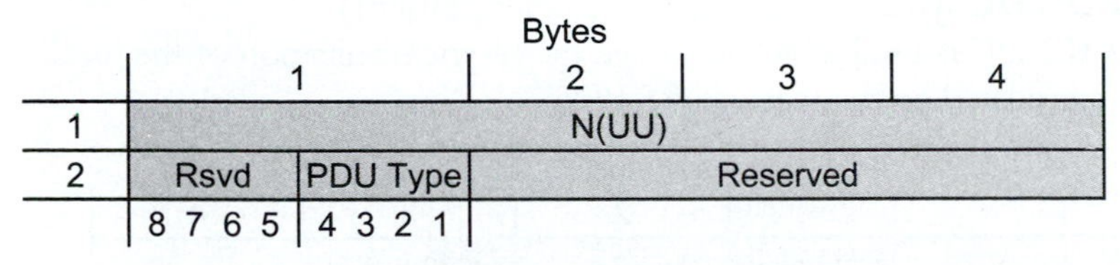

Begin Reject PDU (BGREJ PDU)

END PDU (End)

The END PDU is used to release an SSCOP connection between two peer entities.

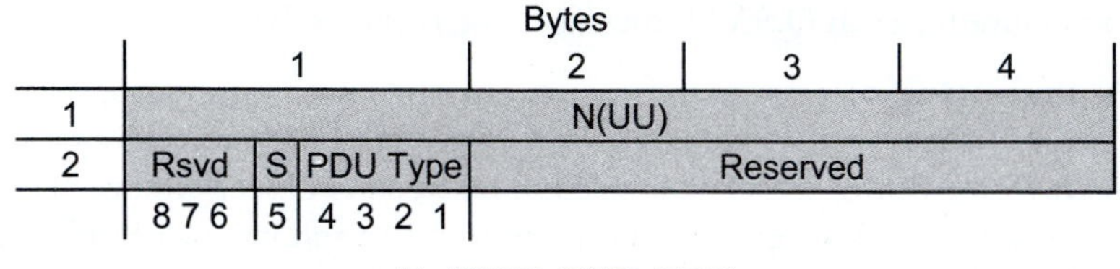

End PDU (END PDU)

ENDAK PDU (End Acknowledge)

The ENDAK PDU is used to confirm the release of an SSCOP connection that was initiated by the peer SSCOP entity.

	Byte 1		Byte 2	Byte 3	Byte 4
1	Rsvd	PDU Type	Reserved		
	8 7 6 5	4 3 2 1			

End Acknowledgement (ENDAK PDU)

RS PDU (Resynchronization Command)

The RS PDU is used to resynchronize the buffers and data transfer state variables in the transmit direction of an SSCOP connection.

	Byte 1		Byte 2	Byte 3	Byte 4
1	N(UU)				
2	Rsvd	PDU Type	Reserved		
	8 7 6 5	4 3 2 1			

Resynchronization PDU (RS PDU)

RSAK PDU (Resynchronization Acknowledgement)

The RSAK PDU is used to acknowledge the resynchronization of the local receiver stimulated by the received RS PDU.

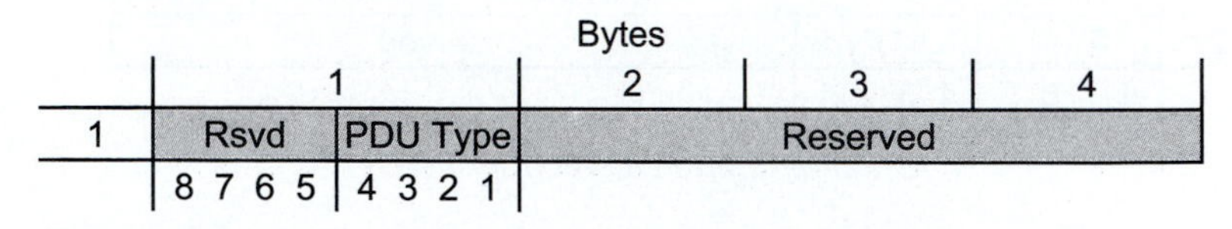

Resynchronization acknowledge PDU (RSAK PDU)

SD PDU (Sequenced Data)
The SD PDU is used to transfer, across an SSCOP connection, sequentially numbered PDUs containing information fields provided by the SSCOP user.

	Bytes 1			2	3	4
1	Information (maximum k bytes)					
. . .				PAD (0-3 Bytes)		
n	PL	Rsd	PDU Type	N(S)		
	8 7	6 5	4 3 2 1			

Sequenced data PDU (SD PDU)

SDP PDU (Sequenced Data with Poll)
The SDP PDU is used to transfer, across an SSCOP connection, sequentially numbered PDUs containing information fields provided by the SSCOP user. The SDP PDU also contains a poll request that is used to stimulate the transmission of a STAT PDU. Therefore, an SDP PDU is the functional concatenation of an SD PDU and a POLL PDU.

	Bytes 1			2	3	4
1	Information (maximum k bytes)					
. . .				PAD (0-3 Bytes)		
	Reserved			N(PS)		
n	PL	Rsd	PDU Type	N(S)		
	8 7	6 5	4 3 2 1			

Sequenced data with poll PDU (SDP PDU)

POLL PDU (Status Request)
The POLL PDU is used to request, across an SSCOP connection, status information about the peer SSCOP entity. It contains a sequence number for use in the retransmission of lost SD or SDP PDUs.

	Bytes 1		2	3	4
1	Reserved		N(PS)		
2	Rsvd	PDU Type	N(S)		
	8 7 6 5	4 3 2 1			

Poll PDU (POLL PDU)

STAT PDU (Solicited Status Response)

The STAT PDU is used to respond to a status request (POLL PDU) received from a peer SSCOP entity. It contains information regarding the reception status of SD or SDP PDUs, credit information for the peer transmitter, and the sequence number of the POLL PDU to which it is in response.

	Bytes 1		2	3	4
1	PAD		List element 1 (a SD PDU N(S))		
2	PAD		List element 2		
...	. . .		. . .		
L	PAD		List element L		
L+1	PAD		N(PS)		
L+2	PAD		N(MR)		
L+3	Rsvd	PDU Type	N(R)		
	8 7 6 5	4 3 2 1			

Solicited status PDU (STAT PDU)

USTAT PDU (Unsolicited Status Response)

The USTAT PDU is used to respond to a detection of a new missing SD or SDP PDU, based on the examination of the sequence number of the SD or SDP PDU. It contains information regarding the reception status of SD or SDP PDUs and credit information for the peer transmitter.

	Bytes 1		2	3	4
1	PAD		List element 1 (a SD PDU N(S))		
2	PAD		List element 2		
3	PAD		N(MR)		
4	Rsvd	PDU Type	N(R)		
	8 7 6 5	4 3 2 1			

Unsolicited status PDU (STAT PDU)

UD PDU (Unnumbered Data)

The UD PDU is used for unassured data transfer between two SSCOP users. When an SSCOP user requests unacknowledged information transfer, the UD PDU is used to send information to the peer without affecting SSCOP status or variables. UD PDUs do not carry a sequence number and therefore, the UD PDU may be lost without notification.

	Bytes 1			2	3	4
1	Information (maximum k bytes)					
. . .				PAD (0-3 Bytes)		
n	PL	Rsd	PDU Type	Reserved		
	8 7	6 5	4 3 2 1			

Unit data PDU (UD PDU) / management data (MD PDU)

MD PDU (Management Data)

The MD PDU is used for unassured management data transfer between two SSCOP entities. When a management entity requests unacknowledged information transfer, the MD PDU is used to send information to the peer management entity without affecting SSCOP status or variables. MD PDUs do not carry a sequence number and therefore, the MD PDU may be lost without notification. (see UD PDU diagram above).

IISP

IISP (Interim Interswitch Signalling Protocol) is a protocol that has been devised to provide signalling between switches from multiple vendors. It has been developed as a stopgap solution, until the more sophisticated PNNI (Private Network-to-Network Interface) specification is complete. There is no migration path from IISP to PNNI.

IISP uses UNI 3.1 signalling procedures as does PNNI. However, the switches using IISP are not peers, i.e., one switch functions as the network node and the other as an end-station. There is no support for dynamic distribution of routing information.

PNNI Signalling and Routing

ATM Forum PNNI 1.0 (af-pnni-0055.000 letter ballot) 1996-03

PNNI (Private Network-to-Network Interface) is a hierarchical, dynamic link-state routing protocol. It is designed to support large-scale ATM networks. The PNNI protocol uses VPI/VCI 0,18 for its messages. In addition, it uses signalling messages to support connection establishment across multiple networks. The signalling is based on UNI 4.0 and Q.2931. Specific information elements were added to UNI 4.0 in order to support the routing process of PNNI.

PNNI Signalling

PNNI Signalling contains the procedure to dynamically establish, maintain and clear ATM connections at the private network to network interface or network-node interface between 2 ATM networks or 2 ATM network nodes. The PNNI signalling protocol is based on the ATM forum UNI specification and on Q.2931.

PNNI messages include:
ALERTING, CALL PROCEEDING, CONNECT, SETUP, RELEASE, RELEASE COMPLETE, NOTIFY, STATUS, STATUS ENQUIRY, RESTART, RESTART ACKNOWLEDGE, STATUS, ADD PARTY, ADD PARTY ACKNOWLEDGE, PARTY ALERTING, ADD PARTY REJECT, DROP PARTY, DROP PARTY ACKNOWLEDGE.

The following messages for ATM call connection control, for the support of 64 Kbits based ISDN circuit code services, are transported without change across the PNNI:
ALERTING, CONNECT PROGRESS, SETUP, RELEASE.

The following message types supported by Q.2931 are not supported by PNNI:
CONNECT ACKNOWLEDGE, SETUP ACKNOWLEDGE, INFORMATION.

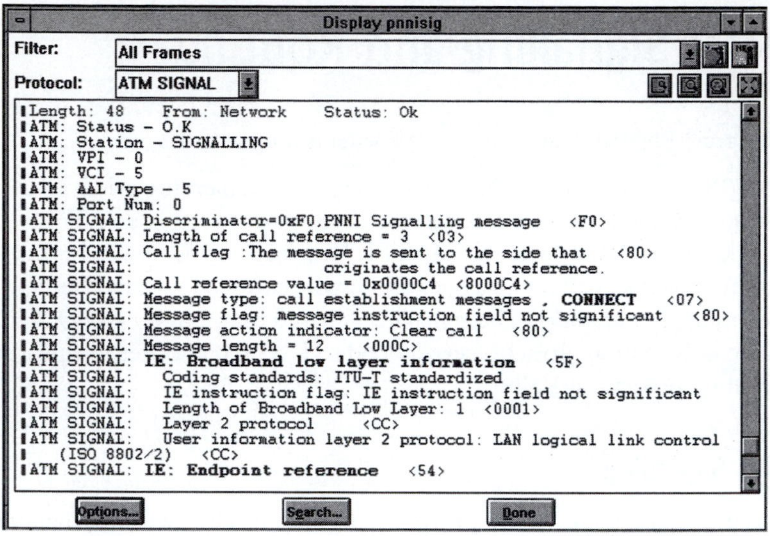

Display pnnisig

Filter: All Frames

Protocol: ATM SIGNAL

```
Length: 48    From: Network    Status: Ok
ATM: Status - O.K
ATM: Station - SIGNALLING
ATM: VPI - 0
ATM: VCI - 5
ATM: AAL Type - 5
ATM: Port Num: 0
ATM SIGNAL: Discriminator=0xF0,PNNI Signalling message   <F0>
ATM SIGNAL: Length of call reference = 3  <03>
ATM SIGNAL: Call flag :The message is sent to the side that   <80>
ATM SIGNAL:                 originates the call reference.
ATM SIGNAL: Call reference value = 0x0000C4  <8000C4>
ATM SIGNAL: Message type: call establishment messages , CONNECT   <07>
ATM SIGNAL: Message flag: message instruction field not significant   <80>
ATM SIGNAL: Message action indicator: Clear call   <80>
ATM SIGNAL: Message length = 12    <000C>
ATM SIGNAL: IE: Broadband low layer information   <5F>
ATM SIGNAL:   Coding standards: ITU-T standardized
ATM SIGNAL:   IE instruction flag: IE instruction field not significant
ATM SIGNAL:   Length of Broadband Low Layer: 1  <0001>
ATM SIGNAL:   Layer 2 protocol     <CC>
ATM SIGNAL:   User information layer 2 protocol: LAN logical link control
    (ISO 8802/2)    <CC>
ATM SIGNAL: IE: Endpoint reference   <54>
```

Options... Search... Done

PNNI signalling decode

PNNI Routing

The structure of the PNNI header is shown in the following illustration:

Packet type	Packet length	Prot ver	Newest ver	Oldest ver	Reserved
2	2	1	1	1	1 bytes

PNNI header structure

Packet type

The following packet types are defined:

Hello	Sent by each node to identify neighbor nodes belonging to the same peer group.
PTSP	PNNI Topology State Packet. Passes topology information between groups.
PTSE	PNNI Topology State Element (Request and Ack). Conveys topology parameters such as active links, their available bandwidth, etc.
Database Summary	Used during the original database exchange between two neighboring peers.

Packet length

The length of the packet.

Prot ver

Protocol version. The version according to which this packet was formatted.

Newest ver / Oldest ver

Newest version supported / oldest version supported. The newest version supported and the oldest version supported fields are included in order for nodes to negotiate the most recent protocol version that can be understood by both nodes exchanging a particular type of packet.

Information groups

The following information groups are found in PNNI packets:

Hello:
Aggregation token
Nodal hierarchy list
Uplink information attribute
LGN horizontal link
Outgoing resource availability
Optional GCAC parameters
System capabilities

PTSP:
PTSE
Nodal state parameters
Nodal information group
Outgoing resource availability
Incoming resource availability
Next higher level binding
Optional GCAC parameters
Internal reachable ATM address
Exterior reachable ATM addresses
Horizontal links
Uplinks
Transit network ID
System capabilities

PTSE Ack:
Nodal PTSE Ack
System capabilities

Database Summary:
Nodal PTSE summaries
System capabilities

PTSE Request:
Requested PTSE header
System capabilities

```
Display PNNI signal example 1
Filter: All Frames
Protocol: Pnni routing
Pnni routing:  PNNI Header
Pnni routing:    Packet type: HELLO    <0001>
Pnni routing:    Packet length: 100  <0064>
Pnni routing:    Version: 172      <AC>
Pnni routing:    Newest version supported: 172  <AC>
Pnni routing:    Oldest version supported: 0  <00>
Pnni routing:   Flags: Reserved     <0180>
Pnni routing:  Sending node id.
Pnni routing:    Parent peer group level: 0x38  <38>
Pnni routing:    Distinguish node identifier: 0xA0  <A0>
Pnni routing:   ATM Address: 0x4700918100000000603E7B170100603E7B170100
   <4700918100000000603E7B170100603E7B170100>
Pnni routing:     ATM Address Details:
Pnni routing:       ATM Format address, ICD   <47>
Pnni routing:       ICD: 0091      <0091>
Pnni routing:       DFI: 0x81      <81>
Pnni routing:       AA: 0x000000   <000000>
Options...    Search...    Done
```

PNNI routing decode

B-ICI

ATM Forum B-ICI Specification version 2.0 12/95

B-ICI, a BISDN Inter Carrier Interface, is an interface connecting two different ATM based public network providers or carriers. This is necessary in order to facilitate end-to-end national and international ATM/BISDN services. The B-ICI specification also includes service specific functions above the ATM layer required to transport, operate and manage a variety of intercarrier services across the B-ICI.

The protocols in this group include: Q.2140 and B-ISUP

B-ISUP

The structure of the B-ISUP header is shown in the following illustration:

Routing label	Type code	Message length	Compatibility	Message	
4	1	2	1	variable	bytes

B-ICI header structure

Routing label

This is the same for each message on a specific ATM virtual connection.

Type code

Defines the function and format of each B-ISDN user part message. Examples of messages are: Address complete, Call progress, Forward transfer, and Release complete.

Message length

Number of octets in the message.

Compatibility

Message compatibility information. Defines the behavior of a switch if a message is not understood.

Message

Contents of the message.

Q.2140

Q.2140 is part of the ATM adaptation layer which supports signalling at the Network Node Interface of the B-ISDN. This protocol implements the Service Specific Coordination Function (SSCF) for signalling at the NNI.

The structure of the Q.2140 header is shown in the following illustration. The Q.2140 contains one field called SSCF status.

Reserved	SSCF status
3 bytes	1 byte

Q.2140 header structure

The SSCF status indicates the status of the sending peer. It can have the following values:

0000 0001	Out of Service.
0000 0010	Processor Outage.
0000 0011	In Service.
0000 0100	Normal.
0000 0101	Emergency.
0000 0111	Alignment Not Successful.
0000 1000	Management Initiated.
0000 1001	Protocol Error.
0000 1010	Proving Not Successful.

SPANS

SPANS UNI release 2.3, NNI release 3.0, UNI release 3.0

SPANS (Simple Protocol for ATM Network Signalling) is a simple signalling protocol, developed by FORE systems and used by FORE and other manufacturers working in cooperation with FORE, for use in ATM networks.

SPANS signalling messages are transferred over a reserved ATM virtual connection, using AAL type 3/4. Currently this connection uses VPI 0 and VCI 15. This channel must be predefined in both directions, on all links used by participants in the signalling protocol.

A null transport layer is used. Retransmission of lost messages and suppression of duplicate messages are performed by the application when necessary.

The first part of this protocol is SPANS UNI which is used in ATM LANs. The protocol specifies the signalling messages that are exchanged between hosts and the ATM network to perform several functions such as opening and closing connections. These functions allow hosts and routers to use an ATM LAN as a subnet of a larger internet. The two classes of messages involved in this protocol are status messages and connection messages.

The second part of the protocol is SPANS NNI, which is a simple signalling protocol for routing and virtual path support in ATM LANs. This part of the protocol specifies the signalling messages that are exchanged between ATM network switches to perform functions such as opening and closing virtual paths. An ATM network switch is a device which is capable of forwarding data over ATM connections from one or more sources to one or more destinations. The two classes of messages involved with this protocol are topology messages and network internal connection messages.

The format of the SPANS header is shown in the following illustration:

8 / 16 / 24	32 bits
Major version	Minor ver
Message type	
Remainder of frame	

SPANS header structure

Major version

The major version of the protocol.

Minor ver(sion)

The minor version of the protocol.

Message type

The types of messages are as follows:

- STATUS messages, which transfer information about the state of a signalling peer. The status messages are as follows:
  Indications - Issued by the network.
  Responses - Issued in response to the network's indications.
  Requests - Issued by hosts when booting.
- CONNECTION messages are used in order to open new connections or close existing ones. The connection messages are as follows:
  Requests - The originating side, either the source or the destination of the connection, sends a request message.
  Indications - In most cases, the request causes an indication message to be issued by the network to the target host.
  Responses - The target host then returns a response message, in response to the request which was received.
  Confirmations - Finally, the network issues a confirmation message to the original source.
- TOPOLOGY messages are exchanged between switches within a network. Through these messages, switches learn of changes in the network topology, e.g., new switches coming on line, switches and links going down, and changing node and link loads.
- NETWORK-INTERNAL CONNECTION messages are exchanged by switches in the network to set up and manage virtual paths. The originating switch issues a request, which may be forwarded as a request by intermediate switches until it reaches the target switch. The target switch produces a reply, which again may be forwarded by intermediate switches on its way back to the originator.

ViVID MPOA

"The ATM Host/Router on Vivid Release 2.x" Draft version 1.0, Newbridge Corporation, 1996-01
"Vivid system forwarding and cache management for MPOA partners" draft version 1.0, 1996-0

ViVID is a proprietary protocol of Newbridge which provides bridged LAN Emulation. and routed LAN Emulation functionality. There are 3 protocols in the ViVID group, BME, ARM and CCP. They share a common header format. All ViVID protocols are LLC/SNAP encapsulated. The Newbridge OUI and a PID value of 0x02 identify them as ViVID.

MPOA

ATM Forum Specification STR-MPOA-MPOA-01.00 04-1997

The Multi Protocol Over ATM (MPOA) deals with the efficient transfer of inter-subnet unicast data in a LANE environment. MPOA integrates LANE and NHRP to preserve the benefits of LAN Emulation, while allowing inter-subnet, internetwork layer protocol communication over ATM VCCs without requiring routers in the data path. MPOA provides a framework for effectively synthesizing bridging and routing with ATM in an environment of diverse protocols, network technologies, and IEEE 802.1 virtual LANs. MPOA is capable of using both routing and bridging information to locate the optimal exit from the ATM cloud.

The format of the header is shown in the following illustration:

8		16		24		32 bits	
ar$afn				ar$pro.type			
ar$pro.snap							
ar$pro.snap		ar$hopcnt		ar$pkstz			
ar$chksum				ar$extoff			
ar$op.version		ar$op.type		ar$shtl		ar$sstl	

MPOA header structure

ar$afn
Defines the type of "link layer" address being carried.

ar$pro.type
This field is a 16 bit unsigned integer.

ar$pro.snap
When ar$pro.type has a value of 0x0080, a snap encoded extension is being used to encode the protocol type. This snap extension is placed in the ar$pro.snap field; otherwise this field should be set to 0.

ar$hopcnt
The hop count. This indicates the maximum number of NHSs that an MPOA packet is allowed to traverse before being discarded.

ar$pktsz

The total length of the MPOA packet in octets.

ar$chksum

The standard IP checksum over the entire MPOA packet.

ar$extoff

This field identifies the existence and location of MPOA extensions.

ar$op.version

This field indicates what version of the generic address mapping and management protocol is represented by this message.

ar$op.type

The MPOA packet type. Possible values for packet types are:

128 MPOA Cache Imposition Request.
129 MPOA Cache Imposition Reply.
130 MPOA Egress Cache Purge Request.
131 MPOA Egress Cache Purge Reply.
132 MPOA Keep-Alive.
133 MPOA Trigger.
134 MPOA Resolution Request.
135 MPOA Resolution Reply.
5 MPOA Data Plane Purge.
6 MPOA Purge Reply.
7 MPOA Error Indication.

ar$shtl

The type and length of the source NBMA address interpreted in the context of the *address family number.*

ar$sstl

The type and length of the source NBMA subaddress interpreted in the context of the *address family number.*

4

ATM Encapsulation Methods

A number of standards have been developed which describe the encapsulation of LAN and WAN protocols over ATM. These standards allow users to integrate ATM into existing WANs and/or LANs, thus providing various cost-effective upgrade routes. This chapter describes the various encapsulation methods in general; specific information concerning the analysis or decoding of particular protocols may be found in the respective protocol chapters.

The following methods are used for encapsulation or transporting LAN or WAN protocols via ATM:

- VC-based Encapsulation: uses predefined VPI/VCI values for specific protocols. Used for WAN or LAN protocols.
- Multiprotocol Encapsulation over ATM Adaptation Layer 5 (IETF RFC1483): covers encapsulation of LAN protocols over ATM with the use of the LLC header.

- Classical IP and ARP over ATM (IETF RFC1577): covers ARP/RARP encapsulation.
- Frame Relay over ATM: describes transmission of Frame Relay via ATM.
- LAN Emulation: emulates Ethernet or Token Ring LANs.

VC-Based (null) Multiplexing

VC-based multiplexing uses one Virtual Channel (VCI/VPI pair) for each protocol. In this way, the protocol is carried directly over the AAL5 PDU. Since no additional information is added to the protocol, this is sometimes called null encapsulation.

For routed protocols e.g., TCP/IP, IPX, the PDU is carried directly in the payload of the AAL5 CPCS PDU. For bridged protocols e.g., Ethernet, Token Ring, FDDI, all fields following the PID field are carried in the payload of the AAL5 CPCS PDU.

Multiprotocol over ATM

IETF RFC 1483 http://www.cis.ohio-state.edu/htbin/rfc/rfc1483.html

One method of transmitting LAN protocols over ATM is by using LLC encapsulation. In this instance, the LLC and SNAP headers within the AAL5 PDU are used to identify the encapsulated protocol. This protocol stack is shown in the following illustration.

Upper-layer applications Upper-layer protocols
Ethernet or TCP/IP
SNAP
802.2 LLC
AAL5
ATM
Physical

LAN protocols encapsulated over ATM

In multiprotocol encapsulation, the PDUs are carried in the Payload field of the Common Part Convergence Sublayer (CPCS) PDU of ATM Adaptation Layer type 5 (AAL5). In this encapsulation method, the LLC header is always 0xAA-AA-03, which indicates that the LLC header is followed by a SNAP header.

DSAP	SSAP	Ctrl
AA	AA	03

LLC header

The format of the SNAP header is as follows:

Organizationally Unique Identifier (OUI)	Protocol Identifier (PID)
3 bytes	2 bytes

SNAP header

Non-ISO PDUs e.g., TCP/IP, IPX, are identified by an OUI value of 0x00-00-00, followed by the PID, which represents the 2-byte EtherType field. For example, an EtherType value of 0x08-00 identifies an IP PDU.

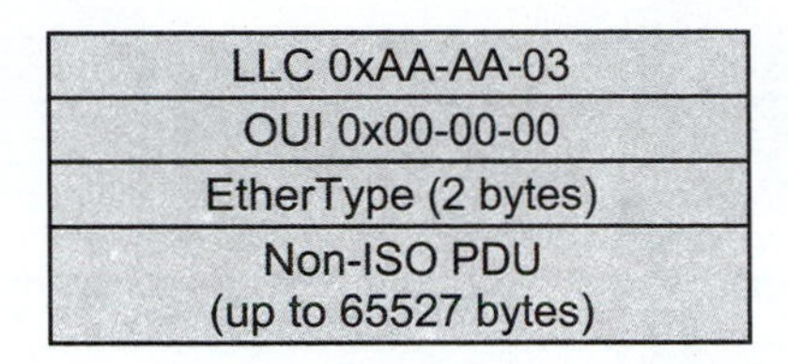

Payload format for routed non-ISO PDUs

Non-routed bridged protocols are identified by an OUI value of 0x00-80-C2 followed by the 2-byte PID which specifies the protocol which follows. The following table lists some possible values for the PID field in this instance.

PID Value . . .	Protocol . . .
0x00-01/0x00-07	Ethernet/802.3
0x00-03/0x00-09	Token Ring/802.5
0x00-04/0x00-0A	FDDI

The following diagrams represent the payload format for the Ethernet, Token Ring, FDDI and SMDS PDUs.

LLC	0xAA-AA-03
OUI	0x00-80-C2
PID	0x00-01 or 0x00-07
PAD	0x00-00
MAC destination address	
remainder of MAC frame	
LAN FCS (if PID is 0x00-01)	

Payload format for bridged Ethernet/802.3 PDUs

LLC	0xAA-AA-03
OUI	0x00-80-C2
PID	0x00-03 or 0x00-09
PAD	0x00-00-xx
Frame Control (1 byte)	
MAC destination address	
remainder of MAC frame	
LAN FCS (if PID is 0x00-03)	

Payload format for bridged Token Ring/802.5 PDUs

LLC 0xAA-AA-03
OUI 0x00-80-C2
PID 0x00-04 or 0x00-0A
PAD 0x00-00-00
Frame Control (1 byte)
MAC destination address
remainder of MAC frame
LAN FCS (if PID is 0x00-01)

Payload format for bridged FDDI PDUs

LLC 0xAA-AA-03		
OUI 0x00-80-C2		
PID 0x00-0B		
Reserved	BEtag	Common PDU Header
BAsize		
MAC destination address		
remainder of MAC frame		
LAN FCS (if PID is 0x00-01)		

Payload format for bridged SMDS PDUs

The following capture display in multi-protocol mode lists the LLC protocol followed by encapsulated IP.

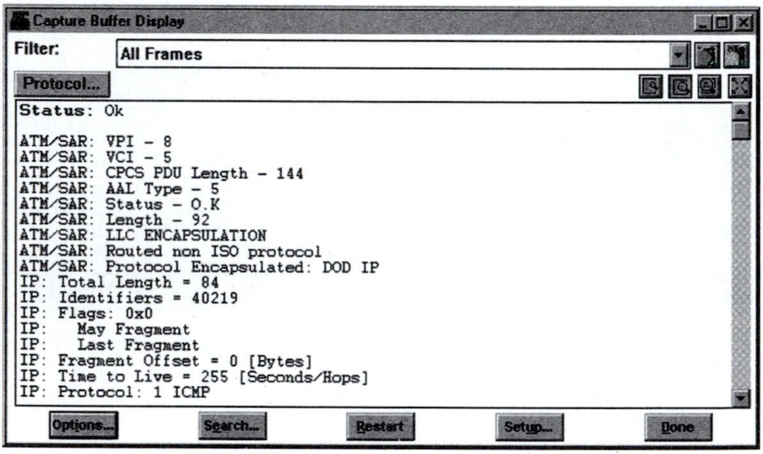

```
Capture Buffer Display
Filter: All Frames
Protocol...
Status: Ok

ATM/SAR: VPI - 8
ATM/SAR: VCI - 5
ATM/SAR: CPCS PDU Length - 144
ATM/SAR: AAL Type - 5
ATM/SAR: Status - O.K
ATM/SAR: Length - 92
ATM/SAR: LLC ENCAPSULATION
ATM/SAR: Routed non ISO protocol
ATM/SAR: Protocol Encapsulated: DOD IP
IP: Total Length = 84
IP: Identifiers = 40219
IP: Flags: 0x0
IP:    May Fragment
IP:    Last Fragment
IP: Fragment Offset = 0 [Bytes]
IP: Time to Live = 255 [Seconds/Hops]
IP: Protocol: 1 ICMP
Options...  Search...  Restart  Setup...  Done
```

Decode of LLC with encapsulated IP

IP Addressing over ATM

IETF RFC 1577 http://www.cis.ohio-state.edu/htbin/rfc/rfc1577.html

ATMARP is a version of ARP and InARP which is appropriate for ATM. In ATMARP the LLC and SNAP headers identify the encapsulated protocol as described above in Multiprotocol Encapsulation over ATM. The LLC header has the value 0xAA-AA-03, which indicates that the LLC header is followed by a SNAP header. The OUI value for ARP is 0x00-00-00, followed by the EtherType field of value 0x08-06.

Internet addresses are assigned independently of ATM addresses. Each host implementation must know its own IP and ATM address(es) and must respond to address resolution requests appropriately. IP members must also use ATMARP and InATMARP to resolve IP addresses to ATM addresses when needed.

The ATMARP PDU is shown in the following illustration.

8	16	24	32 bits
Hardware type		Protocol type	
Type and length of source ATM number	Type and length of source ATM subaddress	Operation	
Length of source protocol address	Type and length of target ATM number	Type and length of target ATM subaddress	Length of target protocol address
Source ATM number (q bytes)		Source ATM subaddress (r bytes)	
Source protocol address (s bytes) (IP=4 bytes)			
Target ATM number (x bytes)		Target ATM subaddress (y bytes)	
Target protocol address (z bytes) (IP=4 bytes)			

Payload format for ATMARP PDU

ATMARP and InATMARP use the same hardware type, protocol type and operation code data formats as ARP and InARP. The location of these fields within the ATMARP packet are in the same byte position as those in ARP and InARP packets. A unique hardware type value has been assigned for ATMARP.

Frame Relay over ATM

Frame Relay/ATM Network Internetworking (FRF.5)

FRF.5 describes network internetworking. This is applicable when 2 distant Frame Relay networks are connected over an ATM backbone. This function encapsulates the Frame Relay frames inside the ATM cells. It requires software that can be added to the ATM switch (or anywhere else between or in the ATM/Frame Relay networks).

Frame Relay/ATM Internetworking (FRF.8)

Mapping between ATM PDUs and Frame Relay PDUs requires examination of the incoming ATM AAL5 CPCS PDU payload header or Frame Relay Q.922 PDU payload header to determine the type, and then to overwrite the incoming header with the outgoing header.

The following diagram shows the translation between the Frame Relay Q.922 PDU payload header and the ATM AAL5 CPCS PDU payload header.

Control 0x03	Pad 0x00
NLPID 0x80	OUI 0x00
0x80-C2	
PID	
Remainder of MAC frame	

Frame Relay payload header

LLC 0xAA-AA	
0x03	OUI 0x00
0x80-C2	
PID	
Pad 0x00-00	
Remainder of MAC Frame	

ATM AAL5 CPCS PDU payload header

Refer to Frame Relay Forum FRF.8 for detailed information concerning the provisions for traffic management between the two protocols.

LAN Emulation

The LAN emulation protocol uses control messages to set up the LAN. LAN Emulation provides for two possible data packet formats: Ethernet and Token Ring. The LAN emulation data frames preserve all the information contained in the original 802.3 or 802.5 frames, but add a 2-byte LEC ID (the source ID), which is unique to each LEC.

Refer to Chapter 22 for a more detailed description of LAN Emulation.

5

Audio/Visual over ATM

A number of standards have been developed to assist in the transfer of audio and video signals over ATM. This chapter covers the following protocols:

- MPEG-2.
- DVB
- AVA
- DSM-CC.
- ATM-Circuit Emulation.

Additional information concerning using Audio Visual over ATM can be found in *ATM* section *AAL2*.

MPEG-2

ISO 13818-1 30/10/94

MPEG-2 is a generic method for compressed representation of video and audio sequences using a common coding syntax defined in the document ISO/IEC 13818 by the International Organization for Standardization. The MPEG-2 Video Standard specifies the coded bit stream for high-quality digital video. As a compatible extension, MPEG-2 Video builds on the completed MPEG-1 Video Standard (ISO/IEC IS 11172-2), by supporting interlaced video formats and a number of other advanced features, including support for applications such as Direct Broadcast Satellite, Cable Television and HDTV.

The ability of ATM to support voice, video and data simultaneously, makes it an excellent candidate for MPEG implementations. In December 1995, the ATM forum issued the Video on Demand (VoD) Specification 1.0, which specifies the implementation of MPEG-2 over ATM. This implementation supports the transport stream MPEG coding, using AAL5 for user data and the Signalling 4.0 stack for call control.

MPEG-2 Transport Stream Header Structure

The structure of the MPEG-2 Transport Stream header is shown in the following illustration.

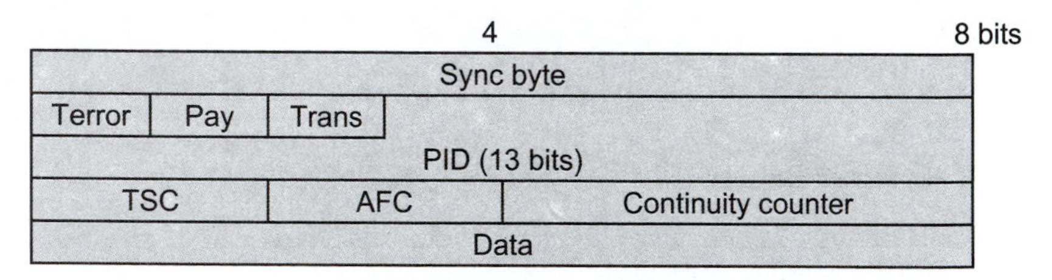

MPEG-2 Transport Stream header

Sync byte

Fixed 8-bit field with the value of 0100 0111.

TError

Transport error indicator. Indicates the presence of at least 1 uncorrectable bit error in the associated transport stream packet.

Pay

Payload unit start indicator. 1-bit flag with normative meaning for transport stream packets.

Trans

Transport priority. 1-bit priority of the packet compared to other packets of the same PID.

PID

13-bit field indicating the type of data stored in the packet payload.

TSC

Transport scrambling control. Indicates the scrambling mode of the Transport stream packet payload.

AFC

Adaptation field control. Indicates whether this transport stream packet header is followed by an adaptation field and/or payload.

Continuity counter

4-bit field incremented with each Transport Stream packet of the same PID.

Data byte

8-bit field containing data.

MPEG-2 Program Stream Header Structure

The structure of the MPEG-2 Program Stream header is shown in the following illustration:

Pack start code	32 bits
01	2 bits
System clock reference base	3 bits
Marker bit	1 bit
System clock reference base	15 bits
Marker bit	1 bit
System clock reference base	15 bits
Marker bit	1 bit
System clock reference	9 bits
Marker bit	1 bit
Program mux rate	22 bits
Marker bit	1 bit
Marker bit	1 bits
Reserved	5 bits
Pack stuffing length	3 bits
Stuffing byte	8 bits

MPEG-2 Program Stream header

Pack start code

The string 0X000001BA identifying the beginning of a pack.

System clock reference base

Indicates the intended time of arrival of the byte. Contains the last bit of the system clock reference base as the input of the program target decoder.

System clock reference extension field

Indicates the number of periods of a 27 MHz clock after a 90 kHz start.

Marker bit

1-bit field with the value 1.

Program mux rate

22 bit integer specifying the rate at which the P-STD receives the program stream during the pack in which it is included. This is measured in units of 50 bytes per second.

Pack stuffing rate

Number of stuff bytes following this field.

Stuffing byte

Fixed value that can be inserted by the encoder to meet the requirements of the channel (for example). It is discarded by the decoder.

DVB

ETS 800 300

Certain implementations suitable for Digital Video Broadcasting (DVB) broadcasting systems are supported by CATV infrastructures. Specifically, implementations of the Return Channel for interactive services are supported by CATV. DVB involves a standard link.

The format of the DVB packet is shown in the following illustration:

Mpegheader (4)	Upstream marker (3)	Slot number (2)	MAC flag control (3)
MAC flags (26)			
Ext. flags (26)			
MAC message (40)			
MAC message (40)			
MAC message (40)			
Rsrvc. (40)			

DVB packet structure

Mpeg header

4 byte Mpeg-2 transport stream header as defined in ISO 13818-1 with a specific PID designated for MAC messages. The value of this PID is 0 x 1C. The transport scrambling control field of the MPEG header is set to 00.

Upstream marker

24 bit field, 3 byte marker that provides upstream QPSK synchronization information. At least one packet with synchronization information must be sent in every period of 3 msec. The definition of the field is as follows:

- **Bit 0: Upstream Marker Enable:**
  Possible valuues
  1 Slot marker pointer is valid.
  0 Slot marker pointer is not valid.

- **Bits 1 - 3: MAC Message Framing** - Bit 1 relates to the first MAC message slot within the MPEG frame, bit 2 to the second MAC message within the MPEG frame, and bit 3 to the last MAC message within the MPEG frame. Possible values:
  0 - A MAC message terminates in this slot.
  1 - A MAC message continues from this slot into the next, or the slot is unused. If the slot is unused, the first two bytes of the slot are 0 x 0000.
- **Bits 4 - 7: Reserved**
- **Bits 8 - 23: Upstream Slot Marker Pointer** - A 16 bit unsigned integer which indicates the number of downstream "symbol" clocks between the next Sync byte and the next 3 msec time marker. Bit 23 is considered the most significant bit of this field.

Slot Number4

A 16 bit field which is defined as follows:

- **Bit 0: Slot Position Register Enable (msb)**
  Possible valuues
  1 Slot marker pointer is valid.
  0 Slot marker pointer is not valid.
- **Bits 1-3: Reserved**
- **Bit 4:** Set to the value '1.' This bit is equivalent to M12 in the case of OOB downstream.
- **Bit 5: Odd Parity** - This bit provides odd parity for upstream slot position register. It is equivalent to M11 in the case of OOB downstream.
- **Bits 6 - 15: Upstream Slot Position Register** - 10 bit counter which counts from 0 to n with bit 6 the msb. These bits are equivalent to M1 - M10 in the case of OOB downstream.

MAC flag control

24 bit field (b0 (msb), b1, b2 . . . b23) that provides control information used in conjunction with the 'MAC Flags' and 'Extension Flags' fields. The definition of the MAC Flag Control field is as follows:

- **b0 - b2** - Channel 0 control field.
- **b3 - b5** - Channel 1 control field.

- **b6 - b8** - Channel 2 control field.
- **b9 - b11** - Channel 3 control field.
- **b12 - b14** - Channel 4 control field.
- **b15 - b17** - Channel 5 control field.
- **b18 - b20** - Channel 6 control field.
- **b21 - b23** - Channel 7 control field.

MAC flags

26 byte field containing 8 slot configuration fields (24 bits each) which contain slot configuration information for the related upstream channels followed by two reserved bytes. The first 3 bytes correspond to MAC Flag Set 1, the second 3 bytes to MAC Flag Set 2, etc.

Ext. flags

A 26 byte field used when one or more 3.088 Mbit/s or 6.176 Mbit/s upstream QPSK links are used. The definition of the Extension Flags field is identical to the definition of the MAC Flags field (above). The Extension Flags field contains the MAC Flags from 9 to 16.

MAC message

The MAC Message field contains a 40 byte message in hexadecimal code.

Reserved field C (Rsrvc.)

Reserved Field C is a 4 byte field reserved for future use.

DSM-CC

ISO 13818-6 6/12/96

The Digital Storage Media Command and Control (DSM-CC) specification is a set of protocols which provides the control functions and operations specific to managing ISO/IEC 11172 (MPEG-1) and ISO/IEC 13818 (MPEG-2) bit streams. The concepts and protocols are, however, considered to apply to more general applications.

The format of the header is shown in the following illustration:

8	16 bits
Protocol discriminator	DSMCC type
Message ID	
Transaction ID (32 bits)	
Download ID (32 bits)	
Reserved	Adaptation length
Message length	

DSM-CC header structure

Protocol discriminator

This field indicates that the message is an MPEG-2 message.

Dsmcc type

MPEG-2 DSMCC type. Possible types are:
UN configuration
UN primitive
UU configuration
UU primitive.

Message ID

The message type.

Transaction ID

A field used for session integrity and error processing.

Download ID

An optional field replacing the transaction ID fields if the message type is a download message.

Reserved

A reserved field, the value of which is always set to zero.

Adaptation length

This field indicates the length of the adaptation part.

Message length

The length of the message including the adaptation part.

ATM Circuit Emulation

Circuit Emulation was developed to facilitate the transmission of constant bit rate (CBR) traffic over ATM networks. Since ATM is a packet- rather than circuit-oriented transmission technology, it must emulate circuit characteristics in order to provide support for CBR traffic. The goal of Circuit Emulation is to connect between CBR equipment across an ATM network, without the CBR equipment realizing it.

6

Banyan Protocols

The Banyan Network, known as VINES (Virtual Networking System), is based on the UNIX operating system. VINES uses UNIX multi-user, multi-tasking characteristics to internetwork LANs and WANs. The Banyan suite includes the following protocols:

- VARP: VINES Address Resolution Protocol.
- VIP: VINES Internet Protocol.
- ICP: Internet Control Protocol.
- RTP: Routing Update Protocol.
- IPC: InterProcess Communications Protocol.
- SPP: Sequenced Packet Protocol.
- NetRPC: NetRemote Procedure Call.
- StreetTalk.

The following diagram illustrates the Banyan protocol suite in relation to the OSI model:

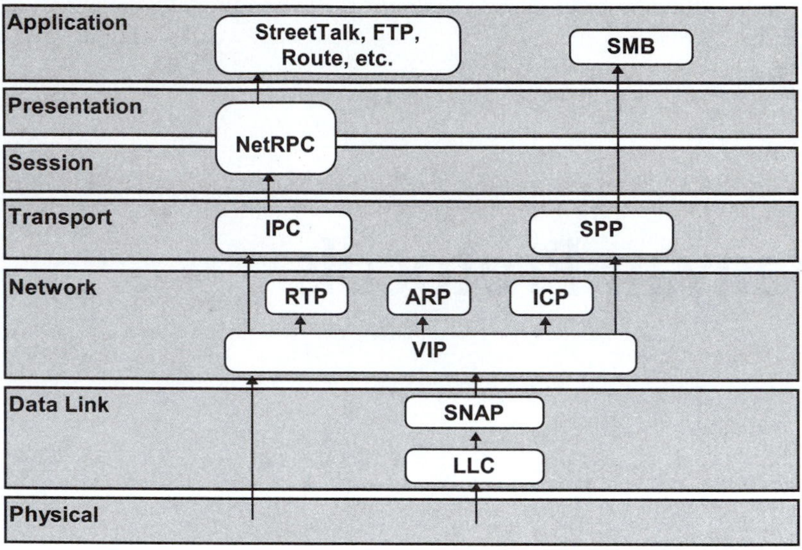

Banyan protocol suite in relation to the OSI model

VARP

The VINES Address Resolution Protocol (VARP), is used for finding node Data Link Control (DLC) addresses from the node IP address.

Frames

VARP frames can be one of the following types:

[service request]	Requests ARP service.
[service reply]	Acknowledges ARP service available.
[assign request]	Requests assignment of IP address.
[assign reply]	Assigns IP address.

Parameters

VARP packets have the following parameters:

Network

Network number of servers responding to ARP requests and the network number assigned to the station requesting an IP address.

Server serial number

Decimal equivalent of the server network number i.e., the server key number.

Subnet

Subnet number assigned to the system requesting a VINES IP address.

VIP

The VINES Internet Protocol (VIP) moves datagrams throughout the network.

Hop count

Maximum number of server hops that the packet can make before a server discards them. VIP decrements the hop count at each server (routing node).

Error flag

Error flag determines the action on routing errors. If set to one, an ICP error frame is generated if a routing error occurs with the packet.

Metric flag

When set to one, the destination server sends an ICP metric frame to report the routing cost to the destination end node.

Broadcast class

VIP uses the broadcast class with the hop count to determine the routing requirements of broadcast packets. Broadcast classes are as follows:

0	All reachable nodes regardless of cost.
1	All nodes reachable at moderate cost.
2	All nodes reachable at low cost.
3	All nodes on the LAN.
4	All reachable servers regardless of cost.
5	All servers reachable at moderate cost.
6	All servers reachable at low cost.
7	All servers on the LAN.

Destination Internet address

The VINES Internet address of the destination node, consists of an 8-digit hexadecimal network number and a 4-digit subnetwork or user number in the form XXXXXXXX.XXXX. VIP uses the subnetwork number 0x0001 for servers. Work stations have subnet numbers starting with 0x8000.

Source Internet address

The VINES Internet address of the source node given in the same form as the destination Internet address.

ICP

The Internet Control Protocol (ICP), is used to notify errors and changes in network topology.

ICP frames may contain the following parameters:

Cost
Routing cost in seconds to reach the specified destination node as given in ICP metric frames.

Communication error
The error message returned by ICP error frames. Possible messages are as follows:

{Invalid socket}	Specified socket invalid.
{Resource in use}	Resource already in use.
{Invalid operation}	Specified operation invalid.
{Bad MemAddr par}	Invalid memory address parameter.
{Dest unreachable}	Destination node unreachable.
{Message overflow}	Message overflow.
{Bad Dest socket}	Invalid destination socket.
{Bad Addr family}	Invalid address family.
{Bad socket type}	Specified socket does not exist.
{Bad protocol}	Protocol does not exist.
{No more sockets}	No more sockets available.
{No more buffers}	No buffer space available.
{Timed out}	Connection time out.
{Bad operation}	Unsupported operation.
{Resource unavail}	Resource unavailable.
{Comm failure}	Internal communication failure.
{H/W Reset failure}	Hardware controller reset failure.
{ARP error}	Internet address resolution error.
{User terminated}	User terminated request.
{Protocol reset}	Protocol reset occurred.
{Protocol discnct}	Protocol disconnect occurred.
{User aborted}	User aborted message.
{Resource discnct}	Resource disconnected.

RTP

The Routing Update Protocol (RTP) is used to distribute network topology.

Packets

RTP packets may be of the following types:

[router update]	Routing update from a router (server).
[endnode update]	Routing update from an end node (workstation).

Frame Parameters

RTP [router update] packets have the following parameters:

Routing table size

Number of entries in the routing table as returned by routing response packets. The routing entry for each known router is given in the form: XXXXXX(CC), where XXXXXXX is the server number and CC is the routing cost to reach the server in units of 0.2 seconds.

Host system type

The host system type may be as follows:

XT, MB	PC-XT class with multi-buffered LAN controller.
AT, SB	PC-AT class with single-buffered LAN controller.
AT, MB	PC-AT class with multi-buffered LAN controller.

Single-buffered LAN controllers use hardware/software that require this protocol to load and transmit each data block one at a time, while multi-buffered LAN controllers are capable of transmitting streams of data.

IPC

The InterProcess Communications (IPC) protocol provides both datagram and reliable message delivery service.

Frames

IPC frames may be one of the following types:

[data]	Bulk data transfer.
[error]	Transport layer error notification.
[detach]	Request to disconnect transport connection.
[probe]	Request for retransmission of missed frame.
[data ack]	Acknowledgment of data transfer.

Frame Parameters

IPC frames have the following parameters:

Source port

Message buffer interface used by the transport layer to access the transport protocol.

Destination port

Local destination port in use by the transport layer.

Sequence number

Numeric index used to track the order of frames transmitted across a virtual connection. Each direction of data flow across the virtual connection uses an independent set of sequence numbers.

Acknowledgement number

Last sequence number received from the other side of the virtual connection. For IPC error packets, the sequence number of the packet causing the error notification.

Source connection ID

Reference code used to identify the sending side of a virtual connection.

Destination connection ID
Reference code used to identify the receiving side of a virtual connection.

SPP

The Sequenced Packet Protocol (SPP) provides a reliable virtual connection service for private connections.

Frames

SPP frames may be one of the following types:

[detach]	Request to disconnect transport connection.
[probe]	Request for retransmission of missed frame.
[data ack]	Acknowledgment of data transfer.

Frame Parameters

SPP frames have the following parameters:

Source port
Message buffer interface used by the transport layer to access the transport protocol.

Destination port
Local destination port in use by the transport layer.

Sequence number
Numeric index used to track the order of frames transmitted across a virtual connection. Each direction of data flow across the virtual connection uses an independent set of sequence numbers.

Acknowledgement number
Last sequence number received from the other side of the virtual connection.

Source connection ID
Reference code used to identify the sending side of a virtual connection.

Destination connection ID
Reference code used to identify the receiving side of a virtual connection.

NetRPC

The NetRemote Procedure Call (NetRPC) protocol is used to access VINES applications such as StreetTalk and VINES Mail. A program number and version identify all VINES applications. Calls to VINES applications must specify the program number, program version, and the specific procedure within the program, where applicable.

Frames

NetRPC frames may be one of the following types:

[request]	Request from a VINES client.
[reply]	Response from a VINES application.

Frame Parameters

NetRPC frames can contain the following parameters:

Transaction ID
Code used to match NetRPC requests with NetRPC replies.

Program number
Code used to refer to the requested application.

Version number
Version number of the requested program.

Procedure number
Procedure number of the requested program.

Error status
Error status of the NetRPC reply.

StreetTalk

StreetTalk maintains a distributed directory of the names of network resources. In VINES, names are global across the Internet and independent of the network topology.

7

Bridge/Router Internetworking Protocols

There are a number of additional protocols which are generally used for bridge/router internetworking. These protocols are located in the Data Link Layer, and may carry encapsulated protocols in higher layers, e.g., IP, IPX, Ethernet and Token Ring.

The following bridge/router protocols are described in this book:

- Cisco Router.
- Cisco SRB.
- Cisco ISL.
- Cisco DRiP.
- CDP: Cisco Discovery Protocol.
- DISL: Dynamic Inter-Switch Link Protocol.
- MAPOS: Multiple Access Protocol over SONET/SDH.
- VTP: VLAN Trunk Protocol.
- RND.

- Wellfleet SRB.
- Wellfleet BOFL.
- BPDU.
- PPP including PPP Multilink, LCP, LQR, PAP, CHAP, IPCP, IPXCP, ATCP, BAP, BACP, BCP, PPP-BPDU, CCP, IPv6CP, SNACP, BVCP, NBFCP, DNCP, L2F, L2TP, ECP, OSINLCP, PPTP, and SDCP (refer to Chapter 25).
- Frame Relay (refer to Chapter 11).
- Cascade (refer to Chapter 11).
- Timeplex (BRE2) (refer to Chapter 11).
- IP over X.25 (refer to Chapter 30
- SMDS/DXI (refer to Chapter 26).

Cisco Router

The Cisco company produces communications equipment such as routers and bridges which use a proprietary protocol header (known as Cisco Router) to transfer LAN protocols via WAN.

Cisco Router's default encapsulation on synchronous serial lines uses HDLC framing with packet contents as defined in the following illustration:

Address	Control	Protocol code	Information
1 byte	1 byte	2 bytes	variable

Cisco router header structure

Address

Specifies the type of packet:

0x0F Unicast packets.

0x8F Broadcast packets.

Control

Always set to zero.

Protocol code

Specifies the encapsulated protocol. The Protocol code is usually Ethernet type codes; however, Cisco has added some codes to support packet types that do not appear in Ethernet.

Standard Ethernet values include:

0x0200 PUP.

0x0600 XNS.

0x0800 IP.

0x0804 Chaos.

0x0806 ARP.

0x0BAD Vines IP.

0x0BAF Vines Echo.

0x6003 DECnet phase IV.

0x8019 Apollo domain.

0x8035 Cisco SLARP.

0x8038 DEC bridge spanning tree protocol.

0x809B Apple EtherTalk.

0x80F3 AppleTalk ARP.

0x8137 Novell IPX.

Cisco-specific values include:

0x0808	Frame Relay ARP.
0x4242	IEEE bridge spanning protocol.
0x6558	Bridged Ethernet/802.3 packet.
0xFEFE	ISO CLNP/ISO ES-IS DSAP/SSAP.

Information

Higher-level protocol data.

Cisco SRB

Cisco uses a proprietary header in order to pass Token Ring information over WAN lines. This is known as Source Routing Bridging (SRB).

Cisco ISL

The Inter-Switch Link or ISL is used to inter-connect two VLAN capable Ethernet switches using the Ethernet MAC and Ethernet media. The packets on the ISL link contain a standard Ethernet, FDDI, or Token Ring frame and the VLAN information associated with that frame. Some additional information is also present in the frame.

The format of the header is shown in the following illustration:

Destination address (5 bytes)	
Frame type (1 byte)	
User type (1 byte)	
Source address (6 bytes)	
Length (2 bytes)	
SNAP LLC (3 bytes)	
HSA (3 bytes)	
Virtual LAN ID (15 bits)	BPDU (1)
Index (2 bytes)	
Reserved (2 bytes)	

Cisco router header structure

Destination address

The destination address field contains a 5 byte destination address.

Frame type

The frame type indicates the type of frame that is encapsulated. In the future, this could be used to indicate alternative encapsulations. The following Type codes are defined:

0000 Ethernet
0001 Token Ring
0010 FDDI
0011 ATM

User type

0 Normal priority.
1 Highest priority.

Source address

Source address of the ISL packet. It should be set to the 802.3 MAC address of the switch port transmitting the frame. It is a 48 bit value.

Length

A 16-bit field containing the length of the packet in bytes, not including the DA, T, U, SA, LEN and CRC fields. The total length of the fields excluded is 18 bytes so the length field is the total length minus 18 bytes.

HSA

High bits of source address field. Contains the upper 3 bytes of the SA field.

Virtual LAN ID

This is the virtual LAN ID of the packet. It is a 15-bit value that is used to distinguish frames on different VLANs. This field is often referred to as the color of the packet.

BPDU and CDP indicator

0 Not forwarded to the CPU for processing.
1 Forwarded to the CPU for processing.

Index

The index field indicates the port index of the source of the packet as it exits the switch. It is used for diagnostic purposes only and may be set to any value by other devices. It is a 16-bit value and ignored in received packets.

Reserved

A reserved field.

DRiP

Cisco Ios Release 11.3(4)T

The Cisco Duplicate Ring Protocol (DRiP) runs on Cisco routers and switches that support VLAN networking and is used to identify active Token Ring VLANs. A VLAN is a logical group of LAN segments with a common set of requirements. DRiP information is used for all-routes explorer filtering and detecting the configuration of duplicate TrCRFs, across routers and switches, which would cause a TrCRF (Token ring Concentrator Relay Function: a logical grouping of ports) to be distributed across ISL trunks. DRiP sends advertisements to a multicast address so the advertisements are received by all neighboring devices. The advertisement includes VLAN information for the source device only. The DRiP database in the router is initialized when TRISL (Cisco's Token ring Inter-Switch Link) encapsulation is configured, at least one TrBRF (Token ring Bridge Relay Function: a logical grouping of TrCRFs) is defined and the interface is configured for SRB (Source Route Bridging) or for routing with RIF.

When a switch receives a DRiP advertisement from a router, it compares the information in the advertisement with its local configuration to determine which TrCRFs have active ports and then denies any configuration that would allow a TrCRF that is already active on another box to be configured on the local switch. If there is a conflict between 2 identical TrCRFs, all ports attached to the conflicting TrCRFs are shut down in the switches and the router's ports remain active. A DRiP advertisement is sent every 30 seconds by the router.

DRiP is assigned the Cisco HDLC protocol type value 0x0102. A Cisco proprietary SNAP value is used. The following fields appear in the structure:

Version
The version number.

Code
The code number.

VLAN info count
The number of VLAN information elements.

VLAN 1... VLAN2...
Various VLAN information elements.

CDP

The Cisco Discovery Protocol (CDP) is a protocol for discovering devices on a network. Each CDP-compatible device sends periodic messages to a well-known multicast address. Devices discover each other by listening at that address.

CDP operation can be enabled or disabled on the FastHub through the object cdpInterfaceEnable. When enabled, the network management module (NMM) SNMP agent discovers neighboring devices and builds its local cache with information about these devices. A management workstation can retrieve this cache by sending SNMP requests to access the CDP MIB.

The format of the header is shown in the following illustration:

Version	TTL	Checksum	TLV	Type
1 byte	2 bytes	2 bytes	variable	2 bytes

CDP header structure

Version
The version of the protocol.

TTL
Time for this frame to live.

Checksum
Checks validity of previous fields.

TLV
Contains a type, length and value field.

Type
1 Device ID.
2 Address.
3 Port ID.
4 Capabilities.
5 Version.
6 Platform.
7 IP Prefix.

DISL

Dynamic Inter-Switch Link protocol (DISL) synchronizes the configuration of two interconnected Fast Ethernet interfaces into an ISL trunk. DISL minimizes VLAN trunk configuration procedures because only one end of a link needs to be configured as a trunk. DISL is a Cisco protocol.

The format of the header is shown in the following illustration.

Version	TLV	Type
1 byte	variable	2 bytes

DISL header structure

Version
The version of the protocol.

TLV
Contains a type, length and value field.

Type
1 Management Domain Name.
2 Status Field.

MAPOS

RFC 2171-6

The MAPOS (Multiple Access Protocol over SONET/SDH) protocol provides multiple access capability over SONET/SDH. It has the scalability of SONET/SDH and also provides a seamless network environment. MAPOS is connectionless, thus well suited for IP traffic. In addition it supports both broadcasts and multicasts. An efficient and simple forwarding mechanism makes it an excellent solution for high-speed networking at all levels: SONET LAN, SONET campus backbone, SONET Internet backbone, and Internet exchange using SONET WAN. MAPOS uses an HDLC-like framing. MAPOS supports a wide range of line rates from 155 Mbps to 10 Gbps, with potential for higher rates in the future.

The fields are transmitted from left to right.

MAPOS frame format

Flag 01111110	Address (16 bits)	Protocol (16 bits)
Information	FCS (16/32 bits)	Flag 01111110

MAPOS frame structure

Flag sequence
Flag sequence is used for frame synchronization. Each frame begins and ends with a flag sequence.

Address
This field contains the destination HDLC address.

Protocol
The protocol field indicates the protocol to which the datagram encapsulated in the information field belongs; for example, 0xFE01 is ARP and 0x0021 is IP.

Information
The information field contains the datagram for the protocol specified in the protocol field.

FCS (frame check sequence)

This is 16 bits long (but may be 32). It is calculated over all bits of the address protocol and information fields.

Note on Interframe fill:

A sending station continuously transmits the flag sequence as Inter-frame fills after the FCS field. The inter-frame flag sequences is silently discarded by the receiving station. When an under-run occurs during DMA in the sending station, it aborts the frame transfer and continuously sends the flag sequence to indicate the error.

VTP

The VLAN Trunk Protocol (VTP) provides for each (router or LAN-switch) device to transmit advertisements in frames on its trunk ports. These advertisement frames are sent to a multicast address to be received by all neighboring devices (but they are not forwarded by normal bridging procedures). Such an advertisement lists the sending device's management domain, its configuration revision number, the VLANs which it knows about, and certain parameters for each known VLAN. By hearing these advertisements, all devices in the same management domain learn about any new VLANs now configured in the transmitting device. Using this method, a new VLAN needs to be created or configured on only one device in the management domain, and the information is automatically learned by all the other devices in the same management domain. VTP is a Cisco protocol.

The format of the VTP packet is shown in the following illustration:

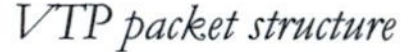

VTP packet structure

Version

The version of the protocol.

Type

1 Summary- Advert Message.
2 Subset- Advert Message.
3 Advert-Request Message.

RND

RND Layer 2 Router Package Description 1993-04 1.00

The RND company produces communications equipment such as routers and bridges. The company uses a proprietary protocol header (known as RND) to transfer LAN protocols via WAN.

The structure of the RND header is shown in the following illustration:

8	16 bits
MPCC	
Destination bridge ID	Destination bridge entity
Source bridge ID	Source bridge entity
Message broadcast ID	Message broadcast bridge
Cost	
Routing flag	Link count
Data length	

RND header structure

MPCC

Specifies normal case or swap bytes case.

Destination bridge ID

Specifies the type of message:

0xF4 IPX router message.
0xF4 DECnet router message.
0xF7 IP router message.
0xF8 TRE management message.
0xF9 ETE management message.
0xFB Routing message bridge ID.
0xFC This bridge entity.
0xFD Channel status message.
0xFE Common LAN bridge ID.
0xFF Broadcast bridge ID.

Destination bridge entity

Value of the entity:

0x0F LAN broadcast entity.
0x64 Smap entity.
0x65 Reml entity.
0x6F C5 reml entity.
0x79 RS232 rem entity.

Source bridge ID

Refer to destination bridge ID above.

Source bridge entity

Refer to destination bridge entity above.

Message broadcast ID

0 Point-to-point.

Message broadcast bridge

0 Point-to-point.

Cost

Accumulated cost.

Routing flag

Routing attributes.

Link count

Count of router hops.

Data length

Length of the data in bytes (swapped).

Wellfleet SRB

Wellfleet, which is known today as Bay Networks, is a manufacturer of routers and bridges. They use a proprietary header in order to pass Token Ring information over WAN lines. This is known as Source Routing Bridging (SRB).

Destination	Source	Route information	LLC
6 bytes	6 bytes	variable & optional	variable

Wellfleet SRB structure

Destination

The address structure is as follows:

I/G	U/L	Address bits

Wellfleet destination address structure

I/G Individual/group address may be:
0 Individual address.
1 Group address.

U/L Universal/local address may be:
0 Universally administered.
1 Locally administered.

Source

The address structure is as follows:

RII	I/G	Address bits

Wellfleet source address structure

RII Routing information indicator:
0 RI absent.
1 RI present.

I/G Individual/group address:
0 Group address.
1 Individual address.

Route information

The structure is as follows:

← RI Field →

← RC Field → ← RD Fields →

RT	LTH	D	LF	r	RD1	RD2	...	RDn	
3	5	1	6	1	16	16		16	bits

← Length in LTH Field →

Wellfleet route information structure

RC Routing control (16 bits).

RDn Route descriptor.

RT Routing type.

LTH Length.

D Direction bit.

LF Largest frame.

r Reserved.

Wellfleet BOFL

The Wellfleet Breath of Life (BOFL) protocol is used as a line sensing protocol on:

- Ethernet LANs to detect transmitter jams.
- Synchronous lines running WFLT STD protocols to determine if the line is up.
- Dial backup PPP lines.

The frame format of Wellfleet BOFL is shown following the Ethernet header in the following illustration:

Destination	Source	8102	PDU	Sequence	Padding	
6	6	2	4	4	n	bytes

←— Ethernet Header —→

Wellfleet BOFL structure

Destination
6-byte destination address.

Source
6-byte source address.

8102
EtherType (0x8102 for Wellfleet BOFL frames).

PDU
PDU field normally equals 0x01010000, but may equal 0x01011111 in some new releases on synchronous links.

Sequence
4-byte sequence field is an incremental counter.

Padding
Padding to fill out the frame to 64 bytes.

BPDU

IEEE 802.3D 1993-07, IEEE 802.1P 199

Bridge Protocol Data Unit (BPDU) is the IEEE 802.1d MAC Bridge Management protocol which is the standard implementation of STP (Spanning Tree Protocol). It uses the STP algorithm to insure that physical loops in the network topology do not result in logical looping of network traffic. Using one bridge configured as root for reference, the BPDU switches one of two bridges forming a network loop into standby mode, so that only one side of a potential loop passes traffic. By examining frequent 802.1d configuration updates, a bridge in the standby mode can switch automatically into the forward mode if the other bridge forming the loop fails.

The structure of the Configuration BPDU is shown in the following illustration:

Field	Octets
Protocol identifier	1-2
Protocol version identifier	3
BPDU type	4
Flags	5
Root identifier	6-13
Root path cost	14-17
Bridge identifier	18-25
Port identifier	26-27
Message age	28-29
Max age	30-31
Hello time	32-33
Forward delay	34-35

Configuration BPDU structure

Protocol identifier

Identifies the spanning tree algorithm and protocol.

Protocol version identifier

Identifies the protocol version.

BPDU type

Identifies the BPDU type: 00000000=Configuration, 10000000=Topology change notification. For the later type, no further fields are present.

Flags

Bit 8 is the Topology Change Acknowledgement flag.
Bit 1 is the Topology Change flag.

Root path cost

Unsigned binary number which is a multiple of arbitrary cost units.

Bridge identifier

Unsigned binary number used for priority designation (lesser number denotes the bridge of the higher priority).

Port identifier

Unsigned binary number used as port priority (lesser number denotes higher priority).

Message age, Max age, Hello time, Forward delay

These are 4 timer values encoded in 2 octets. Each represents an unsigned binary number multiplied by a unit of time of 1/256 of a second. Thus times range from 0 to 256 seconds.

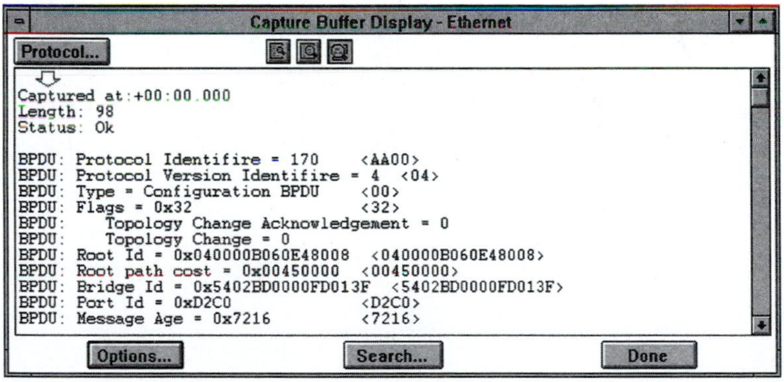

```
Capture Buffer Display - Ethernet
Protocol...
Captured at:+00:00.000
Length: 98
Status: Ok

BPDU: Protocol Identifire = 170      <AA00>
BPDU: Protocol Version Identifire = 4  <04>
BPDU: Type = Configuration BPDU      <00>
BPDU: Flags = 0x32                   <32>
BPDU:    Topology Change Acknowledgement = 0
BPDU:    Topology Change = 0
BPDU: Root Id = 0x040000B060E48008  <040000B060E48008>
BPDU: Root path cost = 0x00450000  <00450000>
BPDU: Bridge Id = 0x5402BD0000FD013F  <5402BD0000FD013F>
BPDU: Port Id = 0xD2C0               <D2C0>
BPDU: Message Age = 0x7216           <7216>
Options...          Search...          Done
```

BPDU decode

8

CDPD Protocols

The basic structuring of the Cellular Digital Packet Data (CDPD) Network is along the lines of the 7-layer OSI model. Each layer within CDPD may be further partitioned into a similar sequence of sub-layers. Each layer or sub-layer in the CDPD network communications architecture is defined with:

- Layer service access points.
- Layer service primitives.
- Layer protocol.
- Layer management entity.

The CDPD network specifications define a number of subprofiles as building blocks that may be selected and combined to define a particular CDPD network element. Subprofiles define the specific multi-layer protocol requirements for a CDPD network element or a CDPD network service. Three major classes of subprofiles are defined:

- Application subprofiles.
- Lower layer subprofiles.
- Subnetwork subprofiles.

The following diagram illustrates the CDPD protocols in relation to the OSI model:

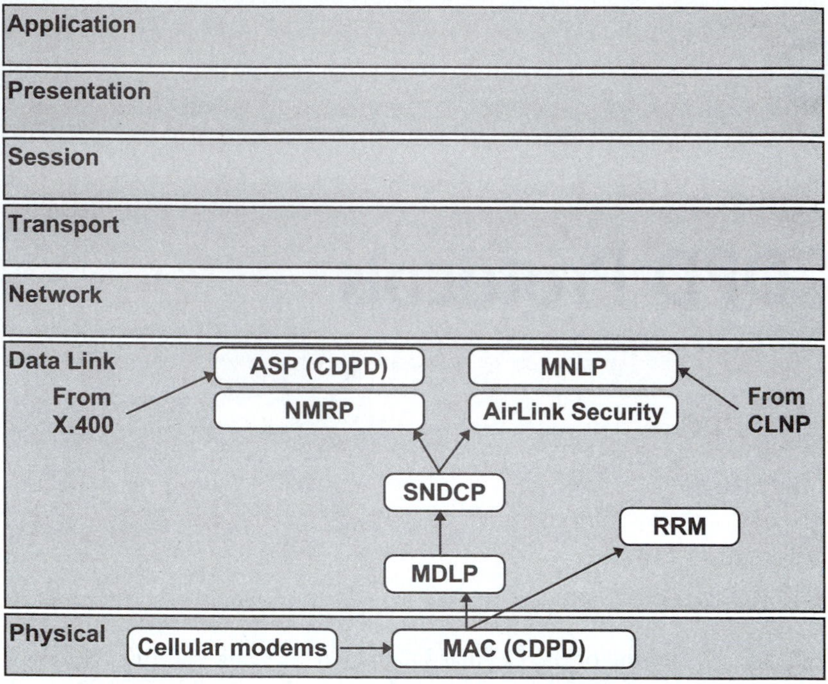

Application
Presentation
Session
Transport
Network
Data Link
ASP (CDPD)
MNLP
From X.400
NMRP
AirLink Security
From CLNP
SNDCP
RRM
MDLP
Physical
Cellular modems
MAC (CDPD)

CDPD protocols in relation to the OSI model

MDLP

CDPD System Specification release 1.1, part 403

The Mobile Data Link Protocol (MDLP) is a protocol that operates within the data link layer of the OSI model to provide logical link control services between Mobile End Systems (M-ESs) and Mobile Data Intermediate Systems (MD-ISs).

MDLP utilizes the services of the CDPD MAC layer to provide access to the physical channel and transparent transfer of link-layer frames between data link layer entities.

The purpose of MDLP is to convey information between network layer entities across the CDPD Airlink interface. It supports multiple M-ESs sharing access to a single channel stream. The channel stream topology is that of a point-to-multipoint subnetwork. In such a subnetwork, direct communication is possible only between the user side and the network side of the channel stream. Direct communication between two M-ESs on the same channel stream is not possible.

The frame format of MDLP is as shown in the following illustration:

Address (1-4 octets)
Control (1-2 octets)
Information (optional)

MDLP frame structure

Address

Variable number of octets as shown in the following illustration:

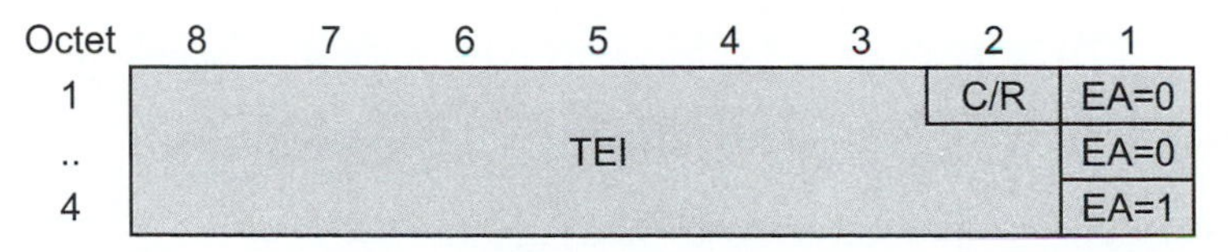

Address field structure

C/R

Command/response field bit identifies a frame as either a command or a response. The user side sends commands with the C/R bit set to 0 and responses with the C/R bit set to 1. The network side does the opposite.

TEI

Temporary equipment identifier. The TEI for a point-to-point data link connection is associated with a single M-ES. An M-ES may contain one TEI used for point-to-point data transfer. The TEI for a broadcast data link connection is associated with all user side data link layer entities. Values are encoded as unsigned binary numbers in a variable length field of a maximum 27 bits in length.

Control

Identifies the type of frame. Possible types are:

I Numbered information transfer
S Supervisory functions
U Unnumbered information transfers and control functions.

Information

Integer number of octets containing the data.

SNDCP

CDPD System Specification release 1.1, part 404

The Subnetwork Dependent Convergence Protocol (SNDCP) provides a number of services to the network layer:

- Connectionless-mode subnetwork service.
- Transparent transfer of a minimum number of octets of user data.
- User data confidentiality.

The SN-Data PDU is conveyed over the acknowledged data link service in the DL-Userdata field of a DL-Data primitive. The format of the SN-Data PDU is as shown in the following illustration:

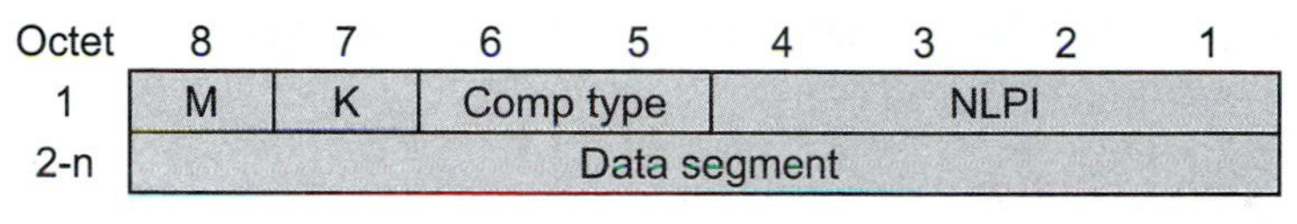

Octet	8	7	6	5	4	3	2	1
1	M	K	Comp type		NLPI			
2-n	Data segment							

SN-Data PDU structure

The SN-Unitdata PDU is conveyed over the unacknowledged data link service in the DL-Userdata field of a DL-Unitdata primitive. The format of the SN-Unitdata PDU is shown in the following illustration:

Octet	8	7	6	5	4	3	2	1
1	M	Reserved			NLPI			
2	Sequence ID				Segment number			
3-n	Data segment							

SN-Unitdata PDU structure

M
More segments bit. When set to 0, the current SN-Data PDU is the last data unit in a complete SN-Data PDU sequence.

K
Key sequence number. Indicates the parity of the encryption/decryption key used to encrypt the data segment field of the SN-Data PDU.

Comp type
Compression type field indicates the Network Layer header compression frame type. This field has meaning only in the first PDUI of a complete SN-

Data PDU sequence, but is copied unchanged into all PDUs in the sequence.

NLPI

Network layer protocol identifier, identifies the associated network layer protocol entities defined as follows:

0 Mobile Network Registration Protocol
1 Security Management Entity
2 CLNP
3 IP
4-15 Reserved for future use

Sequence ID

Identifies the subnetwork service data unit (SNSDU) to which the segment contained in this PDU belongs. All segments belonging to the same SNSDU have the same sequence identifier.

Segment number

Each segment is assigned a segment number, which is sequentially assigned starting from zero. A complete sequence of SN-Unitdata PDUs can consist of 1 to 16 consecutive segments.

Data segment

Exactly one segment of the subnetwork service data unit. The maximum size of a data segment is 128 octets.

9

Cellular Protocols

GSM

In 1989, the European Telecommunication Standards Institute (ETSI) took over responsibility for GSM. Phase I of the GSM specifications were published in 1990, commercial service was started in mid-1991, and by 1993 there were 36 GSM networks in 22 countries. In addition to European countries, South Africa, Australia, and many Middle and Far East countries have chosen GSM. At the beginning of 1994, there were 1.3 million subscribers worldwide. The acronym GSM now aptly stands for Global System for Mobile telecommunications.

GSM was intended to be compatible with ISDN in terms of services offered and control signalling used. However, the standard ISDN bit rate of 64 Kbps could not be practically achieved due to the limitations of the radio link. The digital nature of GSM allows data, both synchronous and asynchronous, to be transported as a bearer service to or from an ISDN terminal. The data rates supported by GSM are 300, 600, 1200, 2400, and 9600 bps.

The most basic teleservice supported by GSM is telephony. A unique feature of GSM compared to older analog systems is the Short Message Service (SMS).

Supplementary services are provided on top of teleservices or bearer services, and include features such as international roaming, caller identification, call forwarding, call waiting, multi-party conversations and barring of outgoing (international) calls, among others.

The following diagram illustrates the structure of the GSM protocol family:

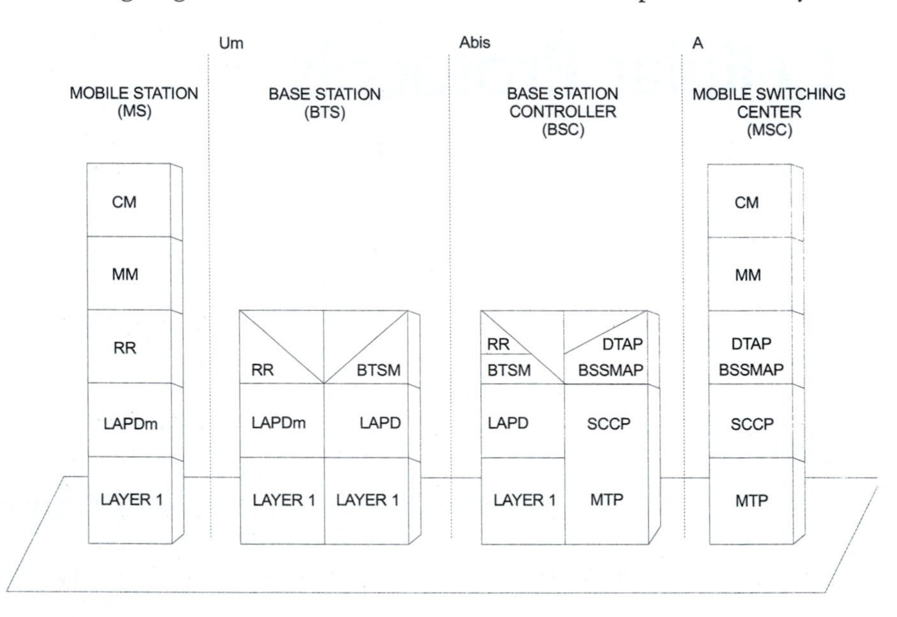

GSM protocol family structure

CDMA

Code Division Multiple Access (CDMA) is a digital air interface standard, claiming eight to fifteen times the capacity of traditional analog cellular systems. It employs a commercial adaptation of a military spread-spectrum technology. Based on spread spectrum theory, it gives essentially the same services and qualities as wireline service. The primary difference is that access to the local exchange carrier (LEC) is provided via a wireless phone.

Though CDMA's application in cellular telephony is relatively new, it is not a new technology. CDMA has been used in many military applications, such as:

- Anti-jamming (because of the spread signal, it is difficult to jam or interfere with a CDMA signal).
- Ranging (measuring the distance of the transmission to know when it will be received).
- Secure communications (the spread spectrum signal is very hard to detect).

CDMA is a spread spectrum technology, which means that it spreads the information contained in a particular signal of interest over a much greater bandwidth than the original signal. With CDMA, unique digital codes, rather than separate RF frequencies or channels, are used to differentiate subscribers. The codes are shared by both the mobile station (cellular phone) and the base station, and are called pseudo-random code sequences. Since each user is separated by a unique code, all users can share the same frequency band (range of radio spectrum). This gives many unique advantages to the CDMA technique over other RF techniques in cellular communication.

CDMA is a digital multiple access technique and this cellular aspect of the protocol is specified by the Telecommunications Industry Association (TIA) as IS-95.

In CDMA, the BSSAP is divided into the DTAP and BSMAP (which corresponds to BSSMAP in GSM). The structure of CDMA is shown in the following illustration:

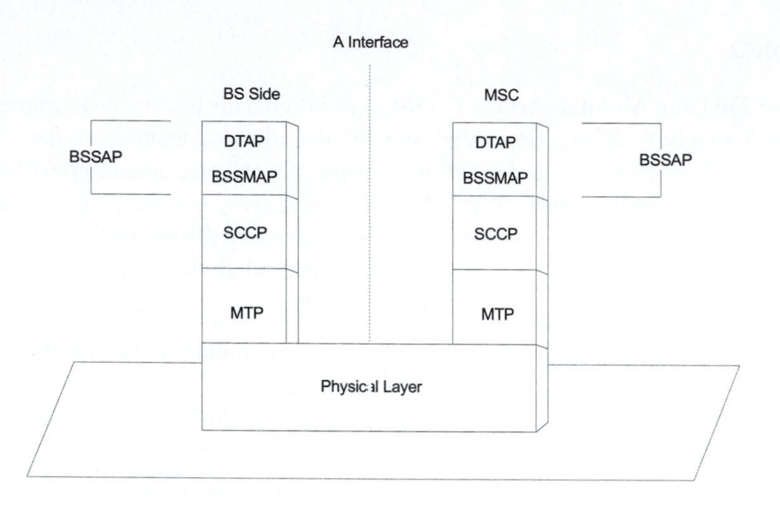

CDMA protocol structure

The following cellular protocols are described in this chapter:

- BSSAP - BSS Application Part.
- BSSMAP - BSS Management Application Part.
- DTAP (GSM) - Direct Transfer Application sub-Part.
- BTSM - Base Station Controller to Base Transceiver Station.
- BSMAP - Base Station Management Application Part.
- DTAP (CDMA) - Direct Transfer Application sub-Part.
- RR - Radio Resource.
- MM - Mobility Management.
- CC - Call Control.
- SMS - Short Message Service.

The following diagram illustrates the GSM and CDMA protocols in relation to other telephony protocols and the OSI model:

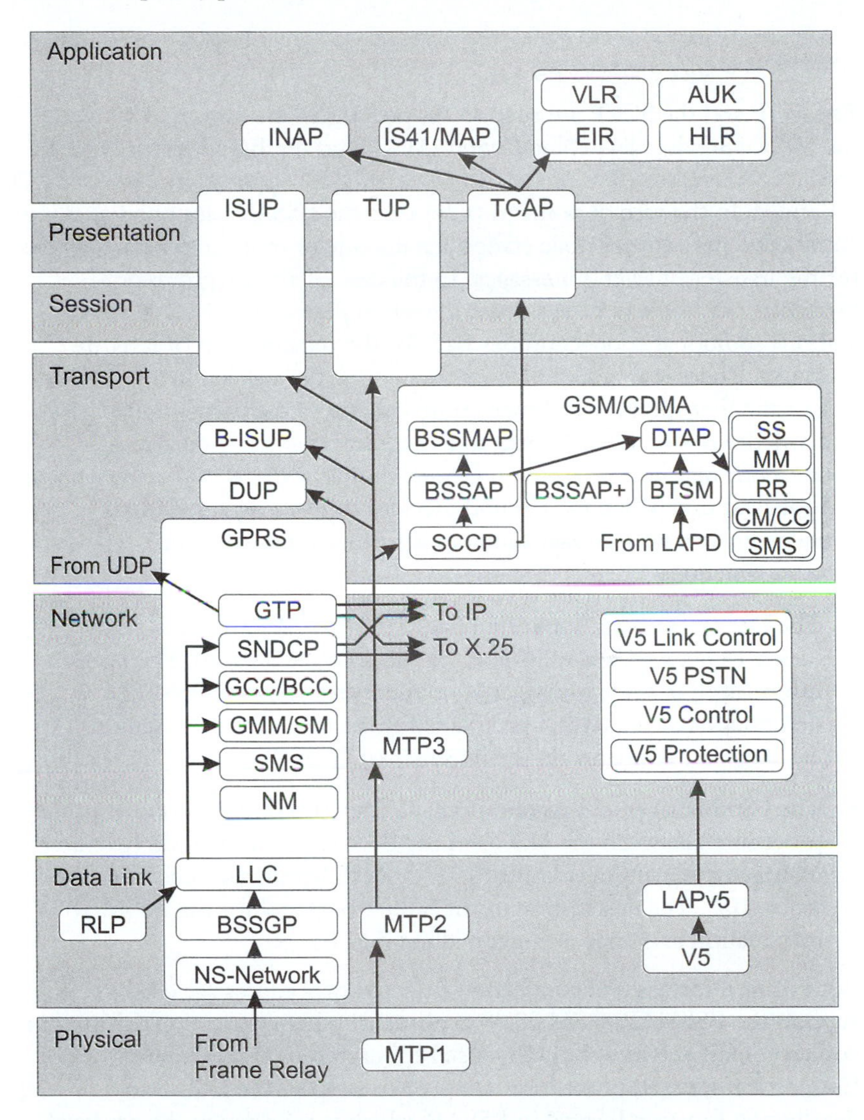

Telephony protocol suite in relation to the OSI model

BSSAP

GSM 08.06 http://www.etsi.org

The MTP and the SCCP are used to support signalling messages between the Mobile Services Switching Center (MSC) and the Base Station System (BSS). One user function of the SCCP, called BSS Application Part (BSSAP) is defined. In the case of point-to-point calls the BSSAP uses one signalling connection per active mobile station having one or more active transactions for the transfer of layer 3 messages. In the case of a voice group or a broadcast call, there is always one connection per cell involved in the call and one additional connection per BSS for the transmission of layer 3 messages. There is an additional connection for the speaker in a broadcast call, or the first speaker in a voice group call, up to the point at which the network decides to transfer them to a common channel. Additional connections may also be required for any mobile stations in the voice group or broadcast call, which the network decides to place on a dedicated connection. The BSSAP user function is further subdivided into two separate functions:

- The Direct Transfer Application sub-Part (DTAP) is used to transfer messages between the MSC and the MS (Mobile Station). The layer-3 information in these messages is not interpreted by the BSS. The descriptions of the layer 3 protocols for the MS-MSC information exchange are contained in the 04 series of GSM Technical Specifications.
- The BSS Management Application sub-Part (BSSMAP) supports other procedures between the MSC and the BSS, related to the MS (resource management, handover control), or to a cell within the BSS, or to the whole BSS. The description of the layer 3 protocol for the BSSMAP information exchange is contained in GSM 08.08.

Both connectionless and connection-oriented procedures are used to support the BSSMAP. GSM 08.08 explains whether connection oriented or connectionless services should be used for each layer 3 procedure. Connection oriented procedures are used to support the DTAP. A distribution function located in BSSAP, which is reflected in the protocol specification by the layer 3 header, performs the discrimination between the data related to those two subparts.

The format of the BSSAP header is shown in the following illustration:

1 byte	1 byte	
Discrimination	DLCI	Length

BSSAP header structure

Discrimination
Discriminates between the 2 sub-protocols: BSSMAP and DTAP.

DLCI
Only used for DTAP. Used in MSC to BSS messages to indicate the type of origination data link connection over the radio interface.

Length
Subsequent layer 3 message parameter length.

BSSMAP

GSM 08.08 http://www.etsi.org

The BSS Management Application Part (BSSMAP) supports all of the procedures between the MSC and the BSS that require interpretation and processing of information related to single call and resource management. Some of the BSSMAP procedures result in, or are triggered by, Radio Resource (RR) management messages defined in GSM 04.08.

The format of the header is shown in the following illustration:

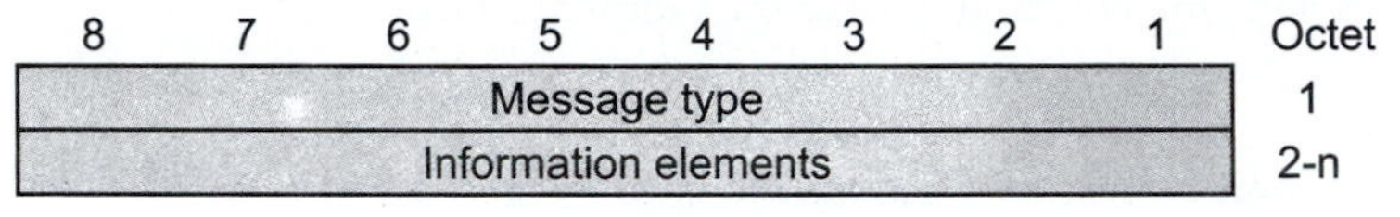

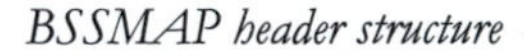
BSSMAP header structure

Message type

Mandatory, one-octet field defining the message type. The message type code uniquely defines the function and format of each BSSMAP message.

Information elements

Each IE has an identifier which is coded as a single octet. The length of an IE may be fixed or variable and may or may not include a length indicator.

DTAP (GSM)

GSM 04.08, 08.06, 08.08 http://www.etsi.org

The Direct Transfer Application Part (DTAP) is used to transfer call control and mobility management messages between the MSC and the MS. The DTAP information in these messages is not interpreted by the BSS. Messages received from the MS are identified as DTAP by the Protocol Discriminator Information Element. The majority of radio interface messages are transferred across the BSS MSC interface by DTAP, except for messages belonging to the Radio Resource (RR) management protocol.

The DTAP function is in charge of transferring layer 3 messages from the MS (or from the MSC) to the MSC (or to the MS) without any analysis of the message contents. The interworking between the layer 2 protocol on the radio side and signalling system 7 at the landside is based on the use of individual SCCP connections for each MS and on the distribution function.

The format of the DTAP header is shown in the following illustration:

8	7	6	5	4	3	2	1	Octet
Protocol discriminator				Transaction / skip				1
0	N(SD)	Message type						2
Information elements								3-n

GSM L3 header structure

Protocol discriminator

Identifies the layer 3 protocol to which the standard layer 3 message belongs. Values may be as follows:

0000 Group call control
0001 Broadcast call control
0010 PDSS1
0011 Call control; call related SS messages
0100 PDSS2
0101 Mobility Management Messages
0110 Radio resources management messages
1001 SMS messages
1011 Non-call related SS messages
1110 Extension of the PD to one octet length
1111 Tests procedures described in TS GSM 11.10

Transaction identifier / skip indicator

Either a transaction identifier, or a skip indictor, depending on the level 3 protocol. The transaction identifier contains the transaction value and flag which identifies who allocated the TI.

N(SD)

For MM and CM, N(SD) is set to the value of the send state variable. In other level 3 messages, bit 7 is set to 0 by the sending side. Messages received with bit 7 set to 1 are ignored.

Message type

Uniquely defines the function and format of each GSM L3 message. The message type is mandatory for all messages. The meaning of the message type is therefore dependent on the protocol (the same value may have different meanings in different protocols) and direction (the same value may have different meanings in the same protocol, when sent from the Mobile Station to the network and when sent from the network to the Mobile Station).

Information elements

The message type may be followed by various information elements depending on the protocol.

BTSM

GSM 08.58 http://www.etsi.org

BTSM is the Base Station Controller to Base Transceiver Station (BSC - BTS) interface protocol (the A-bis interface). BTSM allows sending messages between the Base Station Controller and the Base Transceiver Station. Protocol messages consist of a series of information elements. For each message there are mandatory information elements and optional information elements. BTSM messages are transmitted on the A-bis interface using the I format of LAPD, except for the Measurement Result message which is sent in UI format.

The structure of BTSM messages is shown in the following diagram:

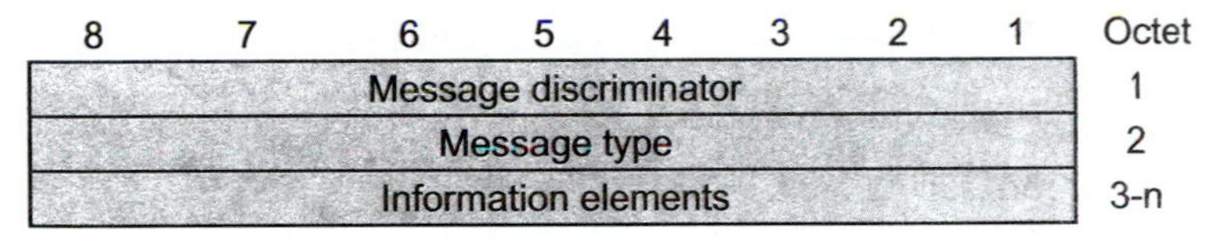

BTSM message structure

Message discriminator

1-octet field used in all messages to discriminate between Transparent and Non-Transparent messages and also between Radio Link Layer Management, Dedicated Channel Management, Common Channel Management and TRX Management messages. The format of the message discriminator is as follows:

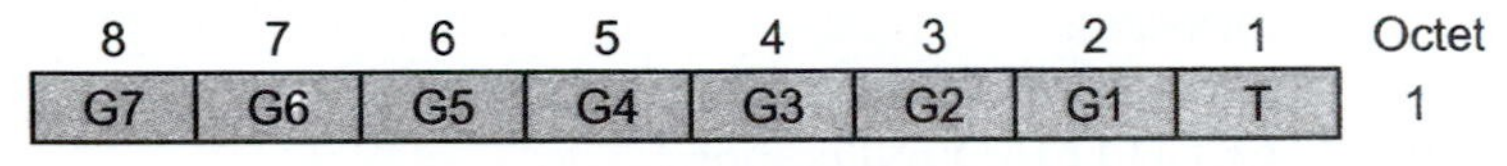

Message discriminator structure

The T-bit is set to 1 to indicate that the message is to be/was considered transparent by BTS. All other messages have the T-bit set to 0. The G-bits are used to group the messages as follows:

G7-G1	*Message group*
0 0 0 0 0 0 0	Reserved
0 0 0 0 0 0 1	Radio Link Layer Management messages
0 0 0 0 1 0 0	Dedicated Channel Management messages

0 0 0 0 1 1 0 Common Channel Management messages
0 0 0 1 0 0 0 TRX Management messages

All other values are reserved for future use.

Message type

Uniquely identifies the function of the message being sent. It is a single octet and coded as follows:

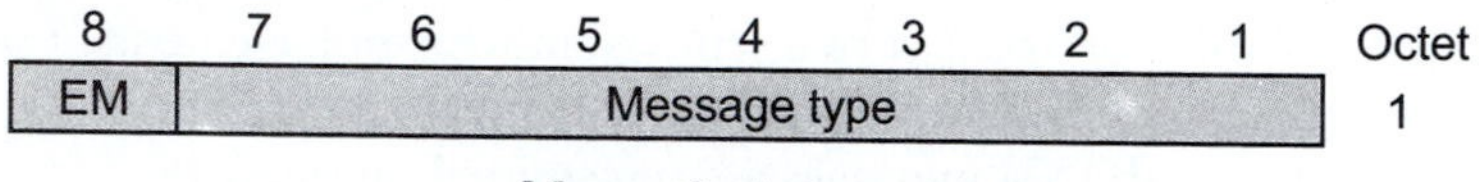

Message discriminator structure

Bit 8 is the extension bit and is reserved for future use. The following message types are used (all other values are reserved):

0000	xxxx	Radio Link Layer Management messages:
	0001	DATA REQuest
	0010	DATA INDication
	0011	ERROR INDication
	0100	ESTablish REQuest
	0101	ESTablish CONFirm
	0110	ESTablish INDication
	0111	RELease REQuest
	1000	RELease CONFirm
	1001	RELease INDication
	1010	UNIT DATA REQuest
	1011	UNIT DATA INDication
0001	xxxx	Common Channel Management/TRX Management messages:
	0001	BCCH INFOrmation
	0010	CCCH LOAD INDication
	0011	CHANnel ReQuireD
	0100	DELETE INDication
	0101	PAGING CoMmanD
	0110	IMMEDIATE ASSIGN COMMAND
	0111	SMS BroadCast REQuest
	1001	RF RESource INDication
	1010	SACCH FILLing
	1011	OVERLOAD
	1100	ERROR REPORT

	1101	SMS BroadCast CoMmanD
	1110	CBCH LOAD INDication
	1111	NOTification CoMmanD
001	xxxxx	Dedicated Channel Management messages:
	00001	CHANnel ACTIVation
	00010	CHANnel ACTIVation ACKnowledge
	00011	CHANnel ACTIVation Negative ACK
	00100	CONNection FAILure
	00101	DEACTIVATE SACCH
	00110	ENCRyption CoMmanD
	00111	HANDOver DETection
	01000	MEASurement RESult
	01001	MODE MODIFY REQuest
	01010	MODE MODIFY ACKnowledge
	01011	MODE MODIFY Negative ACKnowledge
	01100	PHYsical CONTEXT REQuest
	01101	PHYsical CONTEXT CONFirm
	01110	RF CHANnel RELease
	01111	MS POWER CONTROL
	10000	BS POWER CONTROL
	10001	PREPROCess CONFIGure
	10010	PREPROCessed MEASurement RESult
	10011	RF CHANnel RELease ACKnowledge
	10100	SACCH INFO MODIFY
	10101	TALKER DETection
	10110	LISTENER DETection

BSMAP

TIA/EIA/IS-634-A, revision A

The Base Station Management Application Part (BSMAP) supports all Radio Resource Management and Facility Management procedures between the MSC and the BS, or to a cell(s) within the BS. BSMAP messages are not passed to the MS, but are used only to perform functions at the MSC or the BS. A BSMAP message (complete layer 3 information) is also used together with a DTAP message to establish a connection for an MS between the BS and the MSC, in response to the first layer 3 interface message sent by the MS to the BS for each MS system request.

The format of the header is shown in the following illustration:

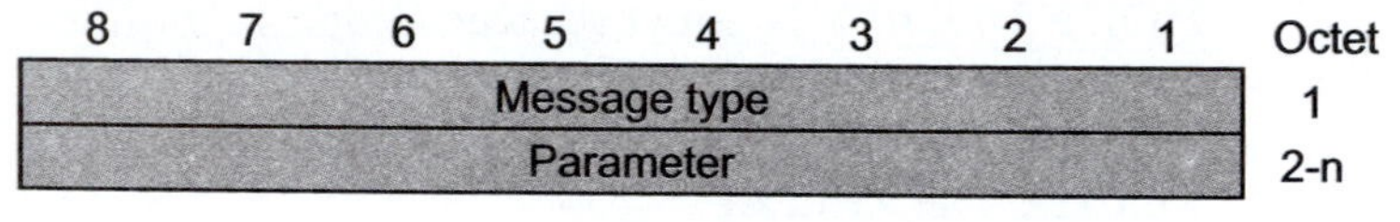

BSMAP header structure

Message type

Mandatory, one-octet field which uniquely defines the function and format of each BSMAP message.

Parameter

Each parameter has a name which is coded as a single octet. The length of a parameter may be fixed or variable and a length indicator for each parameter may be included.

DTAP (CDMA)

TIA/EIA/IS-634-A, revision A

The Direct Transfer Application Part (DTAP) messages are used to transfer call processing and mobility management messages to and from the MS. The BS does not use DTAP messages, but must map messages going to and coming from the MSC into the appropriate air interface signaling protocol. Transaction IDs are used to associate the DTAP messages with a particular MS and the current call.

The format of the header is shown in the following illustration:

8	7	6	5	4	3	2	1	Octet
Transaction identifier				Protocol discriminator				1
Message type								2
Information elements								3-n

DTAP header structure

Protocol discriminator

The protocol discriminator specifies the message being transferred (CC, MM, RR).

Transaction identifier

Distinguishes multiple parallel activities (transactions) within one mobile station. The format of the transaction identifier is as follows:

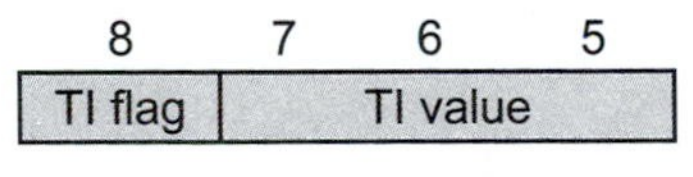

Transaction identifier

TI flag

Identifies who allocated the TI value for this transaction. The purpose of the TI flag is to resolve simultaneous attempts to allocate the same TI value.

TI value

TI values are assigned by the side of the interface initiating a transaction. At the beginning of a transaction, a free TI value is chosen and assigned to this transaction. It then remains fixed for the lifetime of the transaction. After a transaction ends, the associated TI value is free and may be reassigned to a later transaction. Two identical transaction identifier values may be used when each value pertains to a transaction originated at opposite ends of the interface.

Message Type

The message type defines the function of each DTAP message.

Information elements

Each information element has a name which is coded as a single octet. The length of an information element may be fixed or variable and a length indicator for each one may be included.

RR

GSM 04.08 http://www.etsi.org

Radio Resource (RR) management procedures include the functions related to the management of the common transmission resources, e.g., the physical channels and the data link connections on control channels. The general purpose of Radio Resource procedures is to establish, maintain and release RR connections that allow a point-to-point dialogue between the network and a Mobile Station. This includes the cell selection/reselection and the handover procedures. Moreover, Radio Resource management procedures include the reception of the uni-directional BCCH and CCCH when no RR connection is established. This permits automatic cell selection/reselection.

The format of the RR header is shown in the following illustration:

8	7	6	5	4	3	2	1	Octet
Protocol discriminator				Skip indicator				1
Message type								2
Information elements								3-n

RR header structure

Protocol discriminator

0110 identifies the RR Management protocol.

Skip indicator

The value of this field is 0000.

Message type

Uniquely defines the function and format of each RR message. The message type is mandatory for all messages. RR message types may be:

00111 xxx	Channel establishment messages:
011	ADDITIONAL ASSIGNMENT
111	IMMEDIATE ASSIGNMENT
001	IMMEDIATE ASSIGNMENT EXTENDED
010	IMMEDIATE ASSIGNMENT REJECT
00110 xxx	Ciphering messages:
101	CIPHERING MODE COMMAND
010	CIPHERING MODE COMPLETE

00101 xxx	Handover messages:
110	ASSIGNMENT COMMAND
001	ASSIGNMENT COMPLETE
111	ASSIGNMENT FAILURE
011	HANDOVER COMMAND
100	HANDOVER COMPLETE
000	HANDOVER FAILURE
101	PHYSICAL INFORMATION
00001 xxx	Channel release messages:
101	CHANNEL RELEASE
010	PARTIAL RELEASE
111	PARTIAL RELEASE COMPLETE
00100 xxx	Paging messages:
001	PAGING REQUEST TYPE 1
010	PAGING REQUEST TYPE 2
100	PAGING REQUEST TYPE 3
111	PAGING RESPONSE
00011 xxx	System information messages:
000	SYSTEM INFORMATION TYPE 8
001	SYSTEM INFORMATION TYPE 1
010	SYSTEM INFORMATION TYPE 2
011	SYSTEM INFORMATION TYPE 3
100	SYSTEM INFORMATION TYPE 4
101	SYSTEM INFORMATION TYPE 5
110	SYSTEM INFORMATION TYPE 6
111	SYSTEM INFORMATION TYPE 7
00000 xxx	System information messages:
010	SYSTEM INFORMATION TYPE 2bis
011	SYSTEM INFORMATION TYPE 2ter
101	SYSTEM INFORMATION TYPE 5bis
110	SYSTEM INFORMATION TYPE 5ter
00010 xxx	Miscellaneous messages:
000	CHANNEL MODE MODIFY
010	RR STATUS
111	CHANNEL MODE MODIFY ACKNOWLEDGE
100	FREQUENCY REDEFINITION
101	MEASUREMENT REPORT
110	CLASSMARK CHANGE
011	CLASSMARK ENQUIRY

Information elements

The length of an information element may be fixed or variable and a length indicator for each one may be included.

MM

GSM 04.08 http://www.etsi.org

The main function of the Mobility Management (MM) sub-layer is to support the mobility of user terminals, such as informing the network of its present location and providing user identity confidentiality. A further function of the MM sub-layer is to provide connection management services to the different entities of the upper Connection Management (CM) sub-layer.

The format of the header is shown in the following illustration:

8	7	6	5	4	3	2	1	Octet
Protocol discriminator				Skip indicator				1
Message type								2
Information elements								3-n

MM header structure

Protocol discriminator
0101 identifies the MM protocol.

Skip indicator
The value of this field is 0000.

Message type
Uniquely defines the function and format of each MM message. The message type is mandatory for all messages. Bit 8 is reserved for possible future use as an extension bit. Bit 7 is reserved for the send sequence number in messages sent from the mobile station. MM message types may be:

0x00	xxxx	Registration messages:
	0001	IMSI DETACH INDICATION
	0010	LOCATION UPDATING ACCEPT
	0100	LOCATION UPDATING REJECT
	1000	LOCATION UPDATING REQUEST
0x01	xxxx	Security messages:
	0001	AUTHENTICATION REJECT
	0010	AUTHENTICATION REQUEST

	0100	AUTHENTICATION RESPONSE
	1000	IDENTITY REQUEST
	1001	IDENTITY RESPONSE
	1010	TMSI REALLOCATION COMMAND
	1011	TMSI REALLOCATION COMPLETE
0x10	xxxx	Connection management messages:
	0001	CM SERVICE ACCEPT
	0010	CM SERVICE REJECT
	0011	CM SERVICE ABORT
	0100	CM SERVICE REQUEST
	1000	CM REESTABLISHMENT REQUEST
	1001	ABORT
0x11	xxxx	Miscellaneous messages:
	0001	MM STATUS

Information elements

Various information elements.

CC

GSM 04.08 http://www.etsi.org

The Call Control (CC) protocol is one of the protocols of the Connection Management (CM) sub-layer. Every mobile station must support the Call Control protocol. If a mobile station does not support any bearer capability at all, then it must respond to a SETUP message with a RELEASE COMPLETE message. In the Call Control protocol, more than one CC entity is defined. Each CC entity is independent from all others and communicates with the corresponding peer entity using its own MM connection. Different CC entities use different transaction identifiers. Certain sequences of actions of the two peer entities compose elementary procedures. These elementary procedures may be grouped into the following classes:

- Call establishment procedures.
- Call clearing procedures.
- Call information phase procedures.
- Miscellaneous procedures.

The terms *mobile originating* or *mobile originated* (MO) are used to describe a call initiated by the mobile station. The terms *mobile terminating* or *mobile terminated* (MT) are used to describe a call initiated by the network.

The format of the CC header is shown in the following illustration:

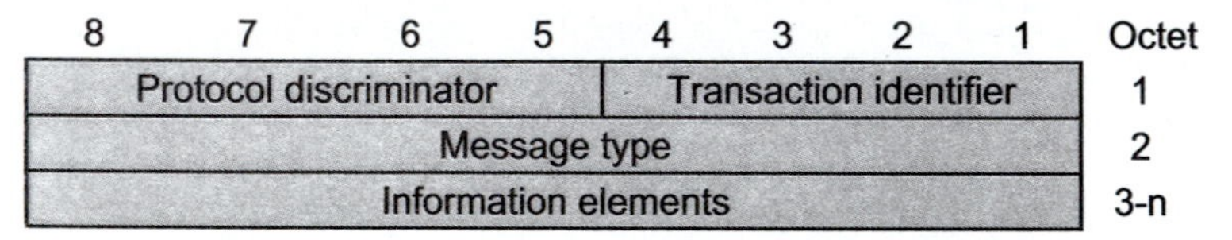

CC header structure

Protocol discriminator

0011 identifies the CC protocol.

Transaction identifier

The transaction identifier distinguishes multiple parallel activities (transactions) within one mobile station. The format of the transaction identifier is as follows:

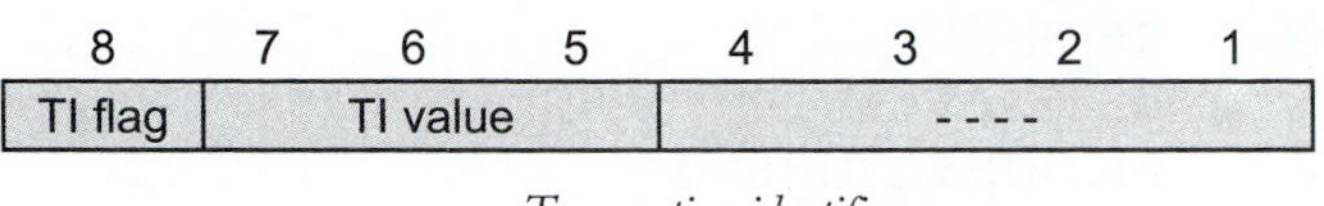

Transaction identifier

TI flag

Identifies who allocated the TI value for this transaction. The purpose of the TI flag is to resolve simultaneous attempts to allocate the same TI value.

TI value

TI values are assigned by the side of the interface initiating a transaction. At the beginning of a transaction, a free TI value is chosen and assigned to this transaction. It then remains fixed for the lifetime of the transaction. After a transaction ends, the associated TI value is free and may be reassigned to a later transaction. Two identical transaction identifier values may be used when each value pertains to a transaction originated at opposite ends of the interface.

Message type

CC message types may be as follows. Bit 8 is reserved for possible future use as an extension bit. Bit 7 is reserved for the send sequence number in messages sent from the mobile station.

0x00	0000	Escape to nationally specific message types
0x00	xxxx	Call establishment messages:
	0001	ALERTING
	1000	CALL CONFIRMED
	0010	CALL PROCEEDING
	0111	CONNECT
	1111	CONNECT ACKNOWLEDGE
	1110	EMERGENCY SETUP
	0011	PROGRESS
	0101	SETUP
0x01	xxxx	Call information phase messages:
	0111	MODIFY
	1111	MODIFY COMPLETE
	0011	MODIFY REJECT
	0000	USER INFORMATION
	1000	HOLD
	1001	HOLD ACKNOWLEDGE
	1010	HOLD REJECT

	1100	RETRIEVE
	1101	RETRIEVE ACKNOWLEDGE
	1110	RETRIEVE REJECT
0x10	xxxx	Call clearing messages:
	0101	DISCONNECT
	1101	RELEASE
	1010	RELEASE COMPLETE
0x11	xxxx	Miscellaneous messages:
	1001	CONGESTION CONTROL
	1110	NOTIFY
	1101	STATUS
	0100	STATUS ENQUIRY
	0101	START DTMF
	0001	STOP DTMF
	0010	STOP DTMF ACKNOWLEDGE
	0110	START DTMF ACKNOWLEDGE
	0111	START DTMF REJECT
	1010	FACILITY

Information elements

Various information elements.

SMS

GSM 04.11 http://www.etsi.org

The Short Message Service (SMS) is used to transfer text messages over mobile networks between a GSM PLMN Mobile Station and a Short Message Entity via a Service Center. The terms MO (Mobile Originating) and MT (Mobile Terminating) are used to indicate the direction in which the short message is sent.

SMS messages can be control or relay messages. The format of the control protocol message header is shown in the following illustration:

8	7	6	5	4	3	2	1	Octet
Protocol discriminator				Transaction identifier				1
Message type								2
Information elements								3-n

SMS control protocol header structure

Protocol discriminator

1001 identifies the SMS protocol.

Transaction identifier

The transaction identifier (TI) distinguishes multiple parallel activities (transactions) within one mobile station. The format of the transaction identifier is as follows:

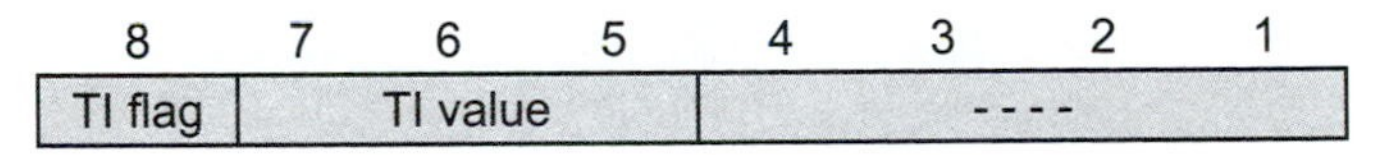

Transaction identifier

TI flag

Identifies who allocated the TI value for this transaction. The purpose of the TI flag is to resolve simultaneous attempts to allocate the same TI value.

TI value

TI values are assigned by the side of the interface initiating a transaction. At the beginning of a transaction, a free TI value is chosen and assigned to this transaction. It then remains fixed for the lifetime of the transaction. After a transaction ends, the associated TI value is free and may be reassigned to a later transaction. Two identical transaction identifier values may be used when each value pertains to a transaction originated at opposite ends of the interface.

Message type

The message type, together with the protocol discriminator, identifies the function of the message being sent. Messages may be of the following:

0000 0001	CP-DATA
0000 0100	CP-ACK
0001 0000	CP-ERROR

Information elements

Each IE has an identifier which is coded as a single octet. The length of an IE may be fixed or variable and may or may not include a length indicator.

The format of the relay protocol message header is shown in the following illustration:

8	7	6	5	4	3	2	1	Octet
0	0	0	0	0	MTI			1
Message reference								2
Information elements								3-n

SMS relay protocol header structure

MTI

Message type indicator. Values are as follows:

Bit Value (3 2 1)	*Direction*	*RP-Message*
0 0 0	ms -> n	RP-DATA
0 0 0	n -> ms	Reserved
0 0 1	ms -> n	Reserved
0 0 1	n -> ms	RP-DATA
0 1 0	ms -> n	RP-ACK
0 1 0	n -> ms	Reserved
0 1 1	ms -> n	Reserved
0 1 1	n -> ms	RP-ACK
1 0 0	ms -> n	RP-ERROR
1 0 0	n -> ms	Reserved
1 0 1	ms -> n	Reserved
1 0 1	n -> ms	RP-ERROR

Bit Value (3 2 1)	*Direction*	*RP-Message*
1 1 0	ms -> n	RP-SMMA
1 1 0	n -> ms	Reserved
1 1 1	ms -> n	Reserved
1 1 1	n -> ms	Reserved

Message reference

Used to link an RP-ACK message or RP-ERROR message to the associated RP-Data or RP-SMMA message transfer attempt.

Information elements

Each IE has an identifier which is coded as a single octet. The length of an IE may be fixed or variable and may or may not include a length indicator.

10

DECnet Protocols

Digital Equipment Corporation (DEC) developed the DECnet protocol to allow high-speed communication between DEC minicomputers across local and wide area networks. The DECnet suite includes the following protocols:

- RP: Routing Protocol.
- MOP: Maintenance Operation Protocol.
- NSP: Network Service Protocol.
- SCP: Session Control Protocol.
- DAP: Data Access Protocol.
- CTERM: Command Terminal.
- LAT: Local Area Transport.
- STP: Spanning Tree Protocol.
- LAVC: Local Area VAX Cluster.

The following diagram illustrates the DECnet protocol suite in relation to the OSI model:

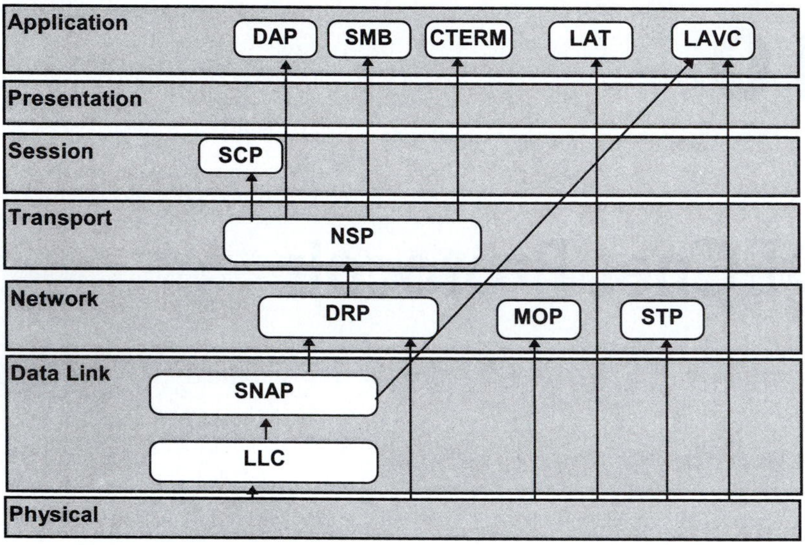

DECnet protocol suite in relation to the OSI model

RP

The Routing Protocol (RP) distributes routing information among DECnet hosts. It defines routing classes into two levels: level 1, which handles routing within a single DECnet routing area; and level 2, which handles routing between areas.

Frames

RP frames can be one of the following types:

[Level 1 hello]	Routing update from a level 1 router.
[Level 2 hello]	Routing update from a level 2 router.
[Endnode hello]	Routing update from an endnode.
[L1 router msg]	Routing status for a local area.
[L2 router msg]	Routing status for other areas.
[Routed data]	Segment of user data.

Frame Parameter

All RP frames have the following parameter:

Node address

Node address. The DECnet area and node given in the decimal dot form: Area.Node (where Area can extend from 1-63 and Node can extend from 1-1023).

Hello Parameters

Hello frames have the following parameters:

Routing priority

Routing priority on a scale of 100 (not used for [endnode hello] frames).

Hello period

Period between routing update hello messages.

Version

The version in use.

Multicast status
Y indicates that this protocol supports multicast traffic on the link; N indicates it does not support multicast traffic.

Maximum
Maximum frame size supported on the link (1500 for Ethernet).

Router Parameters

RP router frames have the following parameters:

Source node
The ID of the sending node.

Number of IDs
Number of IDs contained in the routing table for this level 1 routing message.

Number of areas
Number of areas contained in the routing table for this level 2 message.

Data Parameters

RP [routed data] frames have the following parameters:

Request return
When set to 1, sender is requesting that the other party return the frame. When set to 0, sender is suggesting that the other party discard the frame.

Return path
When set to 1, frame is on return path; when set to 0, frame is on outbound path.

Intra-Ethernet
When set to 1, frame is from a directly connected Ethernet segment; when set to 0, the system forwarded the frame from another segment.

Version
Must be 0.

MOP

The Maintenance Operation Protocol (MOP) is used for utility services such as uploading and downloading system software, remote testing and problem diagnosis.

Frames

MOP frames can be one of the following commands:

[memory load data]	Contains memory load data.
[mem load request]	Request for memory load segment.
[mem load w/addr]	Memory load with transfer address.
[par load w/addr]	Parameter load with transfer address.
[dump service req]	Request for assistance with dump operation.
[mem dump request]	Request for next memory dump segment.
[memory dump data]	Contains memory dump data.
[dump completed]	Acknowledgment of dump completion.
[volunteer assist]	Offer of dump/load/loop assistance.
[request program]	Request for system or loader program.
[rem boot request]	Request for boot program.
[remote ID reqst]	Request for remote console identification.
[remote system ID]	Remote console identification information.
[counters request]	Request for communication information counters.
[counters reply]	Communication information counters.
[reserve console]	Remote console in reserved state.
[release console]	Release of remote console from reserved state.
[rem console poll]	Poll of remote console for status.
[rem console rply]	Response to remote console poll.
[loopback request]	Request to loopback enclosed data.
[loopback reply]	Response to loopback request with data.

Memory Dump and Memory Load Frames

MOP memory dump and memory load frames use the following parameters:

Load number

Data segment sequence number of the current memory data segment.

Load address
Memory load address for storage of the memory data.

Transfer address
Starting memory address of the current segment.

Memory address
Starting physical memory address for the dump.

Count
Number of memory locations to dump.

Version
Protocol format version, currently 1.

Memory size
Size of physical machine memory.

Bits
Generally set to 2 for compatibility reasons.

Buffer size
Local buffer size in bytes.

Communication device
Device type of the requesting system. The following device codes can be used:

Code	Device
DP	DP11-DA
UNA	DEUNA
DU	DU11-DA
CNA	DECNA
DL	DL11-C/E/WA
QNA	DEQNA
DQ	DQ11-DA
CI	Comp. Intercon.
DA	DA11-B/AL
PCL	PCL11-B
DUP	DUP11-DA
DMC	DMC11-DA/FA/MA/MD
DN	DN11-BA/AA

Code	Device
DLV	DLV11-E/F/J
DMP	DMP11
DTE	DTE20 (PDP11-KL10)
DV	DV11-AA/BA
DZ	DZ11-A/B/C/D
KDP	KMC11/DUP11-DA
KDZ	KMC11/DZ11-A/B/C/D
KL	KL8-J
DMV	DMV11
DPV	DPV11
DMF	DMF-32
DMR	DMR11-AA/AB/AC/AE
KMY	KMS11-PX (X.25)
KMX	KMS11-BD/BE (X.25)

Parameter Load Frames

MOP parameter load frames have the following fields:

Load number
The data segment sequence number of the current data segment.

Target name
ASCII system name for target system.

Target address
Hex address of target system.

Host system name
ASCII system name of host.

Host system address
Hex address of host system.

Host system time
Current time for host system.

Request Program Frames

MOP [request program] frames have the following fields:

Communication device

Refer to the values given for dump/load frames above.

Version

Version currently in use.

Type

Type of program that the system requested:

Secondary	A secondary loader program.
Tertiary	A tertiary loader program.
System	An operating system program.

Software ID

Software type that the system requested:

Standard O/S	Standard operating system software.
Maint system	Maintenance system software.

Processor

System processor type:

PDP-11	PDP-11 system.
Comm Srv	Communications server.
Profess	Professional.

Boot Request Frames

Boot request frames have the following fields:

Verification number

A verification code that must match before this protocol can honor a boot request.

Boot server

Boot system device type:

Req	Requesting system.
Def	Default boot server.
<device>	Specified device.

Software ID

The software type as given for the [request program] frame above.

Remote Console Frames

MOP remote console frames have the following fields:

Receipt number

Used to identify a particular request.

Command status

Console command data status given as OK if received, or Lost if not received.

NSP

The Network Services Protocol (NSP) provides reliable virtual connection services with flow control to the network layer Routing Protocol (RP).

Frames

NSP frames can be one of the following commands:

[data segment]	Carries higher level data.
[interrupt]	Carries urgent data.
[data request]	Carries data flow control information.
[interrupt rq]	Carries interrupt flow control information.
[data ackn]	Acknowledges receipt of data.
[control ackn]	Acknowledges receipt of interrupt messages.
[connect ackn]	Acknowledges a [connect init] frame.
[connect init]	Requests a logical link connection.
[connect ackn]	Acknowledges a link connection.
[discnct init]	Requests disconnection of a link.
[discnct ackn]	Acknowledges disconnection of a link.
[no operation]	No operation performed.

Parameters

NSP frames can contain the following fields:

Destination link address
Destination port of the link.

Source link address
Source port of the logical link.

Acknowledge number
The segment number of the last message received successfully or, if followed by {NAK}, the segment number of the message for which the system requests a retransmission.

Acknowledge other
Same as acknowledge number, but used to acknowledge other data.

Segment number
Number of the current data frame.

Flow control
Can indicate the following services:

Seg_reqst	Data segment request count.
SCP_reqst	Session control protocol request count.

Flow control information
The data segment messages can include flow control messages ({send} or {stop}) to indicate the desired action of the receiving system.

BOM/EOM
Beginning of message/end of message. Indicates the start or end of a data segment message.

SCP

The Session Control Protocol (SCP) manages logical links for DECnet connections.

Frames

SCP frames can be one of the following commands:

[connect data]	Transfers connection parameters.
[disconnect]	Supplies disconnect status information.
[reject data]	Supplies reject status information.

Connect Data Parameters

SCP [connect data] frames contain the destination name (Dest) and source name (Src) parameters that can consist of the following fields:

Object type

One of the following object types:

Type	*Description*
(User Process)	General task or end user process.
(Files-DAP 1)	File access through DAP version 1.
(Unit Record)	Unit record service.
(App. TrmSrv)	Application terminal services.
(Cmd. TrmSrv)	Command terminal services.
(RSX-11M TC1)	RSX-11M task control, version 1.
(Op Services)	Operator services interface.
(Node Manage)	Node resource manager.
(3270-BSC GW)	IBM 3270 BSC gateway.
(2780-BSC GW)	IBM 2780 BSC gateway.
(3790-SDLC)	IBM 3790 SDLC gateway.
(TPS Applic.)	TPS application.
(RT-11 DIBOL)	RT-11 DIBOL application.
(TOPS-20 T H)	TOPS-20 terminal handler.
(TOPS-20 R S)	TOPS-20 remote spooler.
(RSX-11M TC2)	RSX-11X task control, version 2.
(TLK Utility)	TLK utility.
(Files-DAP4+)	File access through DAP, version 4+.
(RSX-11S RTL)	RSX-11S remote task Loader.

Type	*Description*
(NICE Proc.)	NICE processor.
(RSTS/E MTP)	RSTS/E media transfer program.
(RSTS/E HCTH)	RSTS/E homogeneous command terminal handler.
(Mail Listen)	Mail listener.
(Host TrmHnd)	Host terminal handler.
(Con. TrmHnd)	Concentrator terminal handler.
(Loop Mirror)	Loopback mirror service.
(Event Rcvr)	Event receiver.
(VAX/VMS PMU)	VAX/VMS personal message utility.
(FTS Service)	FTS service.

Group
Group code identifier.

User
User code identifier.

Descriptor
A user-defined string of data.

Version
The SCP version, as in {SCP 1.0}.

Requestor ID
User name for access verification.

Password
Password for user verification.

Account
Link or service account data.

User data
End user connect data.

Disconnect/Reject Parameters

The decoding for SCP [disconnect] and [reject data] frames lists the reason for disconnection, as follows:

{No error}	Normal disconnect with no error.
{Shutting down}	Source node is deactivating.
{Unknown user}	Destination end user is unknown.
{Invalid username}	Destination end user invalid.
{Dest. overloaded}	Destination out of link resources.
{Unknown error}	Unspecified error.
{Link aborted}	Link aborted by third party.
{User aborted}	Link aborted by end user.
{Host overloaded}	Source is out of link resources.
{Bad ID/password}	Invalid ID or password.
{Bad account info}	Invalid account data.
{Data too long}	Connect data parameters too long.

DAP

The Data Access Protocol (DAP) provides remote file access to systems supporting the DECnet architecture.

Frames

DAP frames can be one of the following commands:

[configuration]	Exchanges information about file systems and supported protocols.
[file attribs]	Provides file attributes.
[open file]	Opens the specified file.
[create file]	Creates the specified file.
[rename file]	Renames the specified file.
[delete file]	Deletes the specified file.
[list dir]	Lists the specified directory.
[submit file]	Submits the specified batch file.
[execute file]	Executes the specified command file.
[control info]	Provides control information about the file system.
[continue]	Continues I/O operation after error.
[acknowledge]	Acknowledges open file and control information commands.
[close file]	Closes file or ends data stream.
[data message]	Carries file I/O data.
[status]	Returns status and error information.
[file index]	Specifies keys for file indexing.
[allocate]	Creates or extends a file.
[summary info]	Returns summary information about a file.
[timestamp]	Specifies time for time-stamped fields.
[protect mode]	Specifies file protection mode.
[file name]	Renames files or lists directories.
[access rights]	Specifies file access rights.

Frame Parameters

DAP commands can contain the following parameters:

Allocation size
The number of blocks allocated to a file.

Attribute
File attribute represented as follows:

Seqnt	Sequential access supported.
Relatv	Relative access supported.
Index	Indexed access supported.
Hashed	Hashed format.

Bit count
Indicates the number of unused bits in the last byte of the data message.

Bits per byte
Number of bits in each byte.

Block size
Physical media block size in bytes.

Bucket size
Bucket size used to access relative, hashed and indexed files.

Checksum
The 16-bit file checksum.

Data type
Type of file data:

ASCII	Standard 7-bit ASCII characters.
Image	Binary data.
EBCDIC	EBCDIC encoded data.
Compr	Compressed data format.
Exec	Executable code.
Privil	Privileged code.
Senstv	Sensitive data, purge after delete.

Device type
Code which indicates the type of device that DAP associates with the file.

File access mode

The open mode for file access which is specified as one of the following:

Put	Put (write) access allowed.
Get	Get (read) access allowed.
Del	Delete access allowed.
Upd	Update access allowed.
Trn	Truncate access allowed.
BIO	Block I/O access allowed.
BRO	Block and Record I/O switching allowed.
FAO	File Access Options. The file access options code.

FilSys

File system, represented as one of the following:

RMS-11	RMS-20	RMS-32
FCS-11	RT-11	None
TOPS-20	TOPS-10	OS-8

Maximum buffer size

The maximum buffer size the sending system can receive.

Operating system type

Operating system type can be of the following:

RT-11	RSTS/E	RSX-11S
RSX-11M	RSX-11D	IAS
VAX/VMS	TOPS-20	TOPS-10
RTS-8	OS-8	RSX-11M+
COPOS/11		

Password

Password required for file access.

Record attributes

Record attributes code.

Record format

Represented as follows:

Undef	Undefined record format.
FixLen	Fixed-length records.
VarLen	Variable-length records.
Var/FC	Variable-length records with fixed control format.
ASCII	ASCII stream format.

Record size

File record size in bytes.

Record number

The record used when accessing file data.

Shared access mode

The open mode for sharing file access, specified as follows:

Put	Put (write) access allowed.
Get	Get (read) access allowed.
Del	Delete access allowed.
Upd	Update access allowed.
MSE	Multi-stream access enabled.
UPI	User-provided interlocking allowed.
Nil	No shared use allowed.

Stream ID

The ID code used to multiplex data streams on one file.

System capabilities

System capabilities code.

System specific information

Information specific to homogeneous systems.

Ver

Version. The DAP version number, the DAP software version number, followed by the user modification number in parenthesis.

Continue Parameters

DAP [continue] frames indicate the recovery action as one of the following:

{try again}	Repeat the attempted operation.
{skip it}	Skip the attempted operation and continue.
{abort transfer}	Abort the I/O transfer.
{resume}	Restart the data stream if suspended.

Status Parameters

DAP [status] frames can report the following status information:

{pending}	Operation in progress.
{OK}	Operation successful.
{bad request}	Specified operation unsupported.
{open error}	Error occurred while opening file.
{I/O error}	Error occurred while transferring data.
{I/O warning}	Non-fatal I/O error occurred.
{close error}	Error occurred while closing file.
{bad format}	Message format invalid.
{sync error}	Message received out of synchronization.

CTERM

The Command Terminal (CTERM) protocol is the terminal emulation protocol of the Digital Network Architecture. CTERM uses DECnet to provide a command terminal connection between DEC terminals and DEC operating systems such as VMS and RSTS/E.

Frames

CTERM frames can be one of the following commands:

[initiate]	Initiates the command terminal connection.
[start read]	Requests a read from the terminal server.
[read data]	Transfers terminal data to the host.
[out-of-band]	Conveys an out-of-band character received by the server.
[abort read]	Requests that the current terminal data read be aborted.
[clear input]	Requests that the input and type-ahead buffers be cleared.
[write data]	Transfers terminal write data and control information.
[write status]	Transfers terminal write status.
[discard stat]	Signals whether to discard terminal output.
[read config]	Requests the current terminal characteristics.
[config data]	Transfers terminal configuration data.
[check input]	Requests the current input character count.
[input count]	Indicates the number of input characters to be read.
[input state]	Indicates the presence of new input characters.

CTERM Parameters

The following are possible CTERM parameters:

Buffer size

Size of the terminal character input buffer.

Character count

Number of characters in the input buffer.

End
Ending character position. Current position of the last character displayed.

EOP
End of prompt. Character position of the first character after the prompt.

Horizontal position
Current horizontal position of the displayed output.

Horizontal position change
Horizontal position change. Horizontal position change since the last read.

Low water mark
Position of the last character not modified.

Maximum receive size
Length of the input character buffer.

Maximum transmit
Maximum transmit buffer size.

Character
Out-of-band character. The out-of-band character received.

Postfix
Postfix new line count. New line postfix count for the current write.

Prefix
Prefix new line count. New line prefix count for the current write.

Software revision
Software revision currently in use.

Start of display
Position of the first character to display.

Termination
Termination set bitmask. The 256-bit termination set for the read.

Time out
Amount of time in seconds before a read request aborts.

Version
Protocol version currently in use.

Vertical position
Current vertical position of the displayed output.

Vertical position change
Change in vertical position since the last read.

Messages

For certain CTERM frames various messages can be displayed as detailed below:

[input state] frames display the current input status as {more characters} or {no more characters}.

[write status] frames can display the write status as {some output lost} or {no output lost}.

[discard stat] frames can display the discard status as {discard} or {no discard}.

[abort read] frames display the abort request as {unconditional} or {if no more input}.

[out-of-band] frames can contain the disposition of the data as {discard}.

LAT

The Local Area Transport (LAT) protocol is designed to handle multiplexed terminal traffic to/from timesharing hosts.

STP

The Spanning Tree Protocol (STP) prevents the formation of logical looping in the network. It is implemented by the 802.1d MAC Bridge Management Protocol, to provide information on bridge topology.

LAVC

The Local Area VAX Cluster (LAVC) protocol communicates between DEC VAX computers in a cluster.

11

Frame Relay

Frame Relay is a protocol standard for LAN internetworking which provides a fast and efficient method of transmitting information from a user device to LAN bridges and routers.

The Frame Relay protocol uses a frame structure similar to that of LAPD, except that the frame header is replaced by a 2-byte Frame Relay header field. The Frame Relay header contains the user-specified DLCI field, which is the destination address of the frame. It also contains congestion and status signals which the network sends to the user.

Virtual Circuits

The Frame Relay frame is transmitted to its destination by way of virtual circuits (logical paths from an originating point in the network) to a destination point. Virtual circuits may be permanent (PVCs) or switched (SVCs). PVCs are set up administratively by the network manager for a dedicated point-to-point connection; SVCs are set up on a call-by-call basis.

Advantages of Frame Relay

Frame Relay offers an attractive alternative to both dedicated lines and X.25 networks for connecting LANs to bridges and routers. The success of the Frame Relay protocol is based on the following two underlying factors:

- Because virtual circuits consume bandwidth only when they transport data, many virtual circuits can exist simultaneously across a given transmission line. In addition, each device can use more of the bandwidth as necessary, and thus operate at higher speeds.
- The improved reliability of communication lines and increased error-handling sophistication at end stations allows the Frame Relay protocol to discard erroneous frames and thus eliminate time-consuming error-handling processing.

These two factors make Frame Relay a desirable choice for data transmission; however, they also necessitate testing to determine that the system works properly and that data is not lost.

Frame Relay Structure

Standards for the Frame Relay protocol have been developed by ANSI and CCITT simultaneously. The separate LMI specification has basically been incorporated into the ANSI specification. The following discussion of the protocol structure includes the major points from these specifications.

The Frame Relay frame structure is based on the LAPD protocol. In the Frame Relay structure, the frame header is altered slightly to contain the Data Link Connection Identifier (DLCI) and congestion bits, in place of the normal address and control fields. This new Frame Relay header is 2 bytes in length and has the following format:

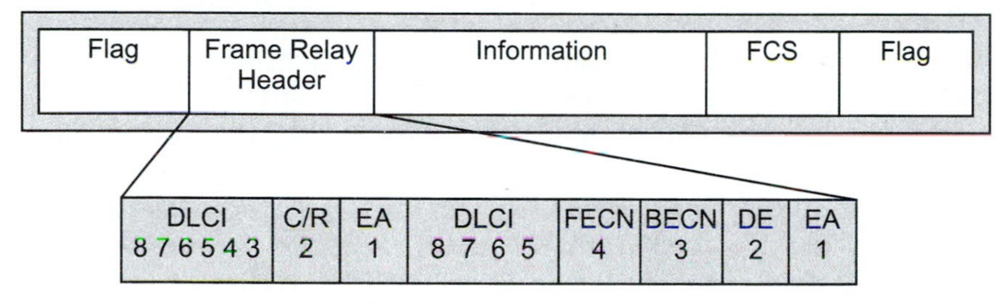

Frame Relay header structure

DLCI

10-bit DLCI field represents the address of the frame and corresponds to a PVC.

C/R

Designates whether the frame is a command or response.

EA

Extended Address field signifies up to two additional bytes in the Frame Relay header, thus greatly expanding the number of possible addresses.

FECN

Forward Explicit Congestion Notification (see ECN below).

BECN

Backward Explicit Congestion Notification (see ECN below).

DE

Discard Eligibility (see DE below).

Information

The Information field may include other protocols within it, such as an X.25, IP or SDLC (SNA) packet.

Explicit Congestion Notification (ECN) Bits

When the network becomes congested to the point that it cannot process new data transmissions, it begins to discard frames. These discarded frames are retransmitted, thus causing more congestion. In an effort to prevent this situation, several mechanisms have been developed to notify user devices at the onset of congestion, so that the offered load may be reduced.

Two bits in the Frame Relay header are used to signal the user device that congestion is occurring on the line: They are the Forward Explicit Congestion Notification (FECN) bit and the Backward Explicit Congestion Notification (BECN) bit. The FECN is changed to 1 as a frame is sent downstream toward the destination location when congestion occurs during data transmission. In this way, all downstream nodes and the attached user device learn about congestion on the line. The BECN is changed to 1 in a frame traveling back toward the source of data transmission on a path where congestion is occurring. Thus the source node is notified to slow down transmission until congestion subsides.

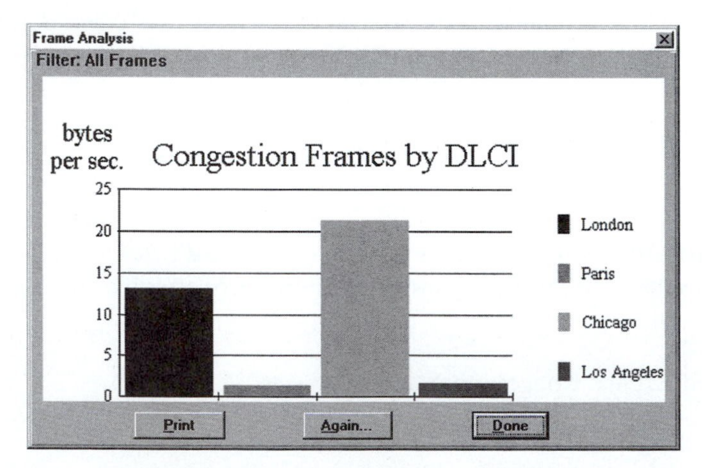

Bytes with congestion bits set according to DLCI value

Consolidated Link Layer Management (CLLM)

It may occur that there are no frames traveling back to the source node which is causing the congestion. In this case, the network will want to send its own message to the problematic source node. The standard, however, does not allow the network to send its own frames with the DLCI of the desired virtual circuit.

To address this problem, ANSI defined the Consolidated Link Layer Management (CLLM). With CLLM, a separate DLCI (number 1023) is reserved for sending link layer control messages from the network to the user device. The ANSI standard (T1.618) defines the format of the CLLM message. It contains a code for the cause of the congestion and a listing of all DLCIs that should act to reduce their data transmission to lower congestion.

Status of Connections (LMI)

Each DLCI corresponds to a PVC (Permanent Virtual Circuit). It is sometimes necessary to transmit information about this connection (e.g., whether the interface is still active) the valid DLCIs for the interface and the status of each PVC. This information is transmitted using the reserved DLCI 1023 or DLCI 0, depending on the standard used.

A multicast status may also be sent with the LMI. Multicasting is where a router sends a frame on a reserved DLCI known as a multicast group. The network then replicates the frame and delivers it to a predefined list of DLCIs, thus broadcasting a single frame to a collection of destinations.

Discard Eligibility (DE)

When there is congestion on the line, the network must decide which frames to discard in order to free the line. Discard Eligibility provides the network with a signal to determine which frames to discard. The network will discard frames with a DE value of 1 before discarding other frames.

The DE bit may be set by the user on some of its lower-priority frames. Alternatively, the network may set the DE bit to indicate to other nodes that a frame should be preferentially selected for discard, if necessary.

Frame Relay Standards

ANSI T1.618

T1.618 describes the protocol supporting the data transfer phase of the Frame Relay bearer service, as defined in ANSI T1.606. T1.618 is based on a subset of ANSI T1.602 (LAPD) called the "Core Aspects" and is used by both switched and permanent virtual calls.

T1.618 also includes the Consolidated Link Layer Mechanism (CLLM). The generation and transport of CLLM is optional. With CLLM, DLCI 1023 is reserved for sending link layer control messages.

T1.618 issues implicit congestion notification from the network to the user device. The congestion notification contains a code indicating the cause of the congestion and lists all DLCIs that have to reduce their traffic in order to lower congestion.

ANSI T1.617

To establish a Switched Virtual Circuit (SVC) connection, Frame Relay users must establish a dialog with the network using the signalling specification in T1.617. The procedure results in the assignment of a DLCI. Once the dialog is established, the T1.618 procedures apply.

To establish a Permanent Virtual Circuit (PVC), a setup protocol is used which is identical to the ISDN D-channel protocol and defined in T1.617.

With ISDN, users can use the D-channel for setup. For non-ISDN callers, there is no D-channel, so the dialog between the user and the network must be separated from the ordinary data transfer procedures. In T1.617, DLCI 0 is reserved.

T1.617 also contains specifications on how Frame Relay service parameters are negotiated.

ANSI LMI

ANSI LMI is a Permanent Virtual Circuit (PVC) management system defined in Annex D of T1.617. ANSI LMI is virtually identical to the Manufacturers' LMI, without the optional extensions. ANSI LMI uses DLCI 0.

Manufacturers' LMI

Manufacturers' LMI is a Frame Relay specification with extension-document number 001-208966, September 18, 1990.

Manufacturers' LMI defines a generic Frame Relay service based on PVCs for interconnecting DTE devices with Frame Relay network equipment. In addition to the ANSI standard, Manufacturers' LMI includes extensions and LMI functions and procedures. Manufacturers' LMI uses DLCI 1023.

Frame Relay NNI PVC (FRF.2)

Network-to-Network (NNI) PVC Frame Relay implementation is described in FRF.2. The NNI interface is concerned with the transfer of C-plane and U-plane information between two network nodes belonging to two different Frame Relay networks.

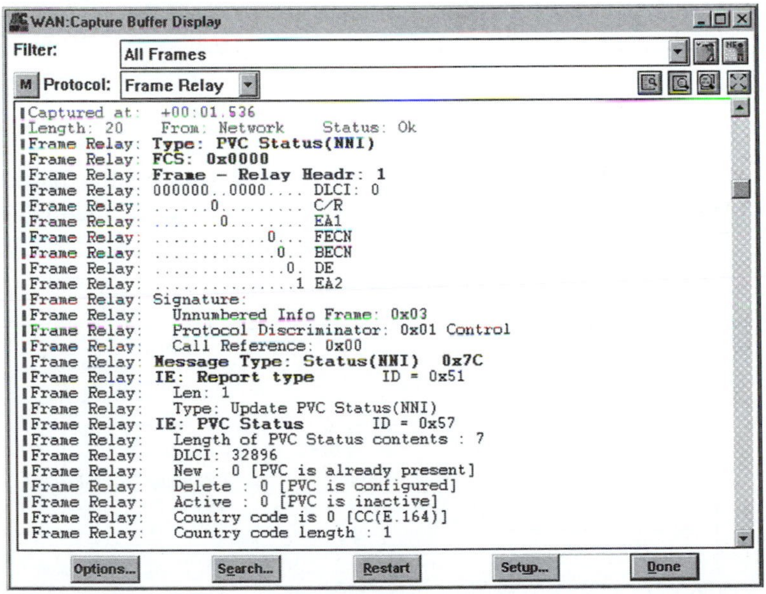

```
WAN:Capture Buffer Display
Filter:    All Frames
M Protocol: Frame Relay
Captured at:   +00:01.536
Length: 20     From: Network     Status: Ok
Frame Relay: Type: PVC Status(NNI)
Frame Relay: FCS: 0x0000
Frame Relay: Frame - Relay Headr: 1
Frame Relay: 000000..0000.... DLCI: 0
Frame Relay: ......0......... C/R
Frame Relay: .......0........ EA1
Frame Relay: ............0... FECN
Frame Relay: .............0.. BECN
Frame Relay: ..............0. DE
Frame Relay: ...............1 EA2
Frame Relay: Signature:
Frame Relay:   Unnumbered Info Frame: 0x03
Frame Relay:   Protocol Discriminator: 0x01 Control
Frame Relay:   Call Reference: 0x00
Frame Relay: Message Type: Status(NNI)  0x7C
Frame Relay: IE: Report type        ID = 0x51
Frame Relay:   Len: 1
Frame Relay:   Type: Update PVC Status(NNI)
Frame Relay: IE: PVC Status        ID = 0x57
Frame Relay:   Length of PVC Status contents : 7
Frame Relay:   DLCI: 32896
Frame Relay:   New : 0 [PVC is already present]
Frame Relay:   Delete : 0 [PVC is configured]
Frame Relay:   Active : 0 [PVC is inactive]
Frame Relay:   Country code is 0 [CC(E.164)]
Frame Relay:   Country code length : 1
Options...   Search...   Restart   Setup...   Done
```

NNI PVC decode

FRF.3

FRF.3 provides multiprotocol encapsulation over Frame Relay within an ANSI T1.618 frame. The structure of such a Frame Relay frame is as follows:

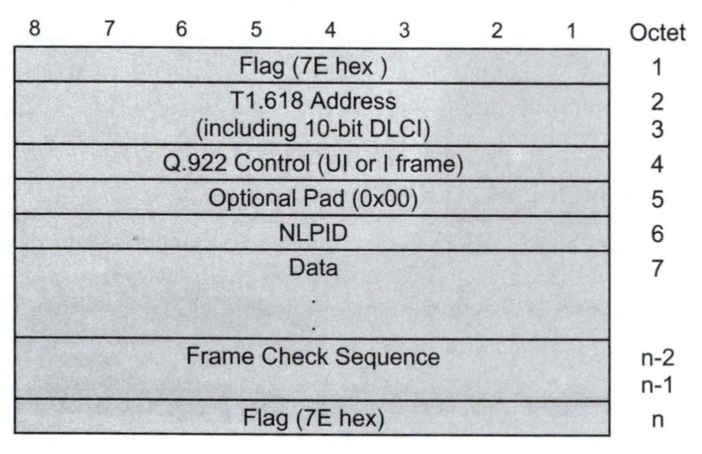

FRF.3 frame structure

The NLPID (Network Level Protocol ID) field designates what encapsulation or what protocol follows. The following diagram details the possible values for NLPID and the protocols which are designated by each value. For example, a value of 0xCC indicates an encapsulated IP frame.

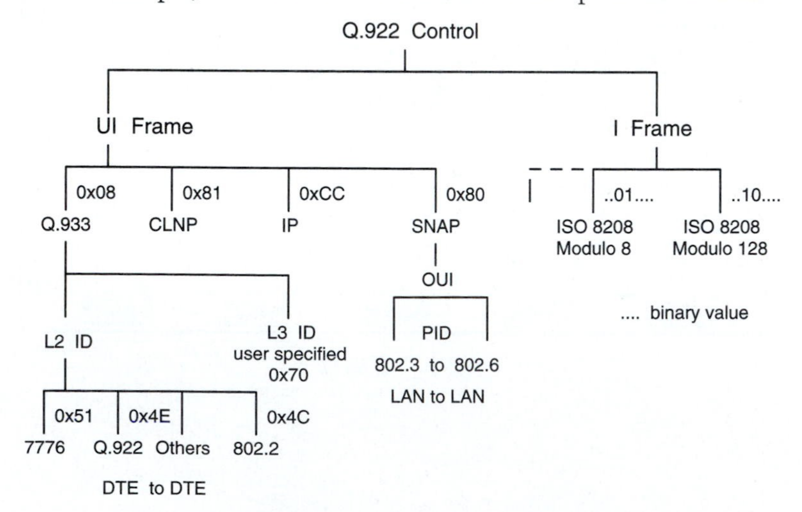

Multiprotocol encapsulation over Frame Relay

UNI SVC (FRF.4)

FRF.4 is Frame Relay switched virtual connection user-to-network interface agreement. It is applied using equipment attached to a non-ISDN Frame Relay network, or to an ISDN network using only case A.

```
WAN:Capture Buffer Display
Filter: All Frames
M Protocol: Frame Relay
Captured at:   +00:01.616
Length: 95      From: Network     Status: Ok
Frame Relay: Type: SVC Message
Frame Relay: FCS: 0x0000
Frame Relay: Frame - Relay Headr: 1
Frame Relay: 000000..0000.... DLCI: 0
Frame Relay: ......0......... C/R
Frame Relay: .......0........ EA1
Frame Relay: ............0... FECN
Frame Relay: .............0.. BECN
Frame Relay: ..............0. DE
Frame Relay: ...............1 EA2
Frame Relay: Signature:
Frame Relay:    Unnumbered Info Frame: 0x03
Frame Relay:    Protocol Discriminator: 0x01 Control
Frame Relay:    Call reference: 0
Frame Relay: Message Type: Setup  0x05
Frame Relay: IE: Bearer capability        ID = 0x4
Frame Relay:    Length of bearer capability contents: 3
Frame Relay:    Coding standard: 0 [CCITT]
Frame Relay:    Information transfer cap: 8 [Unrestricted digital info.]
    [Reserved]
Frame Relay:    Transfer mode: 0 [Reserved]
Frame Relay:    User information layer 2 protocol: 15 [Reserved]
Frame Relay: IE: Data Link Connection identifier        ID = 0x19
Frame Relay:    Length of DLCI contents: 2
Frame Relay:    DLCI: 0
Frame Relay: IE: Connected sub-address        ID = 0x4D
Options...  Search...  Restart  Setup...  Done
```

UNI SVC decode

The following is a list of valid SVC message types:

- Call proceeding.
- Connect.
- Connect Acknowledge.
- Disconnect.
- Progress.
- Release.
- Release complete.
- Setup.
- Status.
- Status enquiry.

FRF.5

FRF.5 defines Network internetworking connecting Frame Relay over ATM. Refer to *Chapter 4* (*Frame Relay over ATM*) for more details.

FRF.8

FRF.8 defines Service internetworking connecting Frame Relay over ATM. Refer to *Chapter 4* (*Frame Relay over ATM*) for more details.

DCP FRF.9

FRF.9
RFC 1661

FRF.9 is the encapsulation of the Data Compression Protocol (DCP) over Frame Relay. It applies to unnumbered information (UI) frames encapsulated using Q.933 Annex E [9] and FRF.3.1 [3] It may be used on Frame Relay connections that are interworked with ATM using FRF.5.

DCP is logically composed of two sublayers: the DCP Control sublayer, and the DCP Function sublayer.

The Data Compression Protocol (DCP) is encapsulated with Multiprotocol Encapsulation over a Frame Relay network using Q.933 Annex E. A DCP PDU is the combination of a DCP Header and DCP Payload or DCPCP PDU.

DCP PROTOCOL DATA UNIT FORMAT

The DCP PDU is used by a DCP entity to communicate data or control information to the remote peer DCP entity. The most significant (earliest sent) octet(s) of the DCP PDU must be the DCP Header. The C/D bit of the DCP Header signals whether the DCP PDU is for control or for data.

DCP Data PDU Format

A DCP data PDU encapsulates compressed or uncompressed user data for transport to the peer DCP entity.

The DCP Header is one or three octets in length.

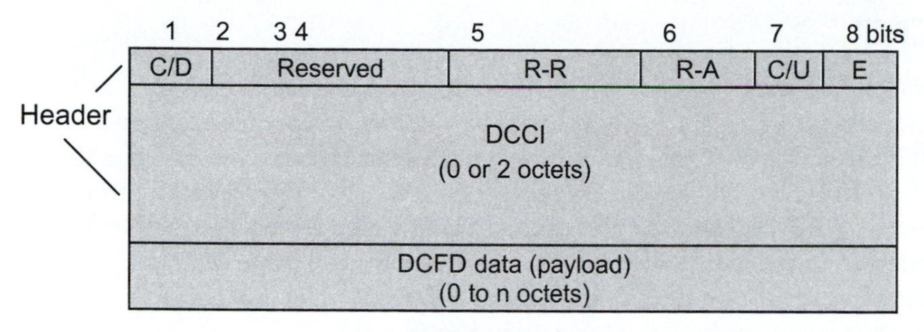

DCP header structure

E

0 extension

1 no extension

The E bit is always 1 for DCP data PDUs.

C/U

0 uncompressed mode

1 for compressed mode

R-A

0 no reset acknowledge

1 reset acknowledge

R-R

0 no reset request

1 reset request

C/D

Shows whether the frame is for control or data.

0 DCP data PDUs

DCCI

Zero or two-octet DC Context Identifier.

Reserved

Reserved for future use. Set to 0.

DCPCP

The DCP Control Protocol (DCPCP) is used to enable, disable, and optionally configure DCP. DCPCP has two modes of operation: Mode-1 and Mode-2. Mode-2 provides full negotiation capabilities to enable, disable and configure DCP using the Point-to-Point Protocol (PPP) Link Control Protocol (LCP) negotiation procedures. Mode-1 uses a subset of the Mode-2 negotiation primitives with simplified procedures to enable and disable DCP with the default DCFD and default parameter values. Mode-1 operation is required; Mode-2 operation is optional.

The length of a DCP Header for DCP control PDUs is one octet.

DCPCP PDUs use the same formats as the PPP LCP as defined in RFC 1661. DCPCP Mode-1 uses a subset of the DCPCP PDU formats (Configure-Request and Configure-Ack with the Mode-1 Configuration Option only).

The format of the DCPCP packet consists of the header followed by the PDU which uses the same format as the PPP LCP as shown below.

Code	Identifier	Length	Data
1 byte	1 byte	2 bytes	Variable

DCPCP PDU structure

Code

Decimal value which indicates the type of packet:

1 Configure-Request.
2 Configure-Ack.
3 Configure-Nak.
4 Configure-Reject.
5 Terminate-Request.
6 Terminate-Ack.
7 Code-Reject.
8 Protocol-Reject.
9 Echo-Request.
10 Echo-Reply.
11 Discard-Request.
12 Link-Quality Report.

Identifier

Decimal value which aids in matching requests and replies.

Length
Length of the packet, including the Code, Identifier, Length and Data fields.

Data
Variable length field which may contain one or more configuration options. The format of the configuration options is as follows:

Type	Length	Revision

DCPCP configuration options

Type
One-byte indication of the type of the configuration option is set to 254 for mode 1 messages.

Length
Length of the configuration option including the Type, Length and Data fields and is set to 3.

Revision
The Revision field is one octet in length and contains the revision number. The current revision is 1.

Mode 1 Request and response messages
The Mode-1 Request message is a DCPCP Configure-Request packet with the Code field set to *1*, the Mode-1 Response message is a DCPCP Configure-Ack packet with the Code field set to *2*.

Mode-2 Formats
Mode-2 formats are the same as the LCP packet formats as shown above, with a unique set of Configurations options. The LCP packets with codes 1 through 7 are required. The other LCP packets specified in RFC 1661 and listed above are optional.

NNI SVC (FRF.10)

Network-to-Network (NNI) SVC Frame Relay implementation is described in FRF.10. This implementation agreement applies to SVCs over Frame Relay NNIs and to SPVCs. It is applicable at NNIs whether both networks are private, both are public, or one is private and the other public. Such frames are automatically recognized by the analyzer and correctly displayed.

FRF.11

Frame Relay is now a major component of many network designs. The protocol provides a minimal set of switching functions to forward variable sized data payloads through a network. The basic frame relay protocol, described in the Frame Relay Forum User to Network (UNI) and Network to Network (NNI) Implementation Agreements, has been augmented by additional agreements which detail techniques for structuring application data over the basic Frame Relay information field. These techniques enabled successful support for data applications such as LAN bridging, IP routing, and SNA.

FRF.11 extends Frame Relay application support to include the transport of digital voice payloads. Specifically FRF.11 addresses the following requirements:

- Transport of compressed voice within the payload of a Frame Relay frame.
- Support a diverse set of voice compression algorithms.
- Effective utilization of low-bit rate Frame Relay connections.
- Multiplexing of up to 255 sub-channels on a single frame relay DLCI.
- Support of multiple voice payloads on the same or different sub-channel within a single frame.
- Support of data sub-channels on a multiplexed Frame Relay DLCI.

FRF.12

FRF.12 is the Frame Relay Fragmentation Implementation Agreement. Fragmentation queuing reduces both delay and delay variation in Frame Relay networks by dividing large data packets into smaller packets and then reassembling the data into the original frames at the destination. This ability is particularly relevant to users who wish to combine voice and other time-sensitive applications, such as SNA mission-critical applications, with non-time-sensitive applications or other data on a single Permanent Virtual Circuit (PVC). The main benefit of fragmentation is the ability to utilize common User to Network Interface (UNI) access lines or Network to Network Interface (NNI) lines and/or PVCs for communications combining large data frames and real-time protocols.

Fragmenting frames enhances the utility and uniformity of Frame Relay networks, reducing delay and delay variation while upgrading application responsiveness, quality and reliability. As a result, multiple types of traffic, such as voice, fax and data, can be transparently combined on a single UNI, NNI and/or PVC.

The Fragmentation Implementation Agreement provides for transmission of Frame Relay Data Terminal Equipment (DTE) and Data Communications Equipment (DCE) with the ability to fragment long data frames into a sequence of shorter frames that are then reassembled into the original frame by the receiving peer DTE or DCE. Frame fragmentation is necessary to control delay and delay variation when real-time traffic, such as voice, is carried across the same interfaces as data traffic. Fragmentation enables the interleaving of delay-sensitive traffic on one PVC with fragments of long data frames on another PVC using the same interface.

FRF.12 supports three fragmentation applications:

1. Locally across a Frame Relay UNI interface between the DTE/DCE peers.
2. Locally across a Frame Relay NNI interface between DCE peers.
3. End-to-End between two Frame Relay DTEs interconnected by one or more Frame Relay networks.

When used end-to-end, the fragmentation procedure is transparent to Frame Relay network(s) between the transmitting and receiving DTEs.

FREther

FREther is a variant of Frame Relay which is comprised of the Frame Relay header followed by the EtherType field. This is an additional form of encapsulation over Frame Relay which is used by some customers.

Timeplex (BRE2)

BRE (Bridge Relay Encapsulation) is a proprietary Ascom Timeplex protocol that extends bridging across WAN links by means of encapsulation. BRE2 is an improved form of the standard, providing better performance due to the fact that it sits directly on the link layer protocol, requires less configuration and provides its own routing protocol. BRE2 was deployed in 4.0 router software and is available in all 4.x and 5.x versions of the software.

The format of the BRE2 frame is as follows:

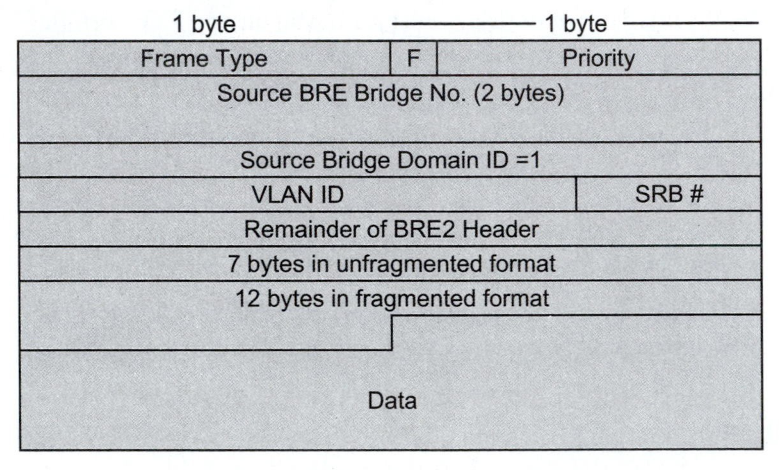

BRE2 frame format

If F is 0 than the frame is unfragmented and the BRE2 header is 17 bytes long. If F is 1, the frame is fragmented and the BRE2 header is 22 bytes long. SRB # is the Source Route Bridge Number (4 bits).

Cascade

In order to provide a Frame Relay service, Regional Bell Operating Companies (RBOCs) deploy Cascade switches in multiple LATAs and interconnect them to provide service to customers across the LATAs, as well as manage the switches in multiple LATAs from a single network management station.

The trunk header format for the Cascade STDX family of switches conform to ANSI T1.618-1991 ISDN Core Aspects of Frame Protocol for use with Frame Relay Bearer Service.

The Cascade trunk header format is as shown below:

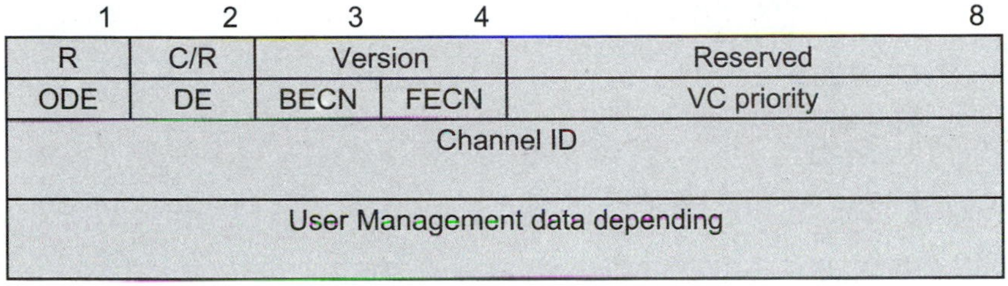

<table>
<tr><th>1</th><th>2</th><th>3</th><th>4</th><th>8</th></tr>
<tr><td>R</td><td>C/R</td><td colspan="2">Version</td><td>Reserved</td></tr>
<tr><td>ODE</td><td>DE</td><td>BECN</td><td>FECN</td><td>VC priority</td></tr>
<tr><td colspan="5">Channel ID</td></tr>
<tr><td colspan="5">User Management data depending</td></tr>
</table>

Cascade trunk header format

R
Reserved.

C/R
Command/Response field bit.

Version
Header version. Defines the version of the trunk header format for the Cascade STDX family of switches. This field is currently set to 0.

ODE
Set to 1 if the ingress rate is greater than the excess burst size.

DE
Discard eligibility indicator which is implemented based on the definitions specified in ANSI T1.618.

BECN

Backward explicit congestion notification according to ANSI T1.618.

FECN

Forward explicit congestion notification according to ANSI T1.618.

VC priority

Virtual circuit priority. Used to differentiate the sensitive traffic from traffic not sensitive to delays, such as file transfers or batch traffic. The priority may be 1, 2 or 3, 1 being the highest.

For management data, a 5th byte contains information concerning the type of PDU information to follow. Values are as follows:

0	Call request PDU
1	Confirmation PDU
2	Rejected PDU
3	Clear PDU
4	Disrupt PDU
5	Hello PDU
6	Hello Acknowledgment PDU
7	Defined Path Hello PDU
8	Defined Path Hello Acknowledgment PDU

LAPF

The purpose of LAPF is to convey data link service data units between DL-service users in the U-plane for frame mode bearer services across the ISDN user-network interface on B-, D- or H-channels. Frame mode bearer connections are established using either procedures specified in Recommendation Q.933 [3] or (for permanent virtual circuits) by subscription. LAPF uses a physical layer service, and allows for statistical multiplexing of one or more frame mode bearer connections over a single ISDN B-, D- or H-channel by use of LAPF and compatible HDLC procedures.

The format of the header is shown in the following illustration:

	8	7	6	5	4	3	2	1
Default address field format (2 octets)	Upper DLCI						C/R	EA 0
	Lower DLCI				FECN (Note)	BECN (Note)	DE (Note)	EA 1

LAPF address format

Control field bits (Modulo 128)	8	7	6	5	4	3	2	1	
I Format	N(S)							0	Octet 4 (Note)
	N(R)							P/F	Octet 5
S Format	X	X	X	X	Su	Su	0	1	Octet 4
	N(R)							P/F	Octet 5
U Format	M	M	M	P/F	M	M	1	1	Octet 4

LAPF format

EA

Address field extension bit.

C/R

Command Response bit.

FECN

Forward Explicit Congestion Notification.

BECN
Backward Explicit Congestion Notification.

DLCI
Data Link Connection Identifier.

DE
Discard Eligibility indicator.

D/C
DLCI or DL-CORE control indicator.

N(S)
Transmitter send sequence number.

N(R)
Transmitter receive sequence number.

P/F
Poll bit when used as a command, final bit when used as a response.

X
Reserved and set to 0.

Su
Supervisory function bit.

M
Modifier function bit.

Multiprotocol over Frame Relay

RFC 1490 http://www.cis.ohio-state.edu/htbin/rfc/rfc1490.html
RFC 2427 http://www.cis.ohio-state.edu/htbin/rfc/rfc2427.html

Multiprotocol over Frame Relay is a method of encapsulating various LAN protocols over Frame Relay. In this case, all protocols encapsulate their packets within a Q.922 Annex A frame. Additionally, frames must contain information necessary to identify the protocol carried within the protocol data unit (PDU), thus allowing the receiver to properly process the incoming packet. The format of such frames is as shown in the following illustration:

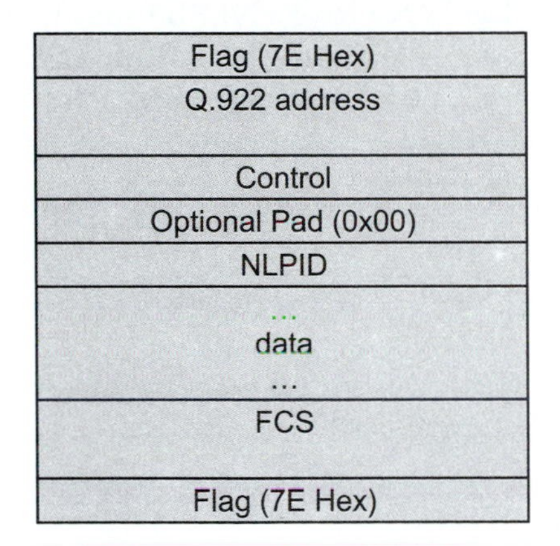

Muliprotocol over Frame Relay frame

Q.922 address

2-octet address field containing the 10-bit DLCI field. On some networks the Q.922 address may optionally contain 3 or 4 octets.

Control

Q.922 control field. The UI value is of 0x03 is used unless negotiated otherwise. The use of XID (0xAF or 0xBF) is permitted.

Pad

Used to align the remainder of the frame to a two octet boundary. There may be 0 or 1 pad octet within the pad field. The value is always 0.

NLPID

Network Level Protocol ID, adminstered by ISO and CCITT. Identifies the encapsulated protocol.

FCS

2-byte frame check sequence.

There are two basic types of data packets that travel within the Frame Relay network: routed packets and bridged packets. These packets have distinct formats and therefore, must contain an indicator that the destination may use to correctly interpret the contents of the frame. This indicator is embedded within the NLPID and SNAP header information.

For those protocols that do not have a NLPID already assigned, it is necessary to provide a mechanism to allow easy protocol identification. There is a NLPID value defined indicating the presence of a SNAP header. The format of the SNAP header is as follows:

Organizationally Unique Identifier (OUI)	Protocol Identifier (PID)
3 bytes	2 bytes

SNAP header

All stations must be able to accept and properly interpret both the NLPID encapsulation and the SNAP header encapsulation for a routed packet.

The three-octet Organizationally Unique Identifier (OUI) identifies an organization which administers the meaning of the Protocol Identifier (PID) which follows. Together they identify a distinct protocol. Note that OUI 0x00-00-00 specifies that the following PID is an Ethertype.

Some protocols have an assigned NLPID, but because the NLPID numbering space is so limited, not all protocols have specific NLPID values assigned to them. When packets of such protocols are routed over Frame Relay networks, they are sent using the NLPID 0x80 followed by SNAP. If the protocol has an Ethertype assigned, the OUI is 0x00-00-00 (which indicates an Ethertype follows), and PID is the Ethertype of the protocol in use. There is one pad octet to align the protocol data on a two octet boundary.

The second type of Frame Relay traffic is bridged packets. These are encapsulated using the NLPID value of 0x80 indicating SNAP. As with other SNAP encapsulated protocols, there is one pad octet to align the data portion of the encapsulated frame. The SNAP header which follows the

NLPID identifies the format of the bridged packet. The OUI value used for this encapsulation is the 802.1 organization code 0x00-80-C2. The PID portion of the SNAP header (the two bytes immediately following the OUI) specifies the form of the MAC header, which immediately follows the SNAP header. Additionally, the PID indicates whether the original FCS is preserved within the bridged frame.

12

FUNI

ATM Forum, Frame Based User to Network Interface Specifications 1995-09

FUNI was developed by the ATM Forum in order to provide users with the ability to connect between ATM networks and existing frame-based equipment (e.g., routers, etc.). FUNI uses a T1/E1 interface and offers a relatively easy and cost-effective method for users to take advantage of ATM infrastructure or an ATM backbone, while not having to replace existing equipment with more expensive ATM equipment.

The frame structure of FUNI is shown in the following illustration:

FUNI PDU

Flag	FUNI header	User SDU	FUNI FCS	Flag
	2	1-n (n<=4096)	2 (up to 4)	1

FUNI frame structure

The FUNI header is as follows:

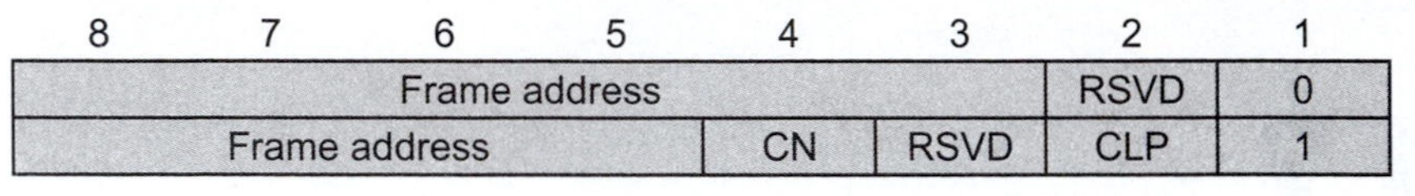

8	7	6	5	4	3	2	1
Frame address						RSVD	0
Frame address				CN	RSVD	CLP	1

FUNI header structure

RSVD

Reserved bits for interface management. These bits are set to 0 unless the frame is used for management.

Frame address

Octet 1, bits 6-3 are mapped to the 4 LSBs of the VPI in the ATM cell header. The 4 MSBs of the VPI are not coded in the address field. Octet 1, bits 8 and 7 and octet 2 bits 8-5 are mapped to the six LSBs of the VCI in the ATM cell header.

FCS

16 bit frame check sequence.

CN

Congestion notification. If the PTI=01x in the last ATM cell composing the FUNI frame, the CN is 1 for the FUNI frame, otherwise it is 0.

CLP

The network equipment copies the CLP bit sent from the user equipment into the CLP bit of all ATM cell headers constituting the FUNI frame. The CLP bit from the network equipment to the user equipment is always set to 0.

13

GPRS Protocols

GPRS (General Packet Radio Service) is used as a data services upgrade to any GSM network. It allows GSM networks to be truly compatible with the Internet. GPRS uses a packet-mode technique to transfer bursty traffic in an efficient manner. It allows transmission bit rates from 9.6 Kbps to more than 150 Kbps per user.

The two key benefits of GPRS are a better use of radio and network resources and completely transparent IP support. GPRS optimizes the use of network and radio resources. It uses radio resources only when there is data to be sent or received. As a true packet technology, it allows end user applications to only occupy the network when a payload is being transferred and so is well adapted to the very bursty nature of data applications.

Another important feature of the GPRS is that it provides immediate connectivity and high throughput. Applications based on standard data protocols such as IP and X.25 are supported. In GPRS, four different quality of service levels are supported. To support data applications GPRS utilizes several new network nodes, in addition to the network nodes in the GSM PLMN. Those nodes are responsible for traffic routing and other interworking functions with external packet-switched data networks, subscriber location, cell selection, roaming and many other functions that any cellular network needs for its operation. Apart from these protocols, GPRS uses the GSM SMS protocol and the GSM MM protocol (which it calls GMM).

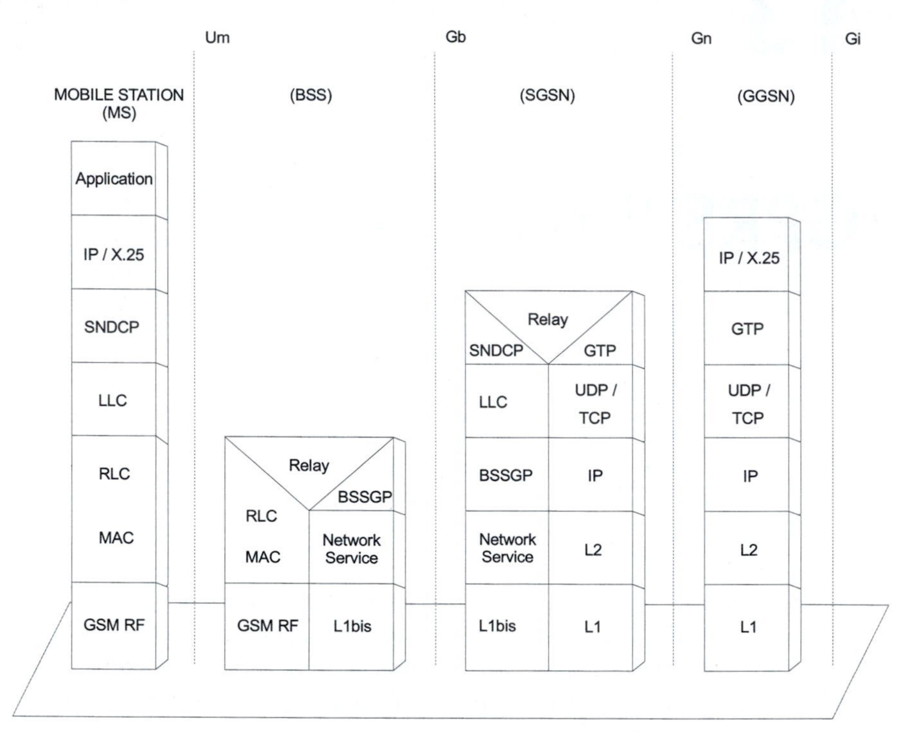

GPRS transmission plan

This chapter describes the following GPRS protocols:

- NS: Network Service.
- BSSGP: Base Station System GPRS Protocol.
- BSSAP+
- BCC: Broadcast Call Control.
- GCC: Group Call Control.
- GMM:: GPRS Mobility Management Protocol.
- GSM: GPRS Session Management.
- GTP: GPRS Tunnelling Protocol.
- LLC: Logical Link Control layer protocol for GPRS.

- RLP: Radio Link Protocol
- SNDCP: Sub-Network Dependant Convergence Protocol.

The following diagram illustrates the GPRS protocols in relation to other telephony protocols and the OSI model:

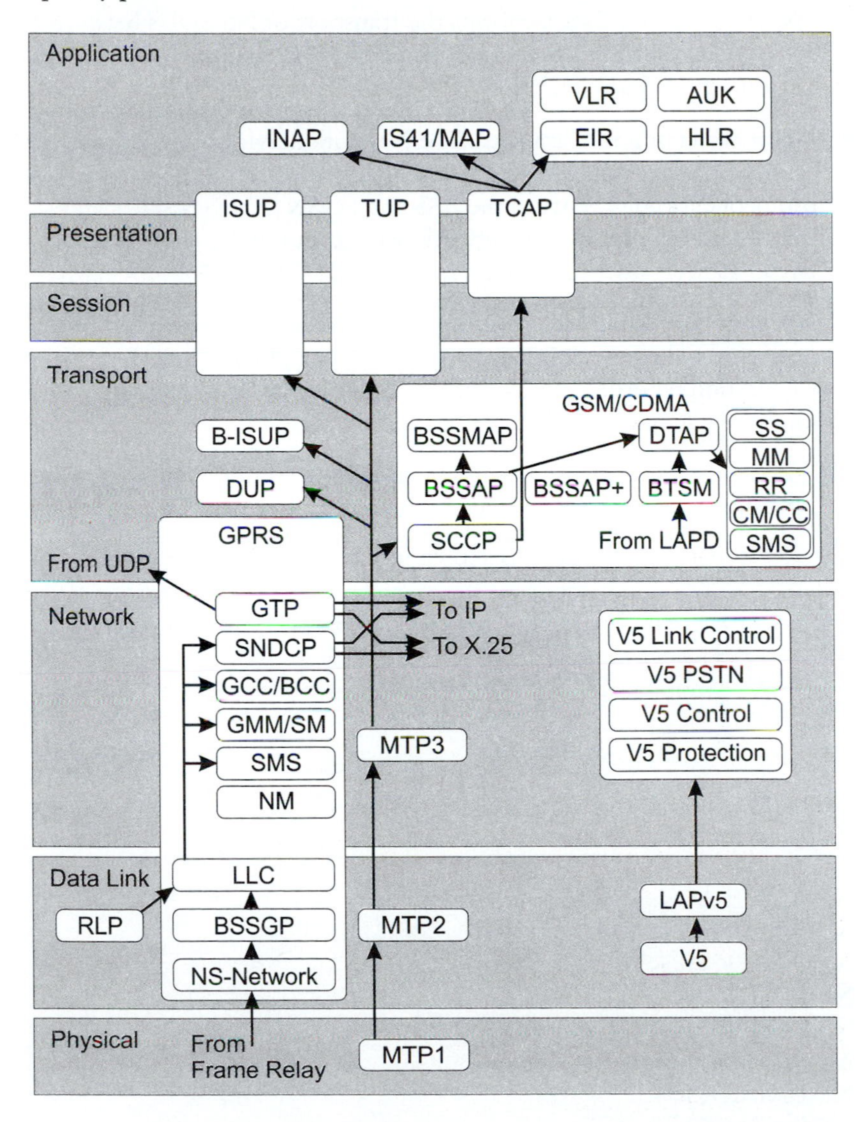

GPRS in relation to the OSI model

NS

GSM 08.16 version 6.1.0 http://www.etsi.fr

The Network Service (NS) performs the transport of NS SDUs between the SGSN (serving GPRS support node) and BSS (base station system). Services provided to the NS user include:

- Network Service SDU transfer. The Network Service provides network service primitives allowing for transmission and reception of upper layer protocol data units between the BSS and SGSN. The NS SDUs are transferred in order by the Network Service, but under exceptional circumstances order may not be maintained.
- Network congestion indication. Congestion recovery control actions may be performed by the Sub-Network Service (e.g., Frame Relay). Congestion reporting mechanisms available in the Sub-Network Service implementation are used by NS to report congestion.
- Status indication. Status indication is used to inform the NS user of the NS affecting events, e.g., change in the available transmission capabilities.

The structure of the NS PDU is shown in the following illustration:

8	7	6	5	4	3	2	1	Octet
PDU type								1
Information elements								2-n

NM header structure

PDU type

PDU type may be:
NS-ALIVE.
NS-ALIVE-ACK.
NS-BLOCK.
NS-BLOCK-ACK.
NS-RESET.
NS-RESET-ACK.
NS-STATUS.
NS-UNBLOCK.
NS-UNBLOCK-ACK.
NS-UNITDATA.

Information elements

The particular IEs present in a PDU depend on the PDU type. The structure of IEs is as shown in the following illustration:

8	7	6	5	4	3	2	1	Octet
Information element ID (IEI)								1
Length indicator								2
Information element value								3

IE structure

Information element ID

The first octet of an information element having the TLV format contains the IEI of the information element. If this octet does not correspond to an IEI known in the PDU, the receiver assumes that the next octet is the first octet of the length indicator field. This rule allows the receiver to skip unknown information elements and to analyze any subsequent information elements.

The following IEs may be present depending on the PDU type:
Cause.
NS-VCI.
NS PDU.
BVCI.
NSEI.

Length indicator

Information elements may be variable in length. The length indicator is one or two octets long, the second octet may be absent. This field consists of the field extension bit, 0/1 ext, and the length of the value field which follows, expressed in octets. The field extension bit enables extension of the length indicator to two octets. Bit 8 of the first octet is reserved for the field extension bit. If the field extension bit is set to 0 (zero), then the second octet of the length indicator is present. If the field extension bit is set to 1 (one), then the first octet is the final octet of the length indicator.

BSSGP

GSM 08.18 version 6.1.0 http://www.etsi.fr

The NS transports BSS (base station system) GPRS protocol PDUs between a BSS and an SGSN (serving GPRS support node). The primary functions of the BSSGP include:

- Provision by an SGSN to a BSS of radio related information used by the RLC/MAC function (in the downlink).
- Provision by a BSS to an SGSN of radio related information derived from the RLC/MAC function (in the uplink).
- Provision of functionality to enable two physically distinct nodes, an SGSN and a BSS, to operate node management control functions.

The structure of BSSGP PDUs is shown in the following illustration:

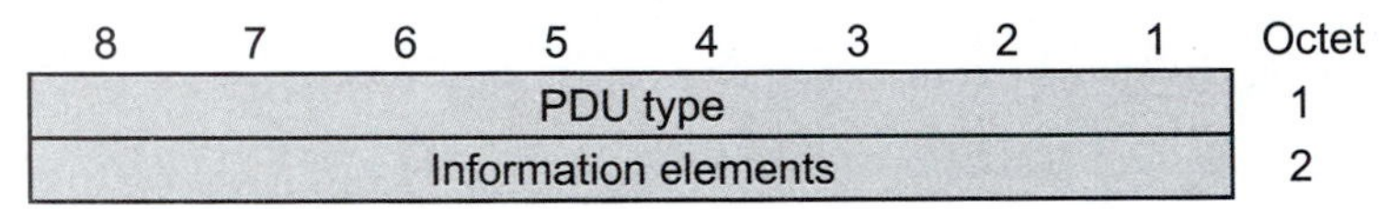

BSSGP header structure

PDU type

PDU type may be:
UL-UNITDATA.
RA-CAPABILITY.
PTM-UNITDATA.
PAGING PS.
PAGING CS.
RA-CAPABILITY-UPDATE.
RA-CAPABILITY-UPDATE-ACK.
RADIO-STATUS.
SUSPEND.
SUSPEND-ACK.
SUSPEND-NACK.
RESUME.
RESUME-ACK.
RESUME-NACK.
FLUSH-LL.

FLUSH-LL-ACK.
LLC-DISCARDED.
FLOW-CONTROL-BVC.
FLOW-CONTROL-BVC-ACK.
FLOW-CONTROL-MS.
FLOW-CONTROL-MS-ACK.
BVC-BLOCK.
BVC-BLOCK-ACK.
BVC-UNBLOCK.
BVC-UNBLOCK-ACK.
BVC-RESET.
BVC-RESET-ACK.
STATUS.
SGSN-INVOKE-TRACE.

Information elements

The following IE types may be present. The hex coding is the value of the IEI field.

0x00 Alignment Octets
0x01 Bmax default MS
0x02 BSS Area Indication
0x03 Bucket Leak Rate
0x04 BVCI
0x05 BVC Bucket Size
0x06 BVC Measurement
0x07 Cause
0x08 Cell Identifier
0x09 Channel needed
0x0a DRX Parameters
0x0b eMLPP-Priority
0x0c Flush Action
0x0d IMSI
0x0e LLC-SDU
0x0f LLC Frames Discarded
0x10 Location Area
0x11 Mobile Id
0x12 MS Bucket Size
0x13 MS Radio Access Capability
0x14 OMC Id
0x15 PDU In Error
0x16 PDU Lifetime

0x17 Priority
0x18 QoS Profile
0x19 Radio Cause
0x1a RA-Cap-UPD-Cause
0x1b Routing Area
0x1c R_default_MS
0x1d Suspend Reference Number
0x1e Tag
0x1f TLLI
0x20 TMSI
0x21 Trace Reference
0x22 Trace Type
0x23 Transaction Id
0x24 Trigger Id
0x25 Number of octets effected
All values not explicitly shown are reserved for future use and are treated by the recipient as an unknown IEI.

BSSAP+

http://www.etsi.org/ GSM 09.18 version 7.1.0 release 1998

BSSAP+ defines use of mobile resources when a mobile station supports both GSM circuit switched services and GSM packet switched services. It defines procedures used on the Serving GPRS Support Node (SGSN) to Visitors Location Register (VLR) for interoperability between circuit and packet switched services. Layer 3 messages on the Gs interface are defined.

BSSAP+		BSSPAP+
SCCP		SCCP
MTP L3		MTP L3
MTP L2		MTP L2
L1		L1
SGSN	*Gs*	*MSC/VLR*

BSSAP+ protocol layer structure over Gs interface

The Gs interface connects the databases in the MSC/VLR and the SGSN. The procedures the of BSSAP+ protocol are used to co-ordinate the location information of MSs that are IMSI attached to both GPRS and non-GPRS services. The Gs interface is also used to convey some circuit switched related procedures via the SGSN.

The basis for the interworking between a VLR and an SGSN is the existence of an association between those entities per MS. An association consists of the SGSN storing the number of the VLR serving the MS for circuit switched services and the VLR storing the number of the SGSN serving the MS for packet switched services. The association is only applicable to MSs in class-A mode of operation and MSs in class-B mode of operation.

All the messages in BSSAP+ use the SCCP class 0 connectionless service.

When the return option in SCCP is used and the sender receives an N_NOTICE indication from SCCP, the sending entity reports to the Operation and Maintenance system (see ITU-T Q.714).

The behaviour of the VLR and the SGSN entities related to the Gs interface are defined by the state of the association for an MS. Individual states per association, i.e. per MS in class-A mode of operation and MS in class-B mode of operation, are held at both the VLR and the SGSN.

8	7	6	5	4	3	2	1	Octet
Message type								1
Information elements								2-n

BSSAP+ header structure

The message type uniquely identifies the message being sent. The following BSSAP+ message types exist:

Value	Message type
00000000	*Unassigned:* treated as an unknown Message type.
00000001	BSSAP+-PAGING-REQUEST.
00000010	BSSAP+-PAGING-REJECT
00000011 to 00001000	*Unassigned:* treated as an unknown Message type.
00001001	BSSAP+-LOCATION-UPDATE-REQUEST.
00001010	BSSAP+-LOCATION-UPDATE-ACCEPT.
00001011	BSSAP+-LOCATION-UPDATE-REJECT.
00001100	BSSAP+-TMSI-REALLOCATION-COMPLETE.
00001101	BSSAP+-ALERT-REQUEST.
00001110	BSSAP+-ALERT-ACK.
00001111	BSSAP+-ALERT-REJECT.
00010000	BSSAP+-MS-ACTIVITY-INDICATION.
00010001	BSSAP+-GPRS-DETACH-INDICATION.
00010010	BSSAP+-GPRS-DETACH-ACK.
00010011	BSSAP+-IMSI-DETACH-INDICATION.
00010100	BSSAP+-IMSI-DETACH-ACK.
00010101	BSSAP+-RESET-INDICATION.
00010110	BSSAP+-RESET-ACK.
00010111	BSSAP+-MS-INFORMATION-REQUEST.
00011000	BSSAP+-MS-INFORMATION-RESPONSE.
00011001	*Unassigned:* treated as an unknown Message type.
00011010	BSSAP+-MM-INFORMATION-REQUEST.
00011101	BSSAP+-MOBILE-STATUS.
00011110	*Unassigned:* treated as an unknown Message type.
00011111	BSSAP+-MS-UNREACHABLE.

Each message type has specific information elements

00000001	IMSI.
00000010	VLR number.
00000011	TMSI.
00000100	Location area identifier.
00000101	Channel Needed.
00000110	eMLPP Priority.
00000111	*Unassigned:* treated as an unknown IEI.
00001000	Gs cause.
00001001	SGSN number.
00001010	GPRS location update type.
00001011	*Unassigned:* treated as an unknown IEI.
00001100	*Unassigned:* treated as an unknown IEI.
00001101	Mobile station classmark 1.
00001110	Mobile identity.
00001111	Reject cause.
00010000	IMSI detach from GPRS service type.
00010001	IMSI detach from non-GPRS service type.
00010010	Information requested.
00010011	PTMSI.
00010100	IMEI.
00010101	IMEISV.
00010110	*Unassigned:* treated as an unknown IEI.
00010111	MM information.
00011000	Cell Global Identity.
00011001	Location information age.
00011010	Mobile station state.
00011011	Erroneous message.
00011100 to 11111111	*Unassigned:* treated as an unknown IEI.

BCC

3G TS 24.069 version 3.1.0

The Broadcast Call Control (BCC) protocol is used by the Voice Group Call Service (VGCS) on the radio interface. It is one of the Connection Management (CM) sublayer protocols (see GSM 04.07).

Generally a number of mobiles stations (MS) participate in a broadcast call. Consequently, there is generally more than one MS with a BCC entity engaged in the same broadcast call, and there is one BCC entity in the network engaged in that broadcast call.

The MS ignores BCC messages sent in unacknowledged mode and which specify as destination a mobile identity which is not a mobile identity of that MS. Higher layers and the MM sub-layer decide when to accept parallel BCC transactions and when/whether to accept BCC transactions in parallel to other CM transactions.

The broadcast call may be initiated by a mobile user or by a dispatcher. The originator of the BCC transaction chooses the Transaction Identifier (TI).

The call control entities are described as communicating finite state machines which exchange messages across the radio interface and communicate internally with other protocol (sub)layers. In particular, the BCC protocol uses the MM and RR sublayer specified in GSM 04.08. The network should apply supervisory functions to verify that the BCC procedures are progressing and if not, take appropriate means to resolve the problems.

The elementary procedures in the BCC include:

- Broadcast call establishment procedures,
- Broadcast call termination procedures
- Broadcast call information phase procedures
- Various miscellaneous procedures.

All messages have the following header:

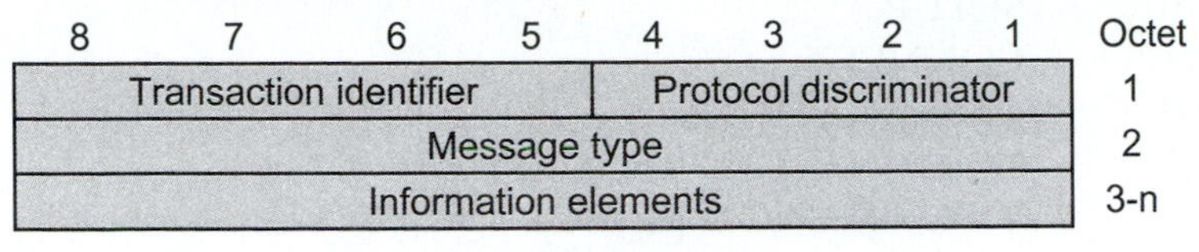

BCC header structure

Protocol discriminator
The protocol discriminator specifies the message being transferred

Transaction identifier
Distinguishes multiple parallel activities (transactions) within one mobile station. The format of the transaction identifier is as follows:

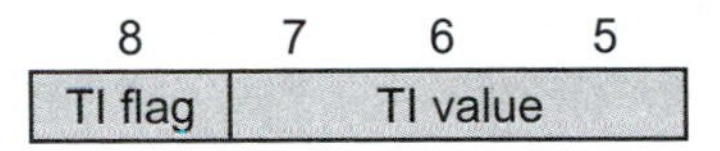

Transaction identifier

TI flag
Identifies who allocated the TI value for this transaction. The purpose of the TI flag is to resolve simultaneous attempts to allocate the same TI value.

TI value
The side of the interface initiating a transaction assigns TI values. At the beginning of a transaction, a free TI value is chosen and assigned to this transaction. It then remains fixed for the lifetime of the transaction. After a transaction ends, the associated TI value is free and may be reassigned to a later transaction. Two identical transaction identifier values may be used when each value pertains to a transaction originated at opposite ends of the interface.

Message type
The message type defines the function of each BCC message. The message type defines the function of each BCC message. The following message types exist:

0x110001	IMMEDIATE SETUP
0x110010	SETUP
0x110011	CONNECT
0x110100	TERMINATION
0x110101	TERMINATION REQUEST
0x110110	TERMINATION REJECT
0x111000	STATUS
0x111001	GET STATUS
0x111010	SET PARAMETER

Information elements

Each information element has a name which is coded as a single octet. The length of an information element may be fixed or variable and a length indicator for each one may be included.

GCC

3G TS 24.068 version 3.1.0

The Group Call Control (GCC) protocol is used by the Voice Group Call Service (VGCS) on the radio interface within the 3GPP system. It is one of the Connection Management (CM) sublayer protocols (see GSM 04.07).

Generally a number of mobiles stations (MS) participate in a group call. Consequently, there is in general more than one MS with a GCC entity engaged in the same group call, and there is one GCC entity in the network engaged in that group call.

The MS ignores GCC messages sent in unacknowledged mode and which specify as destination a mobile identity which is not a mobile identity of that MS. Higher layers and the MM sub-layer decide when to accept parallel GCC transactions and when/whether to accept GCC transactions in parallel to other CM transactions.

The group call may be initiated by a mobile user or by a dispatcher. In certain situations, a MS assumes to be the originator of a group call without being the originator. The originator of the GCC transaction chooses the Transaction Identifier (TI).

The call control entities are described as communicating finite state machines which exchange messages across the radio interface and communicate internally with other protocol (sub) layers. In particular, the GCC protocol uses the MM and RR sublayer specified in GSM 04.08. The network should apply supervisory functions to verify that the GCC procedures are progressing and if not, take appropriate means to resolve the problems.

The elementary procedures in the GCC include:

- Group call establishment procedures,
- Group call termination procedures
- Call information phase procedures
- Various miscellaneous procedures.

All messages have the following header:

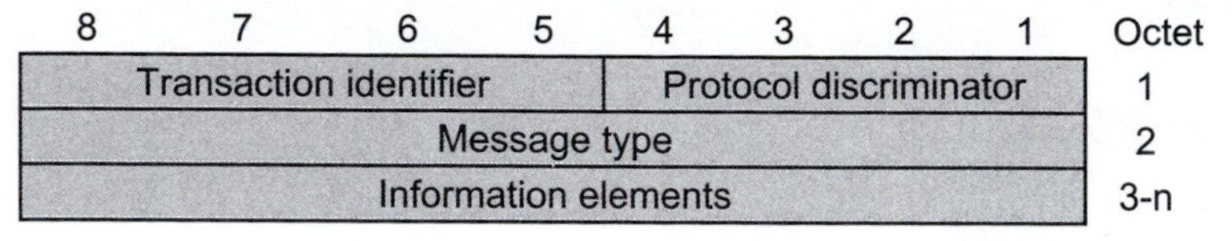

GCC header structure

Protocol discriminator
The protocol discriminator specifies the message being transferred

Transaction identifier
Distinguishes multiple parallel activities (transactions) within one mobile station. The format of the transaction identifier is as follows:

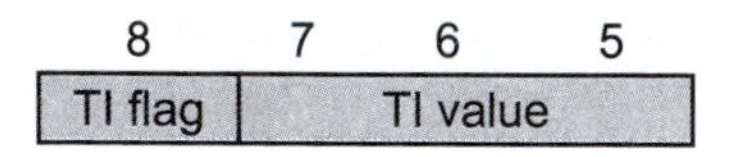

Transaction identifier

TI flag
Identifies who allocated the TI value for this transaction. The purpose of the TI flag is to resolve simultaneous attempts to allocate the same TI value.

TI value
The side of the interface initiating a transaction assigns TI values. At the beginning of a transaction, a free TI value is chosen and assigned to this transaction. It then remains fixed for the lifetime of the transaction. After a transaction ends, the associated TI value is free and may be reassigned to a later transaction. Two identical transaction identifier values may be used when each value pertains to a transaction originated at opposite ends of the interface.

Message type
The message type defines the function of each GCC message. The following message types exist:

0x110001	IMMEDIATE SETUP
0x110010	SETUP
0x110011	CONNECT
0x110100	TERMINATION
0x110101	TERMINATION REQUEST
0x110110	TERMINATION REJECT
0x111000	STATUS
0x111001	GET STATUS
0x111010	SET PARAMETER

Information elements

Each information element has a name which is coded as a single octet. The length of an information element may be fixed or variable and a length indicator for each one may be included.

GMM

GSM 04.08 http://www.etsi.org

GPRS uses the GSM MM (Mobility Management) protocol. Here it is known as the GPRS MM protocol (GMM). The main function of the MM sub-layer is to support the mobility of user terminals, for instance, informing the network of its present location and providing user identity confidentiality. A further function of the GMM sub-layer is to provide connection management services to the different entities of the upper Connection Management (CM) sub-layer.

The format of the header is shown in the following illustration:

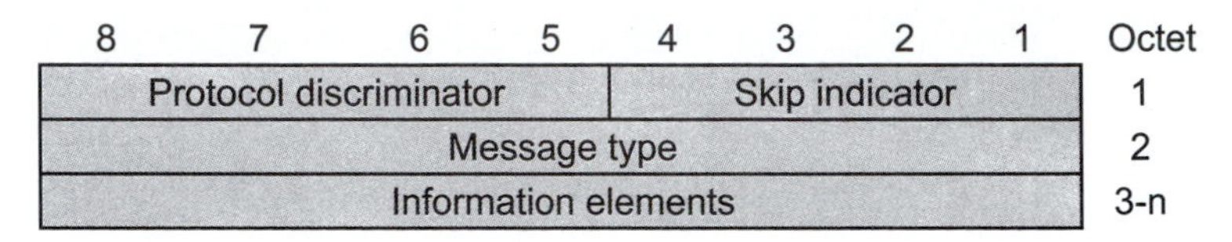

GMM header structure

Protocol discriminator

1000 identifies the GMM protocol.

Skip indicator

The value of this field is 0000.

Message type

Uniquely defines the function and format of each GMM message. The message type is mandatory for all messages. Bit 8 is reserved for possible future use as an extension bit. Bit 7 is reserved for the send sequence number in messages sent from the mobile station. GMM message types may be:

0 0 0 0 0 0 0 1	Attach request
0 0 0 0 0 0 1 0	Attach accept
0 0 0 0 0 0 1 1	Attach complete
0 0 0 0 0 1 0 0	Attach reject
0 0 0 0 0 1 0 1	Detach request
0 0 0 0 0 1 1 0	Detach accept
0 0 0 0 1 0 0 0	Routing area update request

0 0 0 0 1 0 0 1	Routing area update accept
0 0 0 0 1 0 1 0	Routing area update complete
0 0 0 0 1 0 1 1	Routing area update reject
0 0 0 1 0 0 0 0	P-TMSI reallocation command
0 0 0 1 0 0 0 1	P-TMSI reallocation complete
0 0 0 1 0 0 1 0	Authentication and ciphering req
0 0 0 1 0 0 1 1	Authentication and ciphering resp
0 0 0 1 0 1 0 0	Authentication and ciphering rej
0 0 0 1 0 1 0 1	Identity request
0 0 0 1 0 1 1 0	Identity response
0 0 1 0 0 0 0 0	GMM status
0 0 1 0 0 0 0 1	GMM information

Information elements

Various information elements.

GSM

GSM 04.08 http://www.etsi.org

The main function of the GPRS session management (SM) is to support PDP context handling of the user terminal. The SM comprises procedures for: identified PDP context activation, deactivation and modification, and anonymous PDP context activation and deactivation.
The format of the header is shown in the following illustration:

8	7	6	5	4	3	2	1	Octet
Protocol discriminator				Skip indicator				1
Message type								2
Information elements								3-n

GSM header structure

Protocol discriminator

1010 identifies the GSM protocol.

Skip indicator

The value of this field is 0000.

Message type

Uniquely defines the function and format of each GSM message. The message type is mandatory for all messages. Bit 8 is reserved for possible future use as an extension bit. Bit 7 is reserved for the send sequence number in messages sent from the mobile station. GSM message types may be:

0 1 x x x x x x	Session management messages
0 1 0 0 0 0 0 1	Activate PDP context request
0 1 0 0 0 0 1 0	Activate PDP context accept
0 1 0 0 0 0 1 1	Activate PDP context reject
0 1 0 0 0 1 0 0	Request PDP context activation
0 1 0 0 0 1 0 1	Request PDP context activation rej.
0 1 0 0 0 1 1 0	Deactivate PDP context request
0 1 0 0 0 1 1 1	Deactivate PDP context accept
0 1 0 0 1 0 0 0	Modify PDP context request
0 1 0 0 1 0 0 1	Modify PDP context accept

0 1 0 1 0 0 0 0	Activate AA PDP context request
0 1 0 1 0 0 0 1	Activate AA PDP context accept
0 1 0 1 0 0 1 0	Activate AA PDP context reject
0 1 0 1 0 0 1 1	Deactivate AA PDP context request
0 1 0 1 0 1 0 0	Deactivate AA PDP context accept
0 1 0 1 0 1 0 1	SM Status

Information elements

Various information elements.

GTP

GSM 09.60 version 6.1.0 http://www.etsi.fr

GPRS Tunnelling Protocol (GTP) is the protocol between GSN nodes in the GPRS backbone network. GTP is defined both for the Gn interface, i.e. the interface between GSNs within a PLMN, and the Gp interface between GSNs in different PLMNs. GTP is encapsulated within UDP.

GTP allows multiprotocol packets to be tunnelled through the GPRS Backbone between GPRS Support Nodes (GSNs). In the signalling plane, GTP specifies a tunnel control and management protocol which allows the SGSN to provide GPRS network access for an MS. Signalling is used to create, modify and delete tunnels. In the transmission plane, GTP uses a tunnelling mechanism to provide a service for carrying user data packets. The choice of path is dependent on whether the user data to be tunnelled requires a reliable link or not.

The GTP protocol is implemented only by SGSNs and GGSNs. No other systems need to be aware of GTP. GPRS MSs are connected to a SGSN without being aware of GTP. It is assumed that there will be a many-to-many relationship between SGSNs and GGSNs. An SGSN may provide service to many GGSNs. A single GGSN may associate with many SGSNs to deliver traffic to a large number of geographically diverse mobile stations.

The GTP header is a fixed format 16 octet header used for all GTP messages.

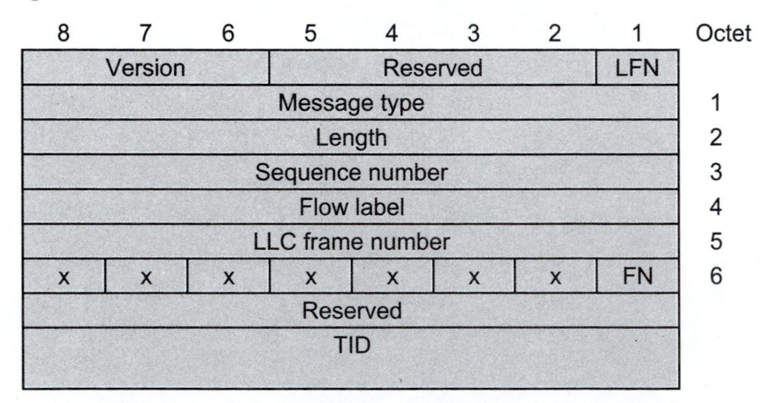

GTP header structure

Version
Set to 0 to indicate the first version of GTP.

Reserved
Reserved bits for future use, set to 1.

LFN
LLC frame number. Flag indicating whether the LLC frame number is included or not, set to 0 in signalling messages.

Message type
Indicates the type of GTP message. In signalling messages, it is set to the unique value that is used for each type of signalling message.

Length
Indicates the length in octets of the GTP message (G-PDU). In signalling messages, this is the length, in octets, of the signalling message including the GTP header.

Sequence number
A transaction identity for signalling messages and an increasing sequence number for tunneled T-PDUs.

Flow label
Identifies unambiguously a GTP flow. In signalling Path Management messages and Location Management messages, the flow label is not used and is set to 0.

LLC frame number
Used at the inter-SGSN routing update procedure to co-ordinate the data transmission on the link layer between the MS and SGSN. Not used for signalling, set to 225 by the sender and ignored by the receiver.

TID
The tunnel identifier that points out MM and PDP contexts in the destination GSN. In signalling messages, it is set to 0 in all V Management messages, Location Management messages and Mobility Management messages. The format of the TID is as follows:

8	7	6	5	4	3	2	1	Octet
MCC digit 2				MCC digit 1				1
MNC digit 1				MCC digit 3				2
MSIN digit 1				MNC digit 2				3
MSIN digit 3				MSIN digit 2				4
MSIN digit 5				MSIN digit 4				5
MSIN digit 7				MSIN digit 6				6
MSIN digit 9				MSIN digit 8				7
NSAPI				MSIN digit 10				8

TID structure

MCC, MNC, MSIN digits
Parts of the IMSI (defined in GMS 04.08).

NSAPI
Network service access point identifier.

LLC

GSM 04.65 version 6.1.0 http://www.etsi.fr

LLC defines the logical link control layer protocol to be used for packet data transfer between the mobile station (MS) and a serving GPRS support node (SGSN). LLC spans from the MS to the SGSN and is intended for use with both acknowledged and unacknowledged data transfer.

The frame formats defined for LLC are based on those defined for LAPD and RLP. However, there are important differences between LLC and other protocols, in particular with regard to frame delimitation methods and transparency mechanisms. These differences are necessary for independence from the radio path.

LLC supports two modes of operation:

- Unacknowledged peer-to-peer operation.
- Acknowledged peer-to-peer operation.

All LLC layer peer-to-peer exchanges are in frames of the following format:

8 7 6 5 4 3 2 1	Octet
Address	1
Control	2
Information	
FCS	

LLC header structure

Address field

The address field contains the SAPI and identifies the DLCI for which a downlink frame is intended and the DLCI transmitting an uplink frame. The length of the address field is 1 byte and it has the following format:

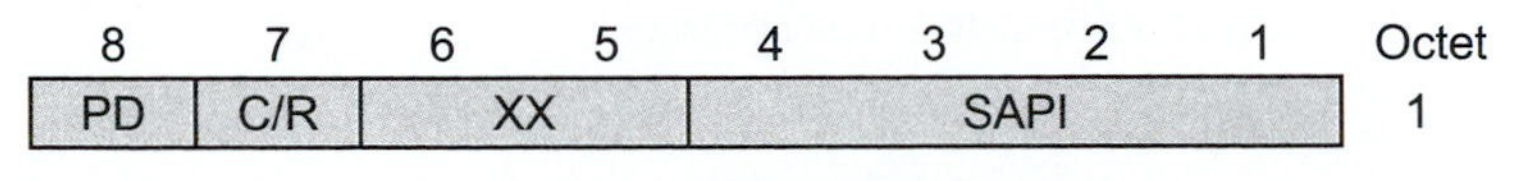

LLC address field structure

PD

Protocol discriminator bit indicates whether a frame is an LLC frame or belongs to a different protocol. LLC frames have the PD bit set to 0. If a frame with the PD bit set to 1 is received, then it is treated as an invalid frame.

C/R

Identifies a frame as either a command or a response. The MS side sends commands with the C/R bit set to 0, and responses with the C/R bit set to 1. The SGSN side does the opposite; i.e., commands are sent with C/R set to 1 and responses are sent with C/R set to 0. The combinations for the SGSN side and MS side are as follows.

Type	*Direction*	*C/R value*
Command	SGSN side to MS side	1
Command	MS side to SGSN side	0
Response	SGSN side to MS side	0
Response	MS side to SGSN side	1

XX

Reserved (2 bits).

SAPI

Service Access Point Identifier identifies a point at which LLC services are provided by an LLE to a layer-3 entity.

Control

Identifies the type of frame. Four types of control field formats are specified:

- Confirmed information transfer (I format)
- Supervisory functions (S format)
- Unconfirmed information transfer (UI format)
- Control functions (U format)

Information

Contains the various commands and responses.

FCS

Frame check sequence consists of a 24 bit CRC code which is used to detect bit errors in the frame header and information fields.

RLP

GSM 04.22 version 7.0.1 http://www.etsi.fr

The Radio Link Protocol (RLP) for data transmission over the GSM PLMN covers the Layer 2 functionality of the ISO OSI reference model. It has been tailored to the needs of digital radio transmission and provides the OSI data link service. RLP spans from the Mobile Station (MS) to the interworking function located at the nearest Mobile Switching Center (MSC) or beyond. Three versions of RLP exist.

Version 0:	Single-link basic version
Version 1:	Single-link extended version
Version 2:	Multi-link version.

The RLP frames have a fixed length of either 240 or 576 bits consisting of a header, information field and an FCS field.

The format of the 240-bit frame is:

Header	Information	FCS
16 bit	200 bit	24 bit
24 bit	192 bit	24 bit

RLP 240-bit frame format

The header is 16 bits in versions 0 and 1 and in version 2 (U frames). It is 24 bits in version 2 (S and I+S frames).

The format of the 576-bit frame is:

Header	Information	FCS
16 bit	536 bit	24 bit
24 bit	528 bit	24 bit

RLP 576-bit frame format

The header is 16 bits in version 1 and version 2 (U frames), and 24 bits in version 2 (S and I+S) frames.

Header

Contains control information of one of the following 3 types: unnumbered protocol control information (U frames), supervisory information (S frames) and user information carrying supervisory information piggybacked (I+S frames).

FCS

This is the Frame Check Sequence field.

The RLP entity will be in the Asynchronous Balanced Mode (ABM), which is the data link operation mode; or Asynchronous Disconnected Mode (ADM), which is the data link non-operational mode.

Header structure of versions 0 and 1

N(S) is a bit 4 low order bit and N(R) is a bit 11 low order bit.

U	C/R	X	X	1	1	1	1	1	1	P/F	M1	M2	M3	M4	M5	X
S	C/R	S1	S2	0	1	1	1	1	1	P/F	N (R)					
I+S	C/R	S1	S2	N (S)						P/F	N (R)					

Bits 1-16

RLP version 0 and 1 header structure

Header structure of version 2

S is a L2R status Bit, N(S) is a bit 1 low order bit, N(R) is a bit 14 low order bit and UP is a UP bit.

U	C/R	X	X	1	1	1	1	1	1	P/F	M1	M2	M3	M4	M5	X		
S	X	X	X	0	1	1	1	1	1	P/F	C/R	S1	S2	N(R)			X	UP
I+S	N(S)									P/F	C/R	S1	S2	N(R)			S	UP

Bits 1-24

RLP version 2 header structure

C/R

The Command Response Bit indicates whether the frame is a command or a response frame. It can have the following values:

1 command
0 response

P/F

The Poll/Final bit marks a special instance of command/response exchange

X

Don't care

Unnumbered Frames (U)

The M1 M2 M3 M4 and M5 bits have the following values in the U frames according to the type of information carried:

SABM	11100
UA	00110
DISC	00010
DM	11000
NULL	11110
UI	00000
XID	11101
TEST	00111
REMAP	10001

SABM 11100

The Set Asynchronous balance mode is used either to initiate a link for numbered information transfer or to reset a link already established.

UA 00110

The Unnumbered Acknowledge is used as a response to acknowledge an SABMM or DISC command.

DISC 00010

The disconnect is used to disestablish a link previously established for information transfer.

DM 11000

The disconnected mode encoding is used as a response message.

NULL 11110

UI 00000

Unnumbered information signifies that the information field is to be interpreted as unnumbered information.

XID 11101

Exchange Identification signifies that the information field is to be interpreted as exchange identification, and is used to negotiate and renegotiate parameters of RLP and layer 2 relay functions.

TEST 00111

The information field of this frame is test information.

REMAP 0001

A remap exchange takes place in ABM following a change of channel coding. If an answer is not received within a specific time, then the mobile end enters ADM.

S and I+S frames

N(S)

The send sequence number contains the number of the I frame.

N(R)

The Receive sequence number is used in ABM to designate the next information frame to be sent and to confirm that all frames up to and including this bit and been received correctly.

S

S represents the L2 status bit.

The S1, S2 bits can have the following significance in the S and I+S frames:
RR 00
REJ 01
RNR 10
SREJ 11

RR

Receive Ready can be used either as a command or a response. It clears any previous busy condition in that area.

REJ

The Reject encoding is used to indicate that in numbered information transfer 1 or more out-of-sequence frames have been received.

RNR

The Receive Not Ready indicates that the entity is not ready to receive numbered information frames.

SREJ

Selective Reject is used to request retransmission of a single frame.

UP

This is used in version 2 to indicate that a service level upgrading will increase the throughput.

SNDCP

GSM 04.65 version 6.1.0 http://www.etsi.fr

The Sub-Network Dependant Convergence Protocol (SNDCP) uses the services provided by the Logical Link Control (LLC) layer and the Session Management (SM) sub-layer. SNDCP splits into either IP or X.25.

The main functions of SNDCP are:

- Multiplexing of several PDPs (packet data protocol).
- Compression/decompression of user data.
- Compression/decompression of protocol control information.
- Segmentation of a network protocol data unit (N-PDU) into Logical Link Control Protocol Data Units (LL-PDUs) and re-assembly of LL-PDUs into a N-PDU.

The SN-DATA PDU is used for acknowledged data transfer. Its format is as follows:

8	7	6	5	4	3	2	1	Octet
X	C	T	M	NSAPI				1
DCOMP				PCOMP				2
Data								3-n

SN-DATA PDU structure

The SN-UNITDATA PDU is used for unacknowledged data transfer. Its format is as follows:

8	7	6	5	4	3	2	1	Octet
X	C	T	M	NSAPI				1
DCOMP				PCOMP				2
Segment offset				N-PDU number				3
E	N-PDU number (continued)							4
N-PDU number (extended)								5
Data								6-n

SN-UNITDATA PDU structure

NSAPI

Network service access point identifier. Values may be:

0 Escape mechanisms for future extensions.
1 Point-to-mutlipoint multicast (PTM-M) information.
2-4 Reserved for future use.
5-15 Dynamically allocated NSAPI value.

M

More bit. Values may be:

0 Last segment of N-PDU.
1 Not the last segment of N-PDU, more segments to follow.

T

SN-PDU type specifies whether the PDU is SN-DATA (0) or SN-UNITDATA (1).

C

Compression indicator. A value of 0 indicates that compression fields, DCOMP and PCOMP, are not included. A value of 1 indicates that these fields are included.

X

Spare bit is set to 0.

DCOMP

Data compression coding, included if C-bit set. Values are as follows:

0 No compression.
1-14 Points to the data compression identifier negotiated dynamically.
15 Reserved for future extensions.

PCOMP

Protocol control information compression coding, included if C-bit set. Values are as follows:

0 No compression.
1-14 Points to the protocol control information compression identifier negotiated dynamically.
15 Reserved for future extensions.

Segment offset

Segment offset from the beginning of the N-PDU in units of 128 octets.

N-PDU number

0-2047 when the extension bit is set to 0.
2048-524287 if the extension bit is set to 1.

E

Extension bit for N-PDU number.

0 Next octet is used for data.
1 Next octet is used for N-PDU number extensions.

14

GR303 (IDLC)

GR-303-CORE Issue 1

The Integrated Digital Loop Carrier (IDLC) is required when a digital loop carrier system is integrated into a local digital link. The link access procedure on the D Channel for ISDN LAPD is the basis for the call processing and operations data-link layer protocol of the IDLC interface. The IDLC is above the LAPD protocol.

The format of the IDLC header is shown in the following illustration:

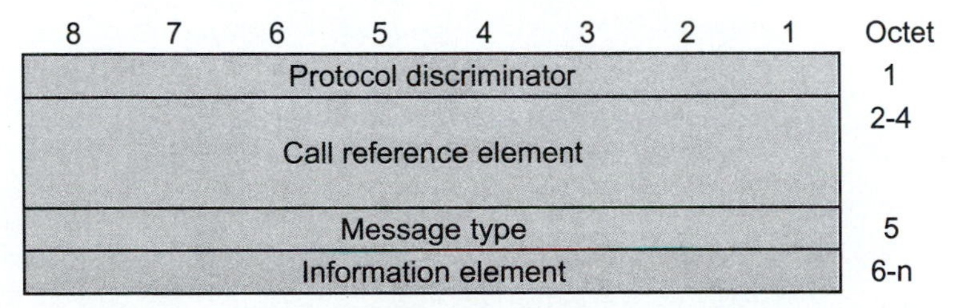

IDLC header structure

Protocol discriminator

Used in both the TMC and CSC reference models to distinguish IDLC call processing messages from standard messages.

Call reference

Used on the TMC and CSC to identify the call (and customer) to which a message applies.

Message type

Used on both the TMC and CSC to identify the message being sent. Available message types include: Alerting, Call Proceeding, Setup, Connect, Setup-Acknowledge, Connect Acknowledge, Disconnect, Release, Release Complete, Notify, Status-Enquiry, Information, Status.

15

H.323 Protocols

The H.323 standard provides a foundation for audio, video, and data communications across IP-based networks, including the Internet. H.323 is an umbrella recommendation from the International Telecommunications Union (ITU) that sets standards for multimedia communications over Local Area Networks (LANs) that do not provide a guaranteed Quality of Service (QoS). These networks dominate today's corporate desktops and include packet-switched TCP/IP and IPX over Ethernet, Fast Ethernet and Token Ring network technologies. Therefore, the H.323 standards are important building blocks for a broad new range of collaborative, LAN-based applications for multimedia communications. They include parts of H.225.0-RAS, Q.931-H.245, RTP/RTCP and audio/video/data codecs, such as the audio codecs (G.711, G.723.1, G.728, etc.), video codecs (H.261, H.263) that compress and decompress media streams and data codecs (T.120).

Media streams are transported on RTP/RTCP. RTP carries the actual media and RTCP carries status and control information. The signalling, with the exception of RAS, is transported reliably over TCP. The following protocols deal with signalling:

- RAS: manages registration, admission and status.

- Q.931: manages call setup and termination.
- H.245: negotiates channel usage and capabilities.

In addition, the following protocols provide optional features within the H.323 framework:

- H.235: security and authentication.
- H.450.x: supplementary services.

The following diagram illustrates the H.323 protocols in relation to the OSI model:

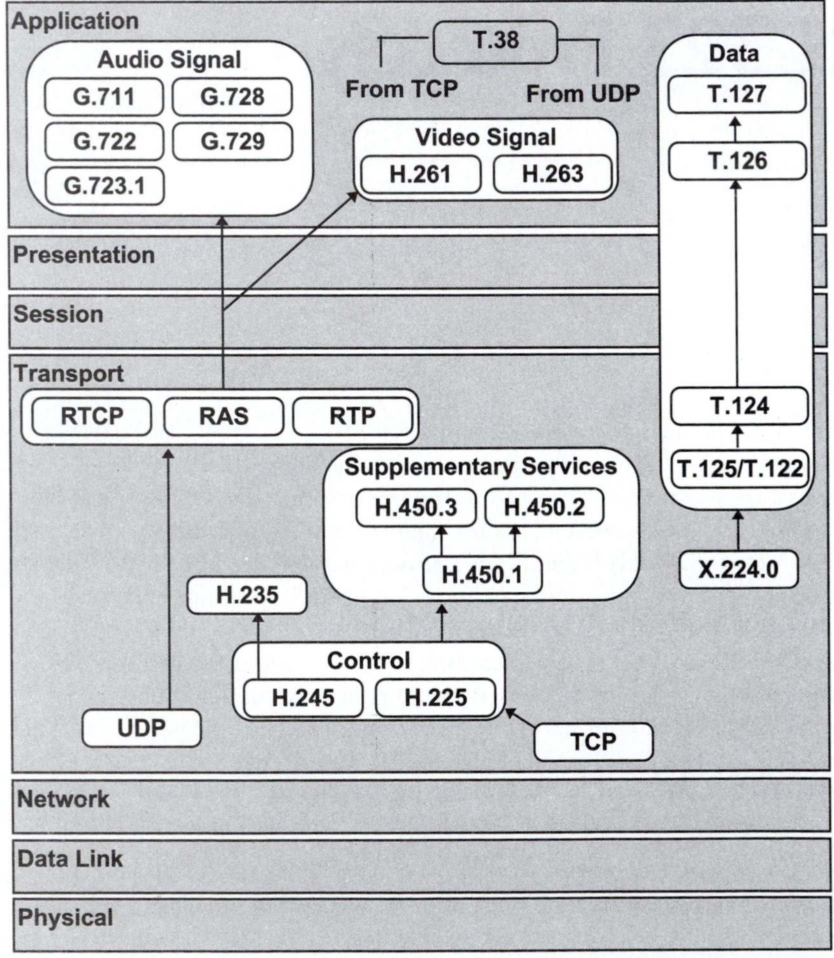

H.323 protocols in relation to the OSI model

RTP

RFC 1889 http://www.cis.ohio-state.edu/htbin/rfc/rfc1889.html

The Real-time Transport (RTP) Protocol provides end-to-end network transport functions suitable for applications transmitting real-time data such as audio, video or simulation data, over multicast or unicast network services. RTP does not address resource reservation and does not guarantee quality-of-service for real-time services. The data transport is augmented by a control protocol (RTCP) to allow monitoring of the data delivery in a manner scalable to large multicast networks, and to provide minimal control and identification functionality. RTP and RTCP are designed to be independent of the underlying transport and network layers. The protocol supports the use of RTP-level translators and mixers.

The format of the RTP Fixed Header Fields is shown in the following illustration:

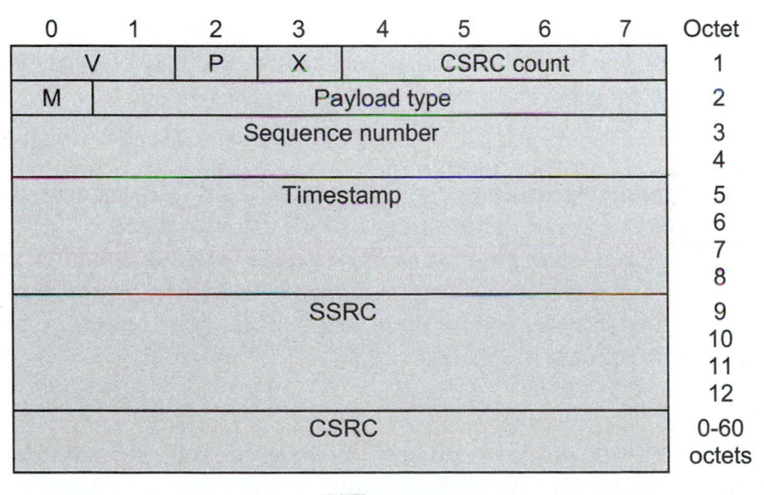

RTP structure

V

Version. Identifies the RTP version.

P

Padding. When set, the packet contains one or more additional padding octets at the end which are not part of the payload.

X
Extension bit. When set, the fixed header is followed by exactly one header extension, with a defined format.

CSRC count
Contains the number of CSRC identifiers that follow the fixed header.

M
Marker. The interpretation of the marker is defined by a profile. It is intended to allow significant events such as frame boundaries to be marked in the packet stream.

Payload type
Identifies the format of the RTP payload and determines its interpretation by the application. A profile specifies a default static mapping of payload type codes to payload formats. Additional payload type codes may be defined dynamically through non-RTP means.

Sequence number
Increments by one for each RTP data packet sent, and may be used by the receiver to detect packet loss and to restore packet sequence.

Timestamp
Reflects the sampling instant of the first octet in the RTP data packet. The sampling instant must be derived from a clock that increments monotonically and linearly in time to allow synchronization and jitter calculations. The resolution of the clock must be sufficient for the desired synchronization accuracy and for measuring packet arrival jitter (one tick per video frame is typically not sufficient).

SSRC
Identifies the synchronization source. This identifier is chosen randomly, with the intent that no two synchronization sources within the same RTP session will have the same SSRC identifier.

CSRC
Contributing source identifiers list. Identifies the contributing sources for the payload contained in this packet.

RTCP

RFC 1889 http://www.cis.ohio-state.edu/htbin/rfc/rfc1889.html

The RTP control protocol (RTCP) is based on the periodic transmission of control packets to all participants in the session, using the same distribution mechanism as the data packets. The underlying protocol must provide multiplexing of the data and control packets, for example using separate port numbers with UDP.

The format of the header is shown in the following illustration:

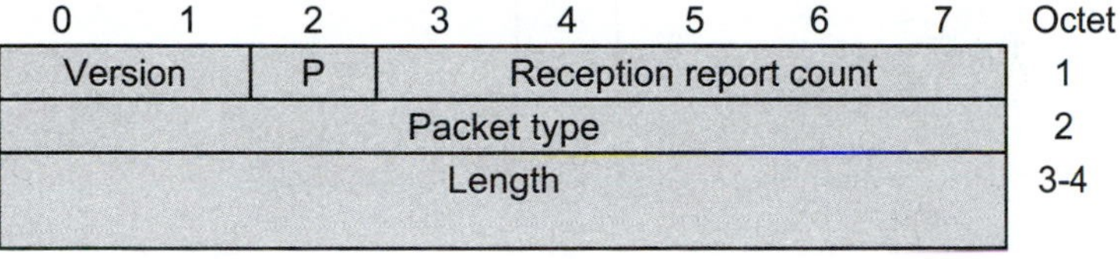

RTCP structure

Version

Identifies the RTP version which is the same in RTCP packets as in RTP data packets. The version defined by this specification is two (2).

P

Padding. When set, this RTCP packet contains some additional padding octets at the end which are not part of the control information. The last octet of the padding is a count of how many padding octets should be ignored. Padding may be needed by some encryption algorithms with fixed block sizes. In a compound RTCP packet, padding should only be required on the last individual packet because the compound packet is encrypted as a whole.

Reception report count

The number of reception report blocks contained in this packet. A value of zero is valid.

Packet type

Contains the constant 200 to identify this as an RTCP SR packet.

Length

The length of this RTCP packet in 32-bit words minus one, including the header and any padding. (The offset of one makes zero a valid length and avoids a possible infinite loop in scanning a compound RTCP packet, while counting 32-bit words avoids a validity check for a multiple of 4.)

RAS

H.225: http://www.itu.int/itudoc/itu-t/rec/h/h225-0.html

Registration, Admission and Status (RAS) channel is used to carry messages used in the gatekeeper discovery and endpoint registration processes which associate an endpoint's alias address with its call signalling channel transport address. Since the RAS channel is an unreliable channel, H.225.0 recommends timeouts and retry counts for various messages. An endpoint or gatekeeper which cannot respond to a request within the specified timeout may use the Request in Progress (RIP) message to indicate that it is still processing the request. An endpoint or gatekeeper receiving the RIP, resets its timeout timer and retry counter.

RAS messages are in ASN.1 syntax. They consist of an exchange of messages.

H.225

H.225: http://www.itu.int/itudoc/itu-t/rec/h/h225-0.html

H.225.0 is a standard which covers narrow-band visual telephone services defined in H.200/AV.120-Series Recommendations. It specifically deals with those situations where the transmission path includes one or more packet based networks, each of which is configured and managed to provide a non-guaranteed QoS, which is not equivalent to that of N-ISDN, such that additional protection or recovery mechanisms beyond those mandated by Rec. H.320 are necessary in the terminals. H.225.0 describes how audio, video, data and control information on a packet based network can be managed to provide conversational services in H.323 equipment.

The structure of H.225 follows the Q.931 standard as shown in the following illustration:

8	7	6	5	4	3	2	1	Octet
Protocol discriminator								1
0	0	0	0	Length of call ref				2
Call reference value								3 (-4)
0	Message type							
Information elements								

H.225 structure

Protocol discriminator
Distinguishes messages for user-network call control from other messages.

Length of call ref
The length of the call reference value.

Call reference value
Identifies the call or facility registration/cancellation request at the local user-network interface to which the particular message applies. May be up to 2 octets in length.

Message type

Identifies the function of the message sent. The following message types are used:

000	xxxxx	Call establishment messages:
	00001	ALERTING
	00010	CALL PROCEEDING
	00111	CONNECT
	01111	CONNECT KNOWLEDGE
	00011	PROGRESS
	00101	SETUP
	01101	SETUP ACKNOWLEDGE
001	xxxxx	Call information phase messages:
	00110	RESUME
	01110	RESUME ACKNOWLEDGE
	00010	RESUME REJECT
	00101	SUSPEND
	01101	SUSPEND ACKNOWLEDGE
	00001	SUSPEND REJECT
	00000	USER INFORMATION
010	xxxxx	Call clearing messages:
	00101	DISCONNECT
	01101	RELEASE
	11010	RELEASE COMPLETE
	00110	RESTART
	01110	RESTART ACKNOWLEDGE
011	xxxxx	Miscellaneous messages:
	00000	SEGMENT
	11001	CONGESTION CONTROL
	11011	INFORMATION
	01110	NOTIFY
	11101	STATUS
	10101	STATUS ENQUIRY

Information elements

Two categories of information elements are defined: single octet information elements and variable length information elements, as shown in the following illustrations.

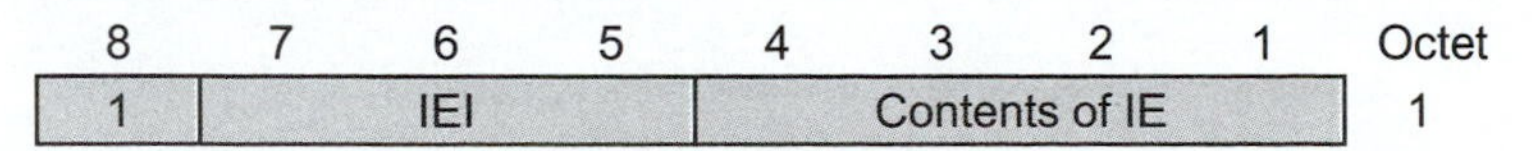

Single octet information element format (type 1)

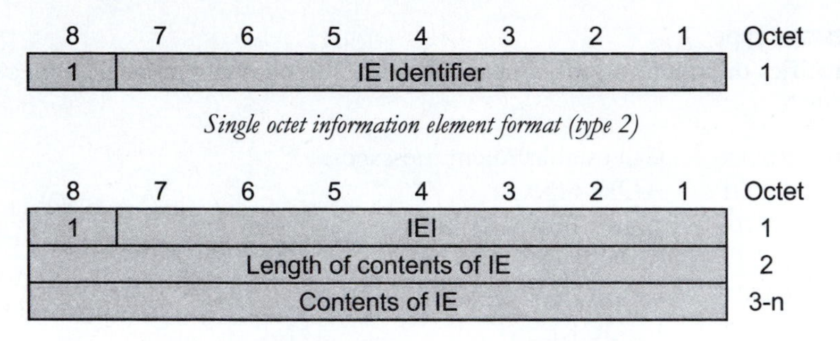

Single octet information element format (type 2)

Variable length information element format

H.245

H.245: http://www.itu.int/itudoc/itu-t/rec/h/h245.html

H.245 is line transmission of non-telephone signals. It includes receiving and transmitting capabilities as well as mode preference from the receiving end, logical channel signalling, and control and indication. Acknowledged signalling procedures are specified to ensure reliable audiovisual and data communication.

H.245 messages are in ASN.1 syntax. They consist of an exchange of messages.

MultimediaSystemControlMessage message types can be defined as request, response, command and indication messages. The following additional message sets are available:

- Master Slave Determination messages.
- Terminal Capability messages.
- Logical Channel signalling messages.
- Multiplex Table signalling messages.
- Request Multiplex Table signalling messages.
- Request Mode messages.
- Round Trip Delay messages.
- Maintenance Loop messages.
- Communication Mode messages.
- Conference Request and Response messages.
- Terminal ID.
- Commands and Indications.

H.261

H.261: http://www.cis.ohio-state.edu/htbin/rfc/rfc2032.html

The H.261 describes a video stream for transport using the real-time transport protocol, RTP, with any of the underlying protocols that carry RTP.

The format of the header is shown in the following illustration:

<table>
<tr><th>0</th><th>1</th><th>2</th><th>3</th><th>4</th><th>5</th><th>6</th><th>7</th><th>Octet</th></tr>
<tr><td colspan="3">SBIT</td><td colspan="3">EBIT</td><td>I</td><td>V</td><td>1</td></tr>
<tr><td colspan="4">GOBN</td><td colspan="4">MBAP</td><td>2</td></tr>
<tr><td>MBAP</td><td colspan="5">QUANT</td><td colspan="2">HMVD</td><td>3</td></tr>
<tr><td colspan="3">HMVD</td><td colspan="5">VMVD</td><td>4</td></tr>
</table>

H.261 header structure

SBIT
Start bit. Number of most significant bits that are to be ignored in the first data octet.

EBIT
End bit. Number of least significant bits that are to be ignored in the last data octet.

I
INTRA-frame encoded data flag. Set to 1 if this stream contains only INTRA-frame coded blocks. Set to 0 if this stream may or may not contain INTRA-frame coded blocks. The sense of this bit may not change during the course of the RTP session.

V
Motion Vector flag. Set to 0 if motion vectors are not used in this stream. This is set to 1 if motion vectors may or may not be used in this stream. The sense of this bit may not change during the course of the session.

GOBN
GOB number. Encodes the GOB number in effect at the start of the packet. This is set to 0 if the packet begins with a GOB header.

MBAP

Macroblock Address Predictor. Encodes the macroblock address predictor (i.e., the last MBA encoded in the previous packet). This predictor ranges from 0-32 (to predict the valid MBAs 1-33), but because the bit stream cannot be fragmented between a GOB header and MB 1, the predictor at the start of the packet can never be 0. Therefore, the range is 1-32, which is biased by -1 to fit in 5 bits. This is set to 0 if the packet begins with a GOB header.

QUANT

Quantizer. Shows the Quantizer value (MQUANT or GQUANT) in effect prior to the start of this packet. Set to 0 if the packet begins with a GOB header.

HMVD

Horizontal Motion Vector Data. Represents the reference horizontal Motion Vector Data (MVD). Set to 0, if V flag is 0 or if the packet begins with a GOB header, or when the MTYPE of the last MB encoded in the previous packet was not MC. HMVD is encoded as a 2's complement number and 10000 corresponding to the value -16 is forbidden (motion vector fields range from +/-15).

VMVD

Vertical Motion Vector Data. Represents the reference vertical Motion Vector Data (MVD). Set to 0 if V flag is 0, or if the packet begins with a GOB header, or when the MTYPE of the last MB encoded in the previous packet was not MC. VMVD is encoded as a 2's complement number and 10000 corresponding to the value -16 is forbidden (motion vector fields range from +/-15).

H.263

RFC 2190 (RTP): http://www.cis.ohio-state.edu/htbin/rfc/rfc2190.html
H.263: http://www.itu.int/itudoc/itu-t/rec/h/h263.html

This protocol specifies the payload format for encapsulating an H.263 bitstream in the Real-time Transport Protocol (RTP). Three modes are defined for the H.263 payload header. An RTP packet can use one of the three modes for H.263 video streams depending on the desired network packet size and H.263 encoding options employed. The shortest H.263 payload header (mode A) supports fragmentation at Group of Block (GOB) boundaries. The long H.263 payload headers (modes B and C) support fragmentation at Macroblock (MB) boundaries.

For each RTP packet, the RTP fixed header is followed by the H.263 payload header, which is followed by the standard H.263 compressed bitstream. The size of the H.263 payload header is variable depending on the modes. The layout of an RTP H.263 video packet is as shown in the following illustration:

4 bytes
RTP header
H.263 payload header
H.263 bitstream

RTP H.263 video packet

In mode A, an H.263 payload header of four bytes is present before an actual compressed H.263 video bitstream in a packet. It allows fragmentation at GOB boundaries. In mode B, an eight byte H.263 payload header is used and each packet starts at MB boundaries, without the PB-frames option. Finally, a twelve byte H.263 payload header is defined in mode C to support fragmentation at MB boundaries for frames that are coded with the PB-frames option.

The mode of each H.263 payload header is indicated by the F and P fields in the header. Packets of different modes can be intermixed. The format of the header for mode A is shown in the following illustration:

1	2	3	4	5	6	7	8	Octet
F	P	SBIT			EBIT			1
SRC			I	U	S	A	R	2
R (cont.)			DBQ		TRB			3
TR								4

H.263 mode A payload header structure

F

Flag bit, indicates the mode of the payload header.

P

Optional PB-frames mode as defined by H.263.

0 Normal I or P frame.

1 PB-frames.

When F=1, P also indicates modes:

0 Mode B.

1 Mode C.

SBIT

Start bit, specifies the number of most significant bits that should be ignored in the first data byte.

EBIT

End bit, specifies the number of least significant bits that should be ignored in the last data byte.

SRC

Source format (bit 6, 7 and 8 in TYPE defined by H.263), specifies the resolution of the current picture.

I

Picture coding type (bit 9 in PTYPE defined by H.263):

0 Intra-coded.

1 Inter-coded.

U
Set to 1 if the Unrestricted Motion Vector option (bit 10 in PTYPE defined by H.263) was set to 1 in the current picture header, otherwise 0.

S
Set to 1 if the Syntax-based Arithmetic Coding option (bit 11 in PTYPE defined by H.263) was set to 1 for current picture header, otherwise 0.

A
Set to 1 if the Advanced Prediction option (bit 12 in PTYPE defined by H.263) was set to 1 for current picture header, otherwise 0.

R
Reserved, must be set to zero.

DBQ
Differential quantization parameter used to calculate the quantizer for the B frame based on the quantizer for the P frame, when the PB-frames option is used. The value should be the same as DBQUANT defined by H.263. Set to zero if the PB-frames option is not used.

TRB
Temporal reference for the B frame as defined by H.263. Set to zero if the PB-frames option is not used.

TR
Temporal reference for the P frame as defined by H.263. Set to zero if the PB-frames option is not used.

The format of the header for mode B is shown in the following illustration:

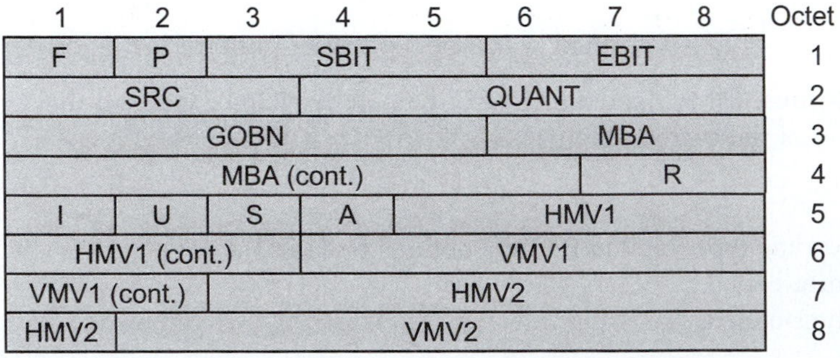

<table>
<tr><th>1</th><th>2</th><th>3</th><th>4</th><th>5</th><th>6</th><th>7</th><th>8</th><th>Octet</th></tr>
<tr><td>F</td><td>P</td><td colspan="3">SBIT</td><td colspan="3">EBIT</td><td>1</td></tr>
<tr><td colspan="3">SRC</td><td colspan="5">QUANT</td><td>2</td></tr>
<tr><td colspan="5">GOBN</td><td colspan="3">MBA</td><td>3</td></tr>
<tr><td colspan="6">MBA (cont.)</td><td colspan="2">R</td><td>4</td></tr>
<tr><td>I</td><td>U</td><td>S</td><td>A</td><td colspan="4">HMV1</td><td>5</td></tr>
<tr><td colspan="3">HMV1 (cont.)</td><td colspan="5">VMV1</td><td>6</td></tr>
<tr><td colspan="2">VMV1 (cont.)</td><td colspan="6">HMV2</td><td>7</td></tr>
<tr><td>HMV2</td><td colspan="7">VMV2</td><td>8</td></tr>
</table>

H.263 mode B payload header structure

F, P, SBIT, EBIT, SRC, I, U, S and A are defined the same as in mode A.

QUANT

Quantization value for the first MB coded at the start of the packet. Set to 0 if the packet begins with a GOB header.

GOBN

GOB number in effect at the start of the packet. GOB number is specified differently for different resolutions.

MBA

The address within the GOB of the first MB in the packet, counting from zero in scan order. For example, the third MB in any GOB is given MBA=2.

R

Reserved, set to zero.

HMV1, VMV1

Horizontal and vertical motion vector predictors for the first MB in this packet. When four motion vectors are used for the current MB with advanced prediction option, they are the motion vector predictors for block number 1 in the MB. Each 7-bit field encodes a motion vector predictor in half pixel resolution as a 2's complement number.

HMV2, VMV2

Horizontal and vertical motion vector predictors for block number 3 in the first MB in this packet when four motion vectors are used with the advanced prediction option. This is needed because block number 3 in the MB needs different motion vector predictors from other blocks in the MB. These two fields are not used when the MB only has one motion vector. Each 7-bit field encodes a motion vector predictor in half pixel resolution as a 2's complement number.

The format of the header for mode C is shown in the following illustration:

<table>
<tr><th>1</th><th>2</th><th>3</th><th>4</th><th>5</th><th>6</th><th>7</th><th>8</th><th>Octet</th></tr>
<tr><td>F</td><td>P</td><td colspan="3">SBIT</td><td colspan="3">EBIT</td><td>1</td></tr>
<tr><td colspan="3">SRC</td><td colspan="5">QUANT</td><td>2</td></tr>
<tr><td colspan="5">GOBN</td><td colspan="3">MBA</td><td>3</td></tr>
<tr><td colspan="6">MBA (cont.)</td><td colspan="2">R</td><td>4</td></tr>
<tr><td>I</td><td>U</td><td>S</td><td>A</td><td colspan="4">HMV1</td><td>5</td></tr>
<tr><td colspan="3">HMV1 (cont.)</td><td colspan="5">VMV1</td><td>6</td></tr>
<tr><td colspan="2">VMV1 (cont.)</td><td colspan="6">HMV2</td><td>7</td></tr>
<tr><td>HMV2</td><td colspan="7">VMV2</td><td>8</td></tr>
<tr><td colspan="8">RR</td><td>9</td></tr>
<tr><td colspan="3">RR (cont.)</td><td colspan="2">DBQ</td><td colspan="3">TRB</td><td>10</td></tr>
<tr><td colspan="8">TR</td><td>11</td></tr>
</table>

H.263 mode B payload header structure

F, P, SBIT, EBIT, SRC, I, U, S, A, DBQ, TRB and TR are defined the same as in mode A. QUANT, GOBN, MBA, HMV1, VMV1, HMV2, VMV2 are defined the same as in mode B.

RR

Reserved, set to zero.

H.235

H.235: http://www.itu.int/itudoc/itu-t/rec/h/h235.html

H.235 provides enhancements within the framework of the H.3xx-Series Recommendations to incorporate security services such as Authentication and Privacy (data encryption). H.235 should work with other H series protocols that utilize H.245 as their control protocol.

All H.235 messages are encrypted as in ASN.1.

H.450.1

The H.45o series defines Supplementary Services for H.323, namely Call Transfer and Call Diversion.

The H.450.1 protocol deals with the procedures and signalling protocol between H.323 entities for the control of supplementary services. This signalling protocol is common to all H.323 supplementary services. The protocol is derived from the generic functional protocol specified in ISO/IEC 11582 for Private Integrated Services Networks (PISN).

The H.450 protocol is used to exchange signalling information to control supplementary services over a LAN. It works together with the H.225 protocol.

This protocol has no header as all messages are in text, in ASN.1 format.

H.450.2

This is a Call Transfer supplementary service for H.323. The H.450.2 protocol describes the procedures and signalling protocol for the call transfer supplementary service in H.323 networks. This supplementary service allows the served user A to transform an existing call (from user A to B) to a new call between user B and a third user C selected by A. User A may or may not have a call established with the third user prior to the call transfer. This is based on H.450.1

This protocol has no header as all messages are in text, in ASN.1 format.

H.450.3

The H.450.3 is a call diversion supplementary service for H.323. It describes the procedures and signalling protocol for the call diversion supplementary service in H.323 networks. This includes the services Call Forwarding Unconditional (SS-CFU), Call Forwarding Busy (SS-CFB), Call Forwarding No Reply (SS-CFNR) and Call Deflection (SS-CD). These are all supplementary services, which apply during call establishment, providing a diversion of an incoming call to another destination endpoint. This is based on H.450.1

This protocol has no header as all messages are in text, in ASN.1 format.

T.38

The T.38 IP-based fax service maps the T.30 fax protocol onto an IP network. Both fax and voiced data are managed through a single gateway. T.38 uses 2 protocols, one for UDP packets and one for TCP packets. Data is encoded using ASN.1 to ensure a standard technique. It allows users to transfer facsimile documents between 2 standard fax terminals over the Internet or other network using IP protocols. H.323 can be used here in the same way that it is used to support Voice over IP.

TCP messages

The T.38 data (Internet Fax Protocol) is contained in the payload of the TCP or UDP messages. The T.38 packet provides an alert for the start of a message. An ASN.1 Application tag identifies it; if this tag is not present the session is aborted.

The following is the format of the TCP Internet Fax Protocol packets.

Type
Data

Type

The type field contains the type of message. It describes the function of and the data of the packet.

Type can be T30_Indicator or T30_Data

Data

The data field is dependent on the type field. It contains the T.30 HDLC control data and the Phase C image (or BFT) data. It contains one or more fields. Each field has 2 parts.

UDP messages

T.38 messages may also be sent over the UDP transport layer. The UDP header is followed by the UDPTL payload which consists of sequence number and a payload.

Sequence number	Mandatory message (primary)	Optional redundant message or optional FEC message

Sequence number

The sequence number is used to identify the sequencing in the payload.

T.125

The T.120 family of protocols describe protocols and services for multipoint Data Conferencing including multilayer protocols which considerably enhance multimedia, MCU and codec control capabilities, permitting greater MCU operational sophistication beyond that described in H.231 and H.243.

T.125 describes the Multipoint Communication Service Protocol (MCS).

It defines:

- Procedures for a single protocol for the transfer of data and control information from one MCS provider to a peer MCS provider.
- The structure and encoding of the MCS protocol data units used for the transfer of data and control information.

Procedures may be:

- Interactions between 2 parallel MCS providers by exchanging MCS protocol data.
- Interactions between MCS providers and users by exchanging MCS primitives.
- Interactions between an MSC provider and a transport service provider by exchanging transport service primitives.

The MCS provider communicates with MCS users through a MCSAP (MCS Service Access Point), by means of MCS primitives defined in T.122. MCSPDU (MCS protocol Data unit) exchanges occur between MCS providers that host the same MCS domain. The MCS provider can have multiple peers; each reached directly by a different MCS connection or indirectly through a peer MCS provider. An MCS connection is composed of either one MAP connection or one or more transport connections. The protocol exchanges are preformed using the transport layer using a pair of TSAPs (Transport Service Access Points).

The MCS PDU is the MCS protocol data unit. This is the information exchanged in the MCS protocol consisting of control information transferred between MCS providers to coordinate their joint operation and possibly data transferred on behalf of MCS users for whom they provide service. Each MCSPDU is transported as one TSDU (Transport service data unit) across a TC (Transport connection) belonging to an MCS connection. Connect MCSPDUs are unlimited in size. Domain MCSPDUs are limited in size by a parameter of the MCS domain.

The structure of Version 2 and Version 3 MCSPDUs is defined in ASN.1 and appears as text or numeric messages.

16

IBM Protocols

The Network Basic I/O System (NetBIOS), TCP/UDP version, was developed for the IBM PC LAN program to support communications between symbolically named stations and transfer of arbitrary data.

The Server Message Block (SMB) is a Microsoft presentation layer protocol providing file and print sharing functions for LAN Manager, VINES and other network operating systems. IBM NetBIOS manages the use of node names and transport layer connections for higher layer protocols such as SMB. The IBM suite includes the following protocols:

- NetBIOS: Network Basic I/O System.
- SMB: Server Message Block.
- SNA: Systems Network Architecture.
- HPR-APPN: High Performance Routing - Advanced Peer to Peer Network.
- NHDR: Network Layer Header.
- THDR: RTP Transport header.
- DLSw: Data Link Switching.

The following diagram illustrates the IBM protocol suite in relation to the OSI model:

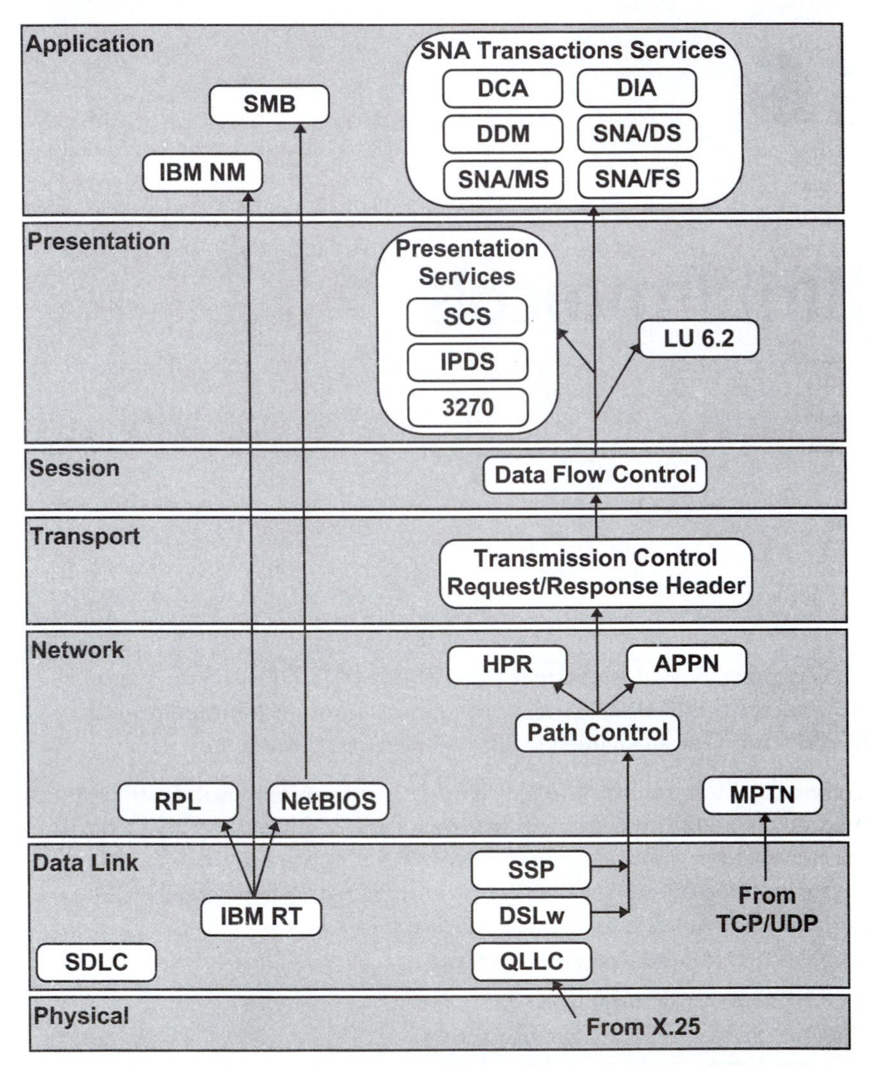

IBM protocol suite in relation to the OSI model

NetBIOS

IBM Local Area Network Technical Reference 1990 4th edition

NetBIOS provides a communication interface between the application program and the attached medium. All communication functions from the physical layer through the session layer are handled by NetBIOS, the adapter support software, and the adapter card. A NetBIOS session is a logical connection between any two names on the network. It is usually encapsulated over LLC.

The format of the header is shown in the following illustration:

Len	XxEFFF	Command	Optional Data 1	Optional Data 2	Xmit/resp correlator	Dest name/ num	Source name/ num

NetBIOS header structure

Len
The length of the NETBIOS header.

XxEFFF
A delimiter indicating that subsequent data is destined for the NetBIOS function.

Command
A specific protocol command that indicates the type of function of the frame.

Data 1
One byte of optional data per specific command.

Data 2
Two bytes of optional data per specific command.

Xmit/response correlator
Used to associate received responses with transmitted requests. Transmit correlator is the value returned in a response to a given query. Response correlator is the value expected when the response to that message is received.

Destination name/num

In non-session frames this field contains the 16-character name. In session frames this field contains a 1 byte destination session number.

Source name/num

In non-session frames this field contains the 16-character source name. In session frames this field contains a 1 byte source session number.

SMB

ftp://ftp.microsoft.com/developr/drg/CIFS/
File Sharing Protocol 1987, SMB File Sharing Protocol Extensions, 1998, SMB File Sharing Protocol 1996

Server Message Block (SMB) is a Microsoft presentation layer protocol providing file and print sharing functions for LAN Manager, Banyan VINES and other network operating systems. IBM NetBIOS manages the use of node names and transport layer connections for higher layer protocols such as SMB.

SMB is used for sharing files, printers, serial ports and communications abstractions, such as named pipes and mail slots between computers. It is a client server request-response protocol. Clients connect to servers using TCP/IP. They can then send commands (SMBs) to the server that allows them to access shares, open files, etc., over the network.

Many protocol variants have been developed. The first protocol variant was the Core Protocol, known also as PC Network Program 1.0. It handled a fairly basic set of operations that included:

- Connecting to and disconnecting from file and print shares.
- Opening and closing files.
- Opening and closing print files.
- Reading and writing files.
- Creating and deleting files and directories.
- Searching directories.
- Getting and setting file attributes.
- Locking and unlocking byte ranges in files.

There are several different versions and sub-versions of this protocol. A particular version is referred to as a dialect. When two machines first come into network contact they negotiate the dialect to be used. Different dialects can include both new messages as well as changes to the fields and semantics of existing messages in other dialects. Each server makes a set of resources available to clients on the network. A resource being shared may be a directory tree, named pipe, printer, etc. So far as clients are concerned, the server has no storage or service dependencies on any other servers; a client considers the server to be the sole provider of the resource being used.

The SMB protocol requires server authentication of users before file accesses are allowed, and each server authenticates its own users. A client system must send authentication information to the server before the server will allow access to its resources.

The general format of the header is shown in the following illustration:

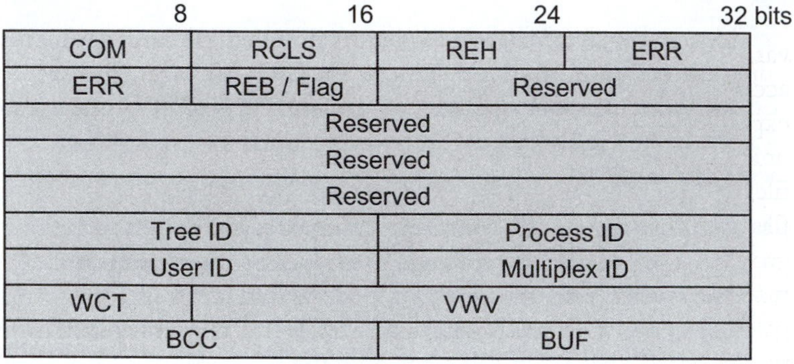

8	16	24	32 bits
COM	RCLS	REH	ERR
ERR	REB / Flag	Reserved	
Reserved			
Reserved			
Reserved			
Tree ID		Process ID	
User ID		Multiplex ID	
WCT	VWV		
BCC		BUF	

SMB frame structure

COM

Protocol commands. The following are possible commands within SMB frames:

Command	*Description.*
[bad command]	Invalid SMB command.
[bind (UNIX)]	Obtain file system address for file.
[cancel forward]	Cancel server recognition of name.
[change/check dir]	Change to directory or check path.
[change group]	Change group association of user.
[change password]	Change password of user.
[close file]	Close file handle and flush buffers.
[close spoolfile]	Close print buffer file.
[consumer logon]	Log on with consumer validation.
[copy file]	Copy file to specified path.
[copy new path]	Copy file to new path name.
[create & bind]	Create file and get file system address.
[create directory]	Create new directory.
[create file]	Create new or open existing file.
[delete dir]	Delete the specified directory.
[delete file]	Delete the specified file.
[echo]	Request echo from server.

Command	*Description.*
[find & close]	Search for file and close directory (UNIX).
[find & close /2]	Search for file and close directory (OS/2).
[find first file]	Find first matching file (OS/2).
[find unique]	Search directory for specified file.
[flush file]	Flush all file buffers to disk.
[fork to PID]	Provide same access rights to new process.
[forward name]	Cause server to accept messages for name.
[get access right]	Get access rights for specified file.
[get exp attribs]	Get expanded attributes for file (OS/2).
[get unix attribs]	Get expanded attributes for file (UNIX).
[get file attribs]	Get attributes for specified file.
[get file queue]	Get print queue listing.
[get group info]	Get logical group associations.
[get machine name]	Get machine name for block messages.
[get pathname]	Get path of specified handle.
[get resources]	Get availability of server resources.
[get server info]	Get total and free space for server disk.
[get user info]	Get logical user associations.
[IOCTL]	Initiate I/O control for DOS-OS/2 devices.
[IOCTL next]	Initiates subsequent I/O control for DOS-OS/2 devices.
[IOCTL (UNIX)]	I/O control for UNIX-Xenix devices.
[link file]	Make an additional path to a file.
[lock and read]	Lock and read byte range.
[lock bytes]	Lock specified byte range.
[lock/unlock & X]	Lock/unlock bytes and execute next command.
[logoff & execute]	Log off and execute next command.
[mail announce]	Query availability of server nodes.
[mailslot message]	Mail slot transaction message.
[make/bind dir]	Make dir and get file system address.
[make temp file]	Make temporary data file.
[make new file]	Make new file only if it does not exist.
[make node]	Make file for use as a device.
[move file]	Move file to specified path (OS/2).
[move new path]	Move file to specified path (UNIX/Xenix).
[multi-block data]	Send data for multi-block message.
[multi-block end]	Terminate multi-block message.
[multi-block hdr]	Send header for multi-block message.
[named pipe call]	Open, write, read, or close named pipe.
[named pipe wait]	Wait for named pipe to become ready.

Command	*Description.*
[named pipe peek]	Look at named pipe data.
[named pipe query]	Query named pipe handle modes.
[named pipe set]	Set named pipe handle modes.
[named pipe attr]	Query named pipe attributes.
[named pipe R/W]	Named pipe read/write transaction.
[named pipe read]	Raw mode named pipe read.
[named pipe write]	Raw mode named pipe write.
[negotiate protoc]	Negotiate SMB protocol version.
[newfile & bind]	Make new file and get file system address.
[notify close]	Close handle used to monitor file changes.
[open file]	Open specified file.
[open & execute]	Open specified file and execute next command.
[open spoolfile]	Open specified print buffer file.
[process exit]	Terminate consumer process.
[read & execute]	Read file and execute next command.
[read and hide]	Read directory ignoring hidden files.
[read block mplex]	Read block data on multiplexed connection.
[read block raw]	Read block data on unique connection.
[read block sec/r]	Read block secondary response.
[read check]	Check file accessibility.
[read from file]	Read from specified file.
[read w/options]	Read from file with specified options.
[rename file]	Rename the specified file to a new name.
[reserve resourcs]	Reserve resources on the server.
[search dir]	Search directory with specified attribute.
[seek]	Set file pointer for handle.
[send broadcast]	Send a one block broadcast message.
[session setup]	Log-in with consumer-based authentication.
[set exp attrib]	Set expanded file attributes (OS/2).
[set unix attribs]	Set expanded file attributes (UNIX/Xenix).
[set file attribs]	Set normal file attributes.
[single block msg]	Send a single block message.
[transaction next]	Subsequent name transaction.
[tree & execute]	Make virtual connection and execute next command.
[tree connect]	Make a virtual connection.
[tree disconect]	Detach a virtual connection.
[unbind]	Discard file system address binding.
[unlock bytes]	Release a locked byte range.
[write & close]	Write to and close specified file handle.

Command	*Description.*
[write & execute]	Write to file and execute next command.
[write & unlock]	Write to and unlock a byte range.
[write block raw]	Write block data on unique connection.
[write block mplx]	Write block data on multiplexed connection.
[write block sec]	Write block secondary request.
[write complete]	Terminate a write block sequence.
[write spoolfile]	Write to the specified print buffer.
[write to file]	Write to the specified file handle.
[X2 open file]	Open file.
[X2 find first]	Find first file.
[X2 find next]	Find next file.
[X2 query FS]	Get file system information.
[X2 set FS info]	Set file system information.
[X2 query path]	Get information on path.
[X2 set path]	Set path information.
[X2 query file]	Get file information.
[X2 set info]	Set file information.
[X2 FS control]	File system control information.
[X2 IOCTL]	I/O control for devices.
[X2 notify]	Monitor file for changes.
[X2 notify next]	Subsequent file monitoring.
[X2 make dir]	Make directory.

RCLS

Error Class. The second field of the decoded SMB protocol contains the error class and error code for each frame as in *E=1/22*, where *1* is the error class and *22* is the error code. The error class identifies the source of the error as shown in the following table:

Error Class	*Name*	*Source of Error*
0		No error, or error was handled by system.
1	ERRDOS	Server operating system.
2	ERRSRV	Server network file manager.
3	ERRHRD	System or device.
4	ERRXOS	Extended operating system.
225-227	ERRRMX	RMX operating system.
255	ERRCMD	Invalid SMB commands.

REH

Reserved.

ERR

Error messages. The following is a list of SMB error messages:

Error Message	*Description*
{Access denied}	Unable to service request.
{Access list full}	Access control list full.
{Bad attrib mode}	Invalid attributes specified.
{Bad disk request}	Disk command invalid.
{Bad drive spec}	Specified drive invalid.
{Bad environment}	Environment invalid.
{Bad EXE file}	Bad executable file format.
{Bad file access}	Invalid access to read only file.
{Bad file ID}	File handle invalid.
{Bad filespec}	Path name invalid.
{Bad format}	Format invalid.
{Bad function}	Function not supported.
{Bad I/O data}	Data invalid on server I/O device.
{Bad math argument}	Math argument invalid.
{Bad media type}	Unknown media type.
{Bad memory block}	Memory block address invalid.
{Bad open mode}	Open mode invalid.
{Bad permissions}	Specified permissions invalid.
{Bad print FID}	Print file ID invalid.
{Bad print request}	Printer device request invalid.
{Bad reqst length}	Bad request structure length.
{Bad semaphore}	Semaphore identifier invalid.
{Bad SMB command}	SMB command invalid.
{Bad Tree ID}	Tree ID invalid.
{Bad User ID}	User ID invalid.
{Bad user/passwrd}	Bad password or user name.
{Bad wait done}	Wait done for unwaited process.
{Continue in MPX}	Continue in block multiplexed mode.
{Can't delete dir}	Cannot delete current directory.
{Can't init net}	Network cannot be initialized.
{Can't mount dev}	Device cannot be mounted.
{Can't RAW, do MPX}	Cannot use raw blocks, use multiplexed.
{Can't ren to vol}	Attempt to rename across volumes failed.
{Can't support RAW}	Cannot support raw block access.
{Can't write dir}	Attempt to write on a directory failed.
{Command not recvd}	Initial command not received.
{CRC data error}	Data CRC error on device.
{Dev out of space}	Device out of space.

Error Message	*Description*
{Device is remote}	Referenced device remote.
{Dir not found}	Directory not found.
{Disk write error}	Disk write fault.
{Disk read error}	Disk read fault.
{Disk seek error}	Disk seek error.
{Drive not ready}	Drive not ready.
{Dup filename}	File name already exists.
{EOF on printer}	End of file found on print queue dump.
{Err buffered}	Error message buffered.
{Err logged}	Error message logged.
{Err displayed}	Error message displayed.
{File not found}	Filespec not found.
{File too big}	Maximum file size exceeded.
{Gen disk failure}	General disk failure.
{Insuf acc rights}	Insufficient access rights.
{Invalid name}	Invalid name supplied on tree connect.
{Invalid pipe}	Invalid pipe specified.
{Lock conflict}	Lock/Unlock conflicts with other locks.
{Memory blks lost}	Memory control blocks destroyed.
{More data coming}	Cannot terminate; more data coming.
{Need block device}	File used where a block device needed.
{Need data file}	Must specify data file.
{No FCBs available}	Out of file control blocks.
{No more files}	No more matching files found.
{No proc to pipe}	No process available to pipe.
{No read process}	Write to a pipe with no read processes.
{No resources}	Server out of resources.
{No room f/message}	No room to buffer message.
{Not a directory}	Must specify directory.
{Not receiving}	Not receiving messages.
{No semaphores}	Semaphore not available.
{OK}	SMB command completed successfully.
{Out of disk space}	Print queue out of disk space.
{Out of handles}	Too many open files.
{Out of memory}	Insufficient memory on server.
{Out of paper}	Printer out of paper.
{Pipe is busy}	Pipe process busy; wait.
{Pipe is closing}	Terminating pipe process.
{Print Q full}	Print file queue table full.
{Proc table full}	Server process table full.

Error Message	*Description*
{Rem I/O error}	Remote I/O error.
{Sector not found}	Sector not found.
{Seek on pipe}	Seek was issued to a pipe.
{Server error}	General server error.
{Server paused}	Server paused.
{Share buffer out}	Share buffer out of space.
{Share conflict}	Share conflicts with existing files.
{Syntax error}	Syntax error in path name.
{Sys call intruptd}	Interrupted system call.
{Table overflow}	Internal table overflow.
{Terminal needed}	Terminal device required.
{Timed out}	Operation has run out of time.
{Too many links}	Too many links.
{Too many names}	Too many remote user names.
{Too many UIDs}	User ID limit exceeded.
{Unit unknown}	Unknown unit.
{Unknown error}	Non-specific error.
{Unknwn process}	No such process.
{Write protected}	Write on write-protected diskette.
{Wrong diskette}	Wrong diskette inserted in drive.

REB/Flag

Reserved. This field is associated with the Core protocol only. The flag field appears in protocol versions later then the Core Protocol.

Tree ID

Uniquely identifies a file sharing connection between consumer and server where this protocol uses a server-based file protection.

Process ID

Identifies a specific consumer process within a virtual connection.

User ID

Used by the server to verify the file access permissions of users where consumer-based file protection is in effect.

Multiplex ID

Reserved for multiplexing multiple messages on a single virtual circuit (VC). A response message will always contain the same value as the corresponding request message. Only one request at a time may be outstanding on any VC.

WCT
Number of parameter words.

VWV
Variable number of words of parameter.

BCC
Number of bytes of data following.

BUF
Variable number of data bytes.

SDLC

IBM SNA Formats GA27-3136-10 1989-06

The SDLC (Synchronous Data Link Control) protocol was developed by IBM to be used as the layer 2 of the SNA hierarchical network. SNA data is carried within the information field of SDLC frames. The format of a standard SDLC frame is as follows:

← Link Header →				← Link Trailer →	
Flag	Address field	Control field	Information	FCS	Flag

SDLC frame format

Flag

The value of the flag is always (0x7E). In order to ensure that the bit pattern of the frame delimiter flag does not appear in the data field of the frame (and therefore cause frame misalignment), a technique known as Bit Stuffing is used by both the transmitter and the receiver.

Address field

The first byte of the frame after the header flag is known as the Address Field. SDLC is used on multipoint lines and it can support as many as 256 terminal control units or secondary stations per line. The address field defines the address of the secondary station which is sending the frame or the destination of the frame sent by the primary station.

Control field

The field following the Address Field is called the Control Field and serves to identify the type of the frame. In addition, it includes sequence numbers, control features and error tracking according to the frame type.

Every frame holds a one bit field called the Poll/Final bit. In SDLC this bit signals which side is 'talking', and provides control over who will speak next and when. When a primary station has finished transmitting a series of frames, it sets the Poll bit, thus giving control to the secondary station. At this time the secondary station may reply to the primary station. When the secondary station finishes transmitting its frames, it sets the Final bit and control returns to the primary station.

Modes of operation

In SDLC there is the notion of primary and secondary stations, defined simply as the initiator of a session and its respondent. The primary station sends commands and the secondary station sends responses.

SDLC operates in Normal Response Mode (NRM). This mode is totally master/slave meaning that only one station may transmit frames at any one time (when permitted to do so). This mode is signified by the SNRM(E) frame. The primary station initiates the session and sends commands. The secondary station sends responses. Full polling is used for all frame transmissions.

FCS

The Frame Check Sequence (FCS) enables a high level of physical error control by allowing the integrity of the transmitted frame data to be checked. The sequence is first calculated by the transmitter using an algorithm based on the values of all the bits in the frame. The receiver then performs the same calculation on the received frame and compares its value to the CRC.

Window size

SDLC supports an extended window size (modulo 128) where the number of possible outstanding frames for acknowledgement is raised from 8 to 128. This extension is generally used for satellite transmissions where the acknowledgement delay is significantly greater than the frame transmission times. The type of the link initialization frame determines the modulo of the session and an "E" is added to the basic frame type name (e.g., SNRM becomes SNRME).

Frame types

The following are the Supervisory Frame Types in SDLC:

RR	Information frame acknowledgement and indication to receive more.
REJ	Request for retransmission of all frames after a given sequence number.
RNR	Indicates a state of temporary occupation of station (e.g., window full).

The following are the Unnumbered Frame Types in SDLC:

DISC	Request disconnection.
UA	Acknowledgement frame.
DM	Response to DISC indicating disconnected mode.

FRMR	Frame reject.
CFGR	Configure.
TEST	Sent from primary to secondary and back again.
BCN	Beacon.
SNRM	Initiator for normal response mode. Full master/slave relationship.
SNRME	SNRM in extended mode.
RD	Request disconnect.
RIM	Secondary station request for initialization after disconnection.
SIM	Set initialization mode.
UP	Unnumbered poll.
UI	Unnumbered information. Sends state information/data.
XID	Identification exchange command.

There is one Information Frame Type in SDLC:

Info Information frame.

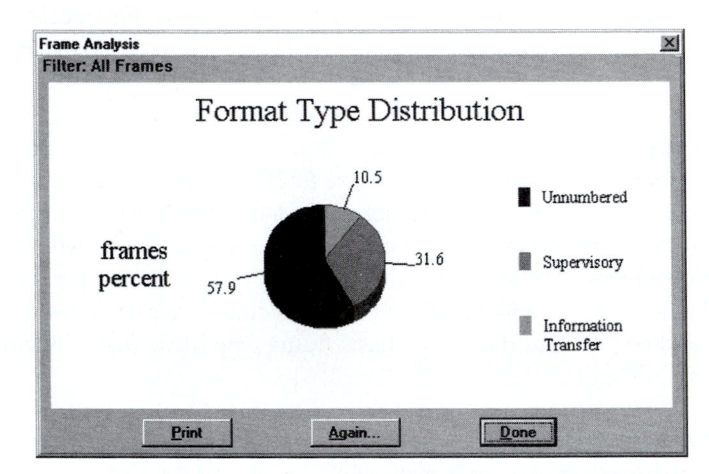

Graph displaying distribution of SDLC frames by format type

QLLC

QLLC is a standard developed for interconnecting SNA LANs over packet switched WANs with X.25. The SDLC header and trailer is stripped off and replaced by similar fields of LAPB before transmission over the network. The standard also defines additional control bytes used to allow the receiving end of the network to reconstruct the original SDLC frame. The SNA information is passed over the network within the X.25 data packet.

The following diagram represents SNA data and QLLC control frames within the X.25 data packet:

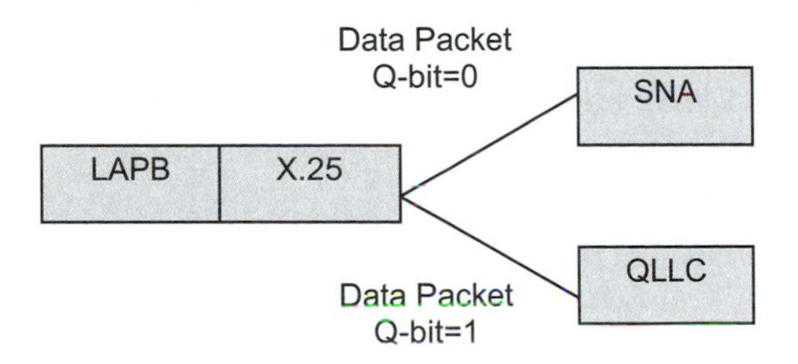

SNA data and QLLC control frames are determined by the value of the Q-bit within the X.25 packet header.

QLLC Frame Types

QRR	Receive Ready.
QDISC	Disconnect.
QUA	Unnumbered Acknowledgement.
QDM	Disconnect Mode.
QFRMR	Frame Reject.
QTEST	Test.
QRD	Request Disconnect.
QXID	Exchange Identification.
QSM	Set Mode.

SNA

The Systems Network Architecture (SNA) was introduced by IBM in 1974 in order to provide a framework for joining together the multitude of mutually incompatible IBM products for distributed processing. SNA was one of the first communications architectures to use a layered model, which later became the basis for the OSI model.

The SNA is a hierarchical network that consists of a collection of machines called nodes. There are four types of nodes; Type 1 (terminals), Type 2 (controllers and machines that manage terminals), Type 4 (front-end processors and machines that take some load off the main CPU) and Type 5 (the main host).

Each node has at least one Network Addressable Unit (NAU). The NAU enables a process to use the network by giving it an address. A process can then reach and be reached by other NAUs.

An NAU can be one of three types; an LU (Logical Unit), a PU (Physical Unit) or an SSCP (System Services Control Point). Usually there is one SSCP for each Type 5 node and none in the other nodes.

SNA distinguishes five different kinds of sessions: SSCP-SSCP, SSCP-PU, SSCP-LU, LU-LU and PU-PU.

The SSCP (PU Type 5) is usually implemented in IBM mainframe machines which use channels to connect to control devices such as disks, tapes and communication controllers. These are high speed communications links (up to 17 Mbps).

The communication controller (the FEP, Front End Processor, PU type 4) is used to connect low speed SDLC lines. All together the SSCPs, FEPs, channels and SDLC lines connecting them create the SNA backbone. Using SDLC, the FEPs also connect Token Ring LAN or X.25 links and other types of SNA devices such as cluster controllers and RJE stations. These are PU type 2/2.1 devices and are used to manage LUs which are the endpoint of SNA network - elements such as the display terminal (the 3270 family).

SNA frames have the following format:

Transmission header (TH)	Request / response header (RH)	Request / response unit (RU)

SNA frame structure

Transmission header

The TH field contains the Format Identifier value (FID). This value corresponds to the type of communication session and the environment in which it is used.

FID2 is the format used between a T4 or T5 node and an adjacent T2.0 or T2.1 node, or between adjacent T2.1 nodes. FID3 is used on links to a PU T1 (such as AS/400 controllers). FID4 is used on links between PU T4s.

The TH field also contains a mapping field (MPF) which indicates whether the frame is a complete SNA frame (containing TH, RH and RU) or just a segment. When the SNA frame is too large to be sent as one frame, it is divided into several segments (first, middle, last or whole). The first segment includes a TH (indicating that it is the first), an RH and the beginning of the RU. Other segments (middle and last) contain a TH (identical to the one of the first except for the MPF field) and the remainder of the RU.

Request/response header

The RH field denotes the SNA category of the frame, the format of the RU, whether requests are chained together, bracket indicators, pacing information and various other SNA frame properties.

Request/response unit

The RU contains the 'user data' that one LU sends to its session partner or a special SNA frame. A field within the RH distinguishes between cases and several classes of SNA frames. There are three categories of SNA frames: NS (function management data), DFC (data flow control) and SC (session control).

SNA TH0 & TH1

SNA TH0 and TH1 correspond to the FID header 0 and 1 respectively.

The format of the packet is shown in the following illustration:

4	6	7	8 bits
FID	MPF	EFI	
DAF (2 bytes)			
OAF (2 bytes)			
SNF (2 bytes)			
DCF (2 bytes)			

SNA TH0, SNA TH1 packet structure

FID

Format Identification: 0=FID 0, 1=FID 1.

MPF

Mapping field:

0 Middle segment of a BIU
1 Last segment of a BIU
2 First segment of a BIU
3 Whole BIU

EFI

Expedited flow indicator:

0 Normal flow
1 Expedited flow

DAF

Destination address field. Network address denoting the BIU's destination network addressable unit (NAU).

OAF

Origin address field. Network address denoting the originating NAU.

SNF

Sequence number field. Numerical identifier for the associated BIU.

DCF

Data count field. A binary count of the number of bytes in the BIU if the BIU segment is associated with the transmission header.

5250 is located in frames with an RH field as may be viewed in the multi-protocol view of the capture buffer.

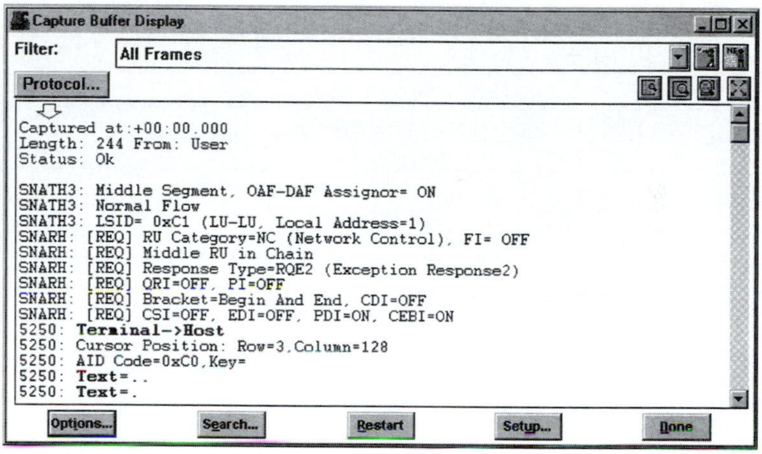

```
Capture Buffer Display
Filter: All Frames
Protocol...

Captured at:+00:00.000
Length: 244 From: User
Status: Ok

SNATH3: Middle Segment, OAF-DAF Assignor= ON
SNATH3: Normal Flow
SNATH3: LSID= 0xC1 (LU-LU, Local Address=1)
SNARH: [REQ] RU Category=NC (Network Control), FI= OFF
SNARH: [REQ] Middle RU in Chain
SNARH: [REQ] Response Type=RQE2 (Exception Response2)
SNARH: [REQ] QRI=OFF, PI=OFF
SNARH: [REQ] Bracket=Begin And End, CDI=OFF
SNARH: [REQ] CSI=OFF, EDI=OFF, PDI=ON, CEBI=ON
5250: Terminal->Host
5250: Cursor Position: Row=3,Column=128
5250: AID Code=0xC0,Key=
5250: Text=..
5250: Text=.

Options...  Search...  Restart  Setup...  Done
```

5250 as viewed in the RH field of the SNA frame

SNA TH5

SNA TH 5 is the FID 5 header.

The format of the packet is shown in the following illustration:

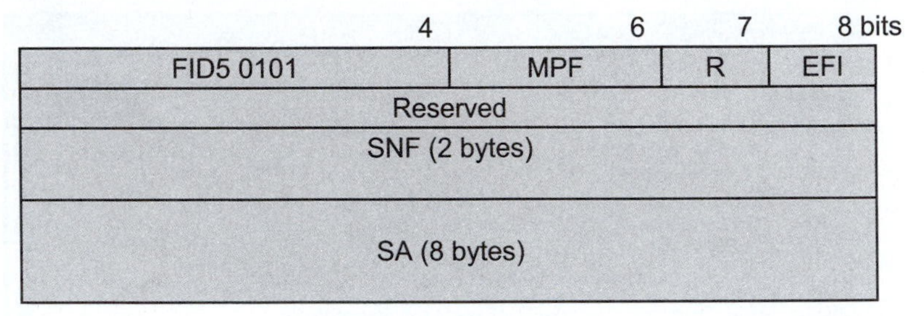

FID 5 header structure

FID 5
The value of this field is 0101.

MPF
Mapping field.

R
Reserved bit.

EFI
Expedited flow indicator (1 bit).

SNF
Sequence number field.

SA
Session address.

HPR-APPN

HPR network is an extension of the SNA network. HPR (High Performance Routing) is an extension of the base-APPN that provides some key advancements. These new functions include:

- Non-disruptive path switching.
- Better utilization of high-speed communication paths.
- An advanced congestion control methodology.
- Additional functionality provided by two new components: Rapid Transport Protocol (RTP) and Automatic Network Routing (ANR). These components provide the added functionality exhibited by HPR nodes.

NHDR

The packet transported along an RTP connection has a specific format. It consists of 3 components. NHDR, THDR and data. The Network Layer Header (NHDR) begins the frame used by RTP (Rapid Transport Protocol) nodes. It provides addressing for the packet as it transverses the HPR network. The components of this header include the transmission priority and the ANR (Automatic Network Routing) labels. NHDR consists of some indicators that identify the packet as a network layer packet.

The format of the header is shown in the following illustration:

1	2	3	4 bits
SM			TPF
TPF	Function type		
Function type	TSP	Slowdown1	Slowdown2
ANR / function routing field (1 or 2 bytes)			

NHDR header structure

SM

Switching mode may have the following values:

5 Function routing.

6 Automatic network routing.

TPF

Transmission priority field may have the following values:

0 Low (L).

1 Medium (M).

2 High (H).

3 Network (N).

Function type (for switching mode 5)

Function type of 1 indicates logical data link control.

TSP

Time-sensitive packet indicator.

Slowdown 1 and 2

This field indicates when ever a minor (slowdown 1) or significant (slowdown 2) congestion condition exists. Possible values are:

0 Does not exist.

1 Exists.

ANR routing field (for SM = 6)

A string of ANR labels 1 or 2 bytes long. The string ends with 0xFF.

Function routing field (for SM = 5)

A 2 byte function routing address (FRA) followed by the value 0xFF.

THDR

THDR is the RTP Transport header. It is used by the RTP endpoints to provide correct processing of the packet. It is used for communication between the endpoints and to identify the RTP connection.

The format of the header is shown in the following illustration:

TCID assignor (7 bytes)
Connection setup (1 bit)
Start-of-message indicator (1 bit)
End-of-message indicator (1 bit)
Status requested indicator (1 bit)
Respond ASAP indicator (1 bit)
Retry indicator (1 bit)
Last message indicator (1 bit)
Connection qualifier field indicator (2 bits)
Optional segments present indicator (1 bit)
Data offset (2 bytes)
Data length (2 bytes)
Byte sequence number (4 bytes)
Control vector 05
Optional segments

THDR header structure

TCID assignor

Transport connection identifier. There are 2 possible values:

0 TCID was assigned by the receiving RTP partner.
1 TCID was assigned by the sending RTP partner.

Connection setup

0 Presented.
1 Not presented.

Start of message indicator

0 Not start of message.

1 Start of message.

End of message indicator

0 Not end of message.

1 End of message.

Status requested indicator

0 Receiver need not reply with a status segment.

1 Receiver must reply with a status segment.

Respond ASAP indicator

1 Sender will retransmit reply ASAP.

Retry indicator

0 Sender will retransmit this packet.

Connection qualifier field indicator

0 None presented.

1 Originator.

Optional segments present indicator

0 Not presented.

1 Presented.

Byte sequence number

Sequence number of the first byte of the data field.

Optional segments

If present the optional segment can contain one or more of the following segments:

0x0E Status segment.

0x0D Connection Setup segment.

0x10 Connection Identifier Exchange segment.

0x14 Switching Information segment.

0x22 Adaptive Rate-Based segment.

0x12 Connection Fault segment.

0x0F Client Out-of-band Bits segment.

The structure of each segment is as follows:

Byte	*Content*
0	Segment length/4.
1	Segment type.
2	Segment data.

Each segment may include control vectors. Supported control vectors are:

0x00	Node identifier Control Vector.
0x03	Network ID Control Vector.
0x05	Network Address Control Vector.
0x06	Cross-Domain Resource Manager Control Vector.
0x09	Activation Request/Response Sequence Identifier Control Vector.
0x0E	Network Name Control Vector.
0x10	Product Set ID Control Vector.
0x13	Gateway Support Capability Control Vector.
0x15	Network-Qualified Address Pair Control Vector.
0x18	SSCP Name Control Vector.
0x22	XID Negotiation Error Control Vector.
0x26	NCE Identifier Control Vector.
0x28	Topic Identifier Control Vector.
0x32	Short-Hold Mode Control Vector.
0x39	NCE Instant Identifier.
0x46	TG Descriptor Control Vector.
0x60	Fully qualified PCID Control Vector.
0x61	HPR Capabilities Control Vector.
0x67	ANR Path Control Vector.
0xFE	Control Vector Keys Not Recognized Control Vector.

DLSw

IETF RFC 1434 http://www.cis.ohio-state.edu/htbin/rfc/rfc1434.html
RFC 1795 http://www.cis.ohio-state.edu/htbin/rfc/rfc1795.html
RFC 2166 http://www.cis.ohio-state.edu/htbin/rfc/rfc2166.html

Data Link Switching (DLSw) is a forwarding mechanism for the IBM SNA (Systems Network Architecture) and IBM NetBIOS (Network Basic Input Output Services) protocols. Over IP networks, DLSw does not provide full routing, but instead provides switching at the SNA Data Link layer (i.e., layer 2 in the SNA architecture) and encapsulation in TCP/IP for transport over the Internet.

A Data Link Switch (abbreviated also as DLSw) can support SNA (Physical Unit (PU) 2, PU 2.1 and PU 4) systems and optionally NetBIOS systems attached to IEEE 802.2 compliant Local Area Networks, as well as SNA (PU 2 (primary or secondary) and PU2.1) systems attached to IBM Synchronous Data Link Control (SDLC) links. For the latter case, the SDLC attached systems are provided with a LAN appearance within the Data Link Switch (each SDLC PU is presented to the SSP protocol as a unique MAC/SAP address pair). For Token Ring LAN attached systems, the Data Link Switch appears as a source-routing bridge. Token Ring Remote systems that are accessed through the Data Link Switch appear as systems attached to an adjacent ring. This ring is a virtual ring that is manifested within each Data Link Switch.

There are two message header formats exchanged between data link switches: Control and Information. These two message formats are as follows:

8	16	Octets
Version number	Header length (=16)	0-1
Message length		2-3
Remote data link correlator		4-7
Remote DLC port ID		8-11
Reserved field		12-13
Message type	Flow control byte	14-15

DLSw information message structure

8	16	Octets
Version number	Header length (=72)	0-1
Message length		2-3
Remote data link correlator		4-7
Remote DLC port ID		8-11
Reserved field		12-13
Message type	Flow control byte	14-15
Protocol ID	Header number	16-17
Reserved		18-19
Largest frame size	SSP flags	20-21
Circuit priority	Message type	22-23
Target MAC address		24-29
Origin MAC address		30-35
Origin link SAP	Target link SAP	36-37
Frame direction	Reserved	38-39
Reserved		40-41
DLC header length		42-43
Origin DLC port ID		44-47
Origin data link correlator		48-51
Origin data link correlator		52-55
Origin transport ID		56-59
Target DLC port ID		60-63
Target data link correlator		64-67
Target Transport ID		68-69
Reserved		70-71

DLSw control message structure

Version number
Set to 0x31 (ASCII 1) to indicate DLSw version 1.

Header length
Set to 0x48 for control messages and 0x10 for information and Independent Flow Control messages.

Message length
Specifies the number of bytes within the data field following the header.

Remote data link correlator / remote DLC port ID

The contents of the DLC and DLC Port ID have local significance only. The values received from a partner DLSw must not be interpreted by the DLSw that receives them and should be echoed as is to a partner DLSw in subsequent messages.

Message type

The following message types are available:

CANUREACH_ex	Can U Reach Station-explorer
CANUREACH_cs	Can U Reach Station-circuit start
ICANREACH_ex	I Can Reach Station-explorer
ICANREACH_cs	I Can Reach Station-circuit start
REACH_ACK	Reach Acknowledgment
DGRMFRAME	Datagram Frame
XIDFRAME	XID Frame
CONTACT	Contact Remote Station
CONTACTED	Remote Station Contacted
RESTART_DL	Restart Data Link
DL_RESTARTED	Data Link Restarted
ENTER_BUSY	Enter Busy
EXIT_BUSY	Exit Busy
INFOFRAME	Information (I) Frame
HALT_DL	Halt Data Link
DL_HALTED	Data Link Halted
NETBIOS_NQ_ex	NETBIOS Name Query-explorer
NETBIOS_NQ_cs	NETBIOS Name Query-circuit setup
NETBIOS_NR_ex	NETBIOS Name Recognized-explorer
NETBIOS_NR_cs	NETBIOS Name Recog-circuit setup
DATAFRAME	Data Frame
HALT_DL_NOACK	Halt Data Link with no Ack
NETBIOS_ANQ	NETBIOS Add Name Query
NETBIOS_ANR	NETBIOS Add Name Response
KEEPALIVE	Transport Keepalive Message
CAP_EXCHANGE	Capabilities Exchange
IFCM	Independent Flow Control Message
TEST_CIRCUIT_REQ	Test Circuit Request
TEST_CIRCUIT_RSP	Test Circuit Response

Flow control byte

Format of the flow control is as follows:

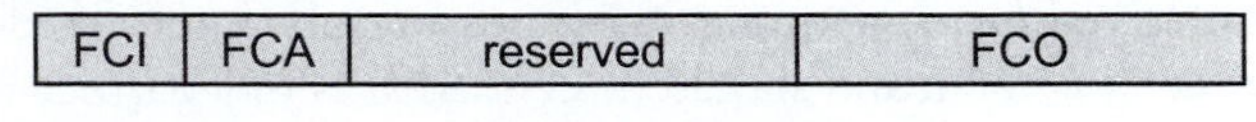

Flow control format

FCI Flow control indicatory.
FCA Flow control ack.
FCO Flow control operator bits.

Protocol ID

Set to 0x42, indicating a decimal value of 66.

Header number

Set to 0x01, indicating a value of one.

Largest frame size

Carries the largest frame size bits across the DLSw connection to ensure that the two end-stations always negotiate a frame size to be used on a circuit that does not require the origin and target DLSw partners to re-segment frames.

SSP flags

Contain additional information related to the SSP message.

Circuit priority

Circuit priority is only valid for CANUREACH_cs, ICANREACH_cs, and REACH_ACK frames.

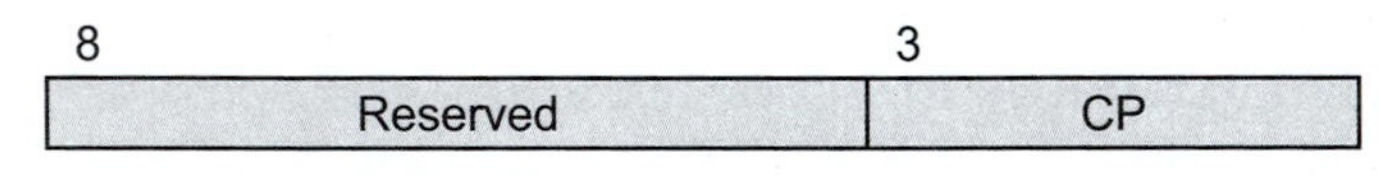

Circuit priority format

The value of the Circuit Priority bits (CP) can be as follows:

000 Unsupported, meaning that the Data Link Switch that originates the circuit does not implement priority.
001 Low Priority.
010 Medium Priority.
011 High Priority.
100 Highest Priority.
101 to 111 are reserved for future use.

Target/Origin MAC address
Origin/Target link SAP

Each attachment address is represented by the concatenation of the MAC address (6 bytes) and the LLC address (1 byte). Each attachment address is classified as either Target, in the context of the destination MAC/SAP addresses of an explorer frame sent in the first frame used to establish a circuit, or Origin, in the context of the source MAC/SAP addresses. All MAC addresses are expressed in non-canonical (Token-Ring) format.

Frame direction

Set to 0x01 for frames sent from the origin DLSw to the target DLSw, and is set to 0x02 for frames sent from the target DLSw to the origin DLSw.

DLC header length

Set to zero for SNA and is set to 0x23 for NetBIOS datagrams, indicating a length of 35 bytes. This includes the Access Control (AC) field, the Frame Control (FC) field, Destination MAC Address (DA), the Source MAC Address (SA), the Routing Information (RI) field (padded to 18 bytes), the Destination link SAP (DSAP), the Source link SAP (SSAP), and the LLC control field (UI).

Origin/Target DLC port ID
Origin/Target data link correlator

An end-to-end circuit is identified by a pair of Circuit IDs. A Circuit ID is a 64 bit number that identifies the DLC circuit within a single DLSw. It consists of a DLC Port ID (4 bytes), and a Data Link Correlator (4 bytes). The Circuit ID must be unique in a single DLSw and is assigned locally. The pair of Circuit IDs along with the Data Link IDs, uniquely identify a single end-to-end circuit. Each DLSw must keep a table of these Circuit ID pairs, one for the local end of the circuit and the other for the remote end of the circuit. In order to identify which Data Link Switch originated the establishment of a circuit, the terms, Origin DLSw and Target DLSw, are used.

Origin/Target transport ID

Used to identify the individual TCP/IP port on a Data Link Switch. The values have only local significance. However, each Data Link Switch is required to reflect the values contained in these two fields, along with the associated values for DLC Port ID and the Data Link Correlator, when returning a message to the other Data Link Switch.

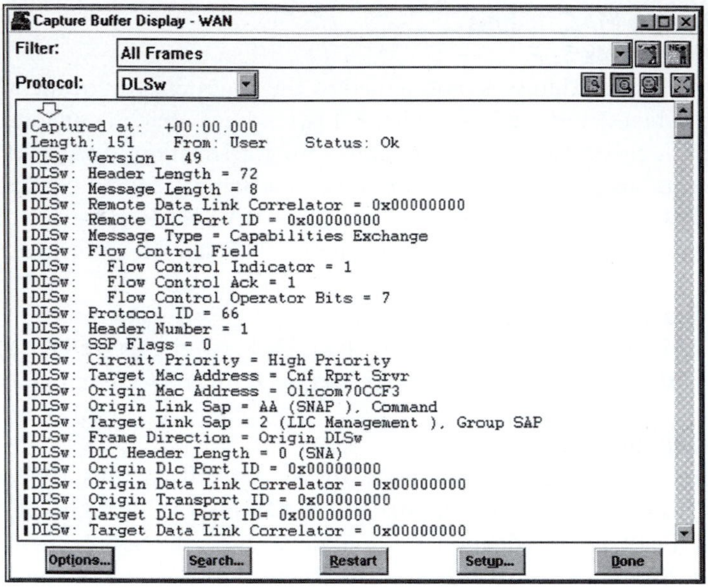

```
Capture Buffer Display - WAN
Filter:   All Frames
Protocol: DLSw

Captured at:  +00:00.000
Length: 151     From: User     Status: Ok
DLSw: Version = 49
DLSw: Header Length = 72
DLSw: Message Length = 8
DLSw: Remote Data Link Correlator = 0x00000000
DLSw: Remote DLC Port ID = 0x00000000
DLSw: Message Type = Capabilities Exchange
DLSw: Flow Control Field
DLSw:   Flow Control Indicator = 1
DLSw:   Flow Control Ack = 1
DLSw:   Flow Control Operator Bits = 7
DLSw: Protocol ID = 66
DLSw: Header Number = 1
DLSw: SSP Flags = 0
DLSw: Circuit Priority = High Priority
DLSw: Target Mac Address = Cnf Rprt Srvr
DLSw: Origin Mac Address = Olicom70CCF3
DLSw: Origin Link Sap = AA (SNAP ), Command
DLSw: Target Link Sap = 2 (LLC Management ), Group SAP
DLSw: Frame Direction = Origin DLSw
DLSw: DLC Header Length = 0 (SNA)
DLSw: Origin Dlc Port ID = 0x00000000
DLSw: Origin Data Link Correlator = 0x00000000
DLSw: Origin Transport ID = 0x00000000
DLSw: Target Dlc Port ID= 0x00000000
DLSw: Target Data Link Correlator = 0x00000000

Options...   Search...   Restart   Setup...   Done
```

DLSw decode

SNA Terminology

Systems Network Architecture (SNA)

The description of the logical structure, formats, protocols and operational sequences for transmitting information units and controlling the configuration and operation of networks.

Network Addressable Unit (NAU)

A logical unit, physical unit or system services control point which is the origin or the destination of information transmitted by the path control network. Each NAU has a network address that represents it to the path control network.

Logical Unit (LU)

A port through which end users access the SNA network in order to communicate with other end users and the functions provided by system services control points (SSCPs). An LU can support at least two sessions (one with an SSCP and one with another LU) and may be capable of supporting many sessions with other logical units.

Physical Unit (PU)

One of the three types of network addressable units (NAUs). Each node of an SNA network contains a physical unit (PU) that manages and monitors the resources (such as attached links) of a node, as requested by a system services control point (SSCP) via an SSCP-PU session. An SSCP activates a session with the PU in order to indirectly manage resources of the node such as attached links through the PU.

System Services Control Point (SSCP)

A focal point within an SNA network for managing the configuration, coordinating network operator/problem determination requests and providing directory support and other session services for network end users. Multiple SSCPs, cooperating as peers, can divide the network into domains of control, with each SSCP having a hierarchical control relationship to the physical and logical units within its domain.

Bracket

One or more chains of request units (RUs) and their responses that are exchanged between the two LU-LU half-sessions and that represent a transaction between them. A bracket must be completed before another bracket can be started.

Data Link Control (DLC) Layer

The layer that consists of the link stations that schedule data transfer over a link between two nodes and perform error control for the link.

Normal Flow

A data flow designated in the transmission header (TH) that is used primarily to carry end-user data. The rate at which requests flow on the normal flow can be regulated by session-level pacing. Normal and expedited flows move in both the primary-to-secondary and secondary-to-primary directions.

Expedited Flow

A data flow designated in the transmission header (TH) that is used to carry network control, session control and various data flow control request/response units (RUs). The expedited flow is separate from the normal flow (which carries primary end-user data) and can be used for commands that affect the normal flow.

Explicit Route (ER)

The path control network elements, including a specific set of one or more transmission groups, that connect two subarea nodes. An explicit route is identified by an origin subarea address, a destination subarea address, an explicit route number and a reverse explicit route number.

LU Type 6.2

LU 6.2 is a particular type of SNA logical unit. It uses SNA-defined interprogram communication protocols and is also referred to as Advanced Program-to-Program Communication (APPC).

Network Services (NS) Header

A 3 byte field in an FMD request/response unit (RU) flowing in an SSCP-LU, SSCP-PU or SSCP-SSCP session. The network services header is used primarily to identify the network services category of the RU and the particular request code within a category.

Node Type

A designation of node according to the protocols it supports and the network addressable units (NAUs) that it can contain. Five types are defined: 1, 2.0, 2.1, 4 and 5. Node types 1, 2.0 and 2.1 are peripheral nodes and types 4 and 5 are subarea nodes.

PU Type 2 (T2)

A network node that can attach to an SNA network as a peripheral node.

PU Type 2.1 (T2.1)

A network node that can attach to an SNA network as a peripheral node using the same protocols as type 2.0 nodes; type 2.1 nodes can be directly attached to one another using SNA low-entry networking.

PU Type 4 (T4)

A network node containing an NCP and that is a subarea node within an SNA network.

PU Type 5 (T5)

A network node containing VTAM and that is a subarea node within an SNA network.

RU Chain

A set of related request/response units (RUs) that are consecutively transmitted on a particular normal or expedited data flow. The request RU chain is the unit of recovery: if one of the RUs in the chain cannot be processed, the entire chain is discarded. Each RU belongs to only one chain, which has a beginning and an end indicated via control bits for request/response headers within the RU chain. Each RU chain can be designated as first-in-chain (FIC), last-in-chain (LIC), middle-in-chain (MIC) or only-in-chain (OIC). Response units and expedited flow request units are always sent as OIC.

Session Control (SC)

One of the components of transmission control. Session control is used to purge data flowing in a session after an unrecoverable error occurs, in order to resynchronize the data flow after such an error and to perform cryptographic verification.

A request unit (RU) category used for requests and responses exchanged between the session control components of a session and for session activation and deactivation requests and responses.

SSCP-LU Session

A session between a system services control point (SSCP) and a logical unit (LU); the session enables the LU to request the SSCP to help initiate LU-LU sessions.

SSCP-PU Session

A session between a system services control point (SSCP) and a physical unit (PU); SSCP-PU sessions allow SSCPs to send requests to, and receive status information from individual nodes in order to control the network configuration.

SSCP-SSCP Session

A session between a system services control point (SSCP) in one domain and the SSCP in another domain. An SSCP-SSCP session is used to initiate and terminate cross-domain LU-LU sessions.

Token Ring

A network with a ring topology that passes tokens from one attaching device to another.

17

ILMI

ATM Forum ILMI specification 4.0 1996-0
IETF RFC 115, May 1990; RFC 1213, March 1991; RFC 1157, May 1990
ATM Forum UNI 3.0; "Managing Internetworks with SNMP" by Mark A. Miller, M&T Books, 1993

The Simple Network Management Protocol (SNMP) and an ATM UNI Management Information Base (MIB) are required to provide any ATM user device with status and configuration information concerning the virtual path and channel connection available at its UNI. In addition, their global operations and network management information may facilitate diagnostic procedures at the UNI.

The Interim Local Management Interface (ILMI) provides bi-directional exchange of management information between UNI Management Entities (UMEs). Both UMEs must contain the same MIB, even though the semantics of some MIB objects may be interpreted differently. Many types of equipment use this ATM UNI ILMI, e.g., high layer switches, workstations, computers with ATM interface, ATM network switches and more.

MIB Names

The following is a list, in tree form, of MIB names.

enterprises
353 atmForum
    1 atmForumAdmin
        2 atmfTransmissionTypes
            1 atmfUnknownType
            2 atmfSonetSTS3c
            3 atmfDs3
            4 atmf4B5B
            5 atmf8B10B
        3 atmfMediaTypes
            1 atmfMediaUnknownType
            2 atmfMediaCoaxCable
            3 atmfMediaSingleMode
            4 atmfMediaMultiMode
            5 atmfMediaStp
            6 atmfMediaUtp
        4 atmTrafficDescrTypes
            1 atmfNoDescriptor
            2 atmfPeakRate
            3 atmfNoClpNoScr
            4 atmfClpNoTaggingNoScr
            5 atmfClpTaggingNoScr
            6 atmfNoClpScr
            7 atmfClpNoTaggingScr
            8 atmfClpTaggingScr
        5 atmfSrvcRegTypes
            1 atmfSrvcRegLecs
    2 atmForumUni
        1 atmfPhysicalGroup
            1 atmfPortTable
                1 atmfPortEntry
                    1 atmfPortIndex
                    2 atmfPortAddress
                    3 atmfPortTransmissionType
                    4 atmfPortMediaType
                    5 atmfPortOperStatus

6 atmfPortSpecific
2 atmfAtmLayerGroup
1 atmfAtmLayerTable
1 atmfAtmLayerEntry
1 atmfAtmLayerIndex
2 atmfAtmLayerMaxVPCs
3 atmfAtmLayerMaxVCCs
4 atmfAtmLayerConfiguredVPCs
5 atmfAtmLayerConfiguredVCCs
6 atmfAtmLayerMaxVpiBits
7 atmfAtmLayerMaxVciBits
8 atmfAtmLayerUniType
3 atmfAtmStatsGroup
1 atmfAtmStatsTable
1 atmfAtmStatsEntry
1 atmfAtmStatsIndex
2 atmfAtmStatsReceivedCells
3 atmfAtmStatsDroppedReceivedCells
4 atmfAtmStatsTransmittedCells
4 atmfVpcGroup
1 atmVpcTable
1 atmVpcEntry
1 atmVpcPortIndex
2 atmfVpcVpi
3 atmfVpcOperStatus
4 atmfVpcTransmitTrafficDescriptorType
5 atmfVpcTransmitTrafficDescriptorParam1
6 atmfVpcTransmitTrafficDescriptorParam2
7 atmfVpcTransmitTrafficDescriptorParam3
8 atmfVpcTransmitTrafficDescriptorParam4
9 atmfVpcTransmitTrafficDescriptorParam5
10 atmfVpcReceiveTrafficDescriptorType
11 atmfVpcReceiveTrafficDescriptorParam1
12 atmfVpcReceiveTrafficDescriptorParam2
13 atmfVpcReceiveTrafficDescriptorParam3
14 atmfVpcReceiveTrafficDescriptorParam4
15 atmfVpcReceiveTrafficDescriptorParam5
16 atmfVpcQoSCategory
17 atmfVpcTransmitQoSClass
18 atmfVpcReceiveQoSClass
5 atmfVccGroup

```
1 atmfVccTable
    1 atmfVccEntry
        1 atmVccPortIndex
        2 atmfVccVpi
        3 atmfVccVci
        4 atmfVccOperStatus
        5 atmfVccTransmitTrafficDescriptorType
        6 atmfVccTransmitTrafficDescriptorParam1
        7 atmfVccTransmitTrafficDescriptorParam2
        8 atmfVccTransmitTrafficDescriptorParam3
        9 atmfVccTransmitTrafficDescriptorParam4
        10 atmfVccTransmitTrafficDescriptorParam5
        11 atmfVccReceiveTrafficDescriptorType
        12 atmfVccReceiveTrafficDescriptorParam1
        13 atmfVccReceiveTrafficDescriptorParam2
        14 atmfVccReceiveTrafficDescriptorParam3
        15 atmfVccReceiveTrafficDescriptorParam4
        16 atmfVccReceiveTrafficDescriptorParam5
        17 atmfVccQoSCategory
        18 atmfVccTransmitQoSClass
        19 atmfVccReceiveQoSClass
8 atmfSrvcRegistryGroup
    1 atmfSrvcRegTable
        1 atmfSrvcRegEntry
            1 atmfSrvcRegPort
            1 atmfSrvcRegServiceID
            1 atmfSrvcRegATMAddress
            1 atmfSrvcRegAddressIndex
```

SNMP

RFC 1157: http://www.cis.ohio-state.edu/htbin/rfc/rfc1157.html

ILMI uses SNMP, which is designed to be simple and has a very straightforward architecture. The SNMP message is divided into two sections: a version identifier plus community name and a PDU.

The version identifier and community name are sometimes referred to as the authentication header. The version number assures that both manager and agent are using the same version of SNMP. Messages between manager and agent containing different version numbers are discarded without further processing. The community name authenticates the manager before allowing access to the agent. The community name, along with the manager's IP address, is stored in the agent's community profile. If there is a difference between the manager and agent values for the community name, the agent will send an authentication failure trap message to the manager.

GetRequest and GetResponse PDUs

The manager uses the GetRequest PDU to retrieve the value of one or more object(s) from an agent. Under error-free conditions, the agent generates a GetResponse PDU. On both Request and Response PDUs there is a Request Index field that correlates the manager's request to the agent's response, an Error Status field which is set to noError, and an Error Index field which is set to zero. In this process, four errors are possible:

1. If a variable does not exactly match an available object, the agent returns a GetResponse PDU with the Error Status set to NoSuchName and the Error Index set the same as the index of the variable in question.
2. If a variable is of aggregate type, the Response is the same as above.
3. If the size of the appropriate GetResponse PDU would exceed a local limitation, the agent returns a GetResponse PDU of identical form, where the value of the Error Status is set to tooBig and the Error Index is set to 0.
4. If the value of a requested variable cannot be retrieved for any other reason, then the agent returns a GetResponse PDU, with the Error Status set to genErr and the Error Index set the same as the index of the variable in question.

The following is an example of a GetRequest PDU decode:

```
Capture Buffer Display
Protocol: ILMI

Captured at:+00:00.000
Length: 96 From: User
Status: Ok

ILMI: Message::= SEQUENCE {                          <30>
ILMI: Length: 43
ILMI:     Version::= INTEGER {                      <02>
ILMI:     Length: 1
ILMI:     Version: 00                    <00>
ILMI:     } .
ILMI:     Community::= STRING {                      <04>
ILMI:     Length: 4
ILMI:     Community: ILMI                <494C4D49>
ILMI:     } .
ILMI:     Command ::= Get-Request     <A0>
ILMI:     Length : 32
ILMI:     RequestID ::= INTEGER {                  <02>
ILMI:     Length: 2
ILMI:     RequestId: 17E3                <17E3>
ILMI:     } .
ILMI:     ErrorStatus::= INTEGER {                    <02>
ILMI:     Length: 1
ILMI:     Error Status: No Error
ILMI:     } .

Options...    Search...    Done
```

```
Capture Buffer Display
Protocol: ILMI

ILMI:     } .
ILMI:     ErrorIndex::= INTEGER {                      <02>
ILMI:     Length: 1
ILMI:     Error Index: 00                <00>
ILMI:     } .
ILMI:     VarBindList ::= SEQUENCE {                    <30>
ILMI:     Length : 20
ILMI:        Length error checking : OK
ILMI:        Mib No 1
ILMI:        VarBind ::= SEQUENCE {                <30>
ILMI:        Length : 18
ILMI:           Name ::= Object ID {                  <06>
ILMI:           Length : 14
ILMI:           Mib Tree :1.3.6.1.4.1.45.1.4.2.10.3.2.1.0.
ILMI:           } .
ILMI:           Data ::= Null {              <05>
ILMI:           Length : 0
ILMI:        } }.
ILMI:     } .
ILMI:           Total Object length check : OK
ILMI: } .
ILMI: SNMP ended correctly
User Data
OFFST DATA                       ASCII
0000: 00 00 00 00 00 00 00 00   ........

Options...    Search...    Done
```

GetRequest PDU

GetNextRequest PDU

The GetNextRequest PDU is used to retrieve one or more objects from an agent. Under error-free conditions, the agent generates a GetResponse PDU, with the same Request Index. The Variable Bindings contain the name and value associated with the lexicographic successor of each of the object identifiers (OIDs) noted in the GetNextRequest PDU. The main difference between GetRequest and GetNextRequest PDUs is that the GetNextRequest PDU retrieves the value of the next object within the agent's MIB view. Three possible errors may occur in this process:

1. If a variable in the Variable Bindings field does not lexicographically proceed the name of an object that may be retrieved, the agent returns a GetResponse with the Error Status set to noSuchName and the Error Index set to the same as the variable in question.

2. If the size of the appropriate GetResponse PDU would exceed a local limitation, the agent returns a GetResponse PDU of identical form, with the Error Status set to tooBig and the Error Index set to zero.

3. If the value of the lexicographic successor to a requested variable cannot be retrieved for any other reason, the agent returns the GetResponse PDU, with the Error Status set to genErr and the Error Index set the same as the index of the variable in question.

SetRequest PDU

The SetRequest PDU is used to assign a value to an object residing in the agent. When the agent receives the SetRequest PDU, it alters the values of the named objects to the values in the variable binding. Under error-free conditions, the agent generates a GetResponse PDU of identical form, except that the assigned PDU type is 2. Four different errors may occur in this process:

1. If a variable is not available for set operations within the relevant MIB view, the agent returns a GetResponse PDU with the Error Status set to NoSuchName (or readOnly) and the Error Index set the same as the index of the variable in question.
2. If a variable does not conform to the ASN.1 type, length and value, the agent returns a GetResponse with the Error Status set to badValue and the same Error Index.
3. If the size of the appropriate GetResponse PDU exceeds a local limitation, the agent returns a GetResponse PDU of identical form, with the Error Status set to tooBig and the Error Index set to zero.
4. If the value of a requested variable cannot be altered for any other reason, the agent returns a GetResponse PDU, with the Error Status set to genErr and the Error Index set the same as the index of the variable in question.

Trap PDU

The last PDU type is the Trap PDU which has a different format from the other four PDUs. It contains the following fields:

- Enterprise field, which identifies the management enterprise under whose registration authority the trap was defined.
- Generic trap type, which provides more specific information on the event being reported. There are seven unique values for this field: coldStart, warmStart, linkDown, linkUp, authenticationFailure, egpNeighborLoss, and enterpriseSpecific.
- Specific Trap Type field, which identifies the specific Trap.
- Timestamp field, which represents the amount of time elapsed between the last initialization of the agent and the generation of that Trap.
- Variable bindings.

The following is an example of the Trap decode:

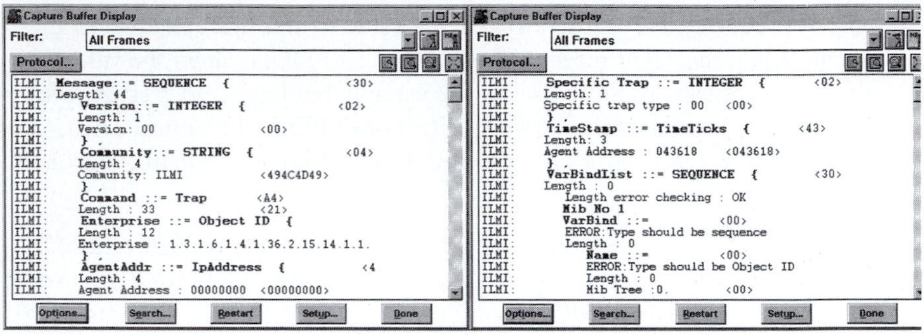

```
Capture Buffer Display
Filter: All Frames
Protocol...
ILMI: Message::= SEQUENCE  {                    <30>
ILMI: Length: 44
ILMI:    Version::= INTEGER  {                 <02>
ILMI:    Length: 1
ILMI:    Version: 00               <00>
ILMI:    } .
ILMI:    Community::= STRING  {                <04>
ILMI:    Length: 4
ILMI:    Community: ILMI           <494C4D49>
ILMI:    } .
ILMI:    Command ::= Trap         <A4>
ILMI:    Length : 33               <21>
ILMI:    Enterprise ::= Object ID  {
ILMI:    Length : 12
ILMI:    Enterprise : 1.3.1.6.1.4.1.36.2.15.14.1.1.
ILMI:    } .
ILMI:    AgentAddr ::= IpAddress  {             <4
ILMI:    Length: 4
ILMI:    Agent Address : 00000000  <00000000>
Options...  Search...  Restart  Setup...  Done
```

```
Capture Buffer Display
Filter: All Frames
Protocol...
ILMI:      Specific Trap ::= INTEGER  {      <02>
ILMI:      Length: 1
ILMI:      Specific trap type : 00   <00>
ILMI:      } .
ILMI:      TimeStamp ::= TimeTicks  {      <43>
ILMI:      Length: 3
ILMI:      Agent Address : 043618    <043618>
ILMI:      } .
ILMI:      VarBindList ::= SEQUENCE  {      <30>
ILMI:      Length : 0
ILMI:        Length error checking : OK
ILMI:        Mib No 1
ILMI:        VarBind ::=            <00>
ILMI:        ERROR:Type should be sequence
ILMI:        Length : 0
ILMI:          Name ::=             <00>
ILMI:          ERROR:Type should be Object ID
ILMI:          Length : 0
ILMI:          Mib Tree :0.          <00>
Options...  Search...  Restart  Setup...  Done
```

Trap PDU

SMI

General

SMI (Structure of Management Information) is the standard used for defining the rules of managed object identification. The SMI organizes, names, and describes information so that logical access can occur. The SMI states that each managed object must have a name, a syntax , and an encoding. The name or OID uniquely identifies the object. The syntax defines the data type, such as an integer or a string of octets. The encoding describes how the information associated with the managed object is serialized for transmission between machines.

SMI defines the syntax that retrieves and communicates information, controls the way information is placed into logical groups, and the naming mechanism, known as the object identifiers, that identify each managed object. This can be extended to include MIBs, which store management information. Managed objects are accessed via an MIB. Objects in the MIB are defined using Abstract Syntax Notation One (ASN.1). Each type of object (termed an object type) has a name, a syntax, and an encoding. The name is represented uniquely as an OBJECT IDENTIFIER, which is an administratively assigned name. The syntax defines the abstract data structure corresponding to that object type. For example, the structure of a given object type might be an INTEGER or OCTET STRING. The encoding of an object type is simply how instances of that object type are represented using the object's type syntax.

Object Identifier

An object identifier is a sequence of integers which traverse a global tree. The tree consists of a root connected to a number of labeled nodes via edges. Each node may, in turn, have children of its own which are labeled. In this case we may term the node a subtree. This process may continue to an arbitrary level of depth.

The root node is unlabeled, but has at least three children directly under it; one node is administrated by the International Organization for Standardization, with label iso(1); another is administrated by the International Telegraph and Telephone Consultative Committee, with label ccitt(0); and the third is jointly administered by the ISO and CCITT, joint-iso-ccitt(2). Under the iso(1) node, the ISO has designated one subtree for use by other (inter)national organizations, org(3). Of the children nodes present, two have been assigned to the US National Institute of Standards and Technology. One of these subtrees has been transferred by the NIST to the US Department of Defense, dod(6). DoD will allocate a node to the Internet community, to be administered by the Internet Activities Board (IAB) as follows:

internet OBJECT IDENTIFIER::={iso org(3) dod(6) 1} -> 1.3.6.1

In this subtree four nodes are present:

directory OBJECT IDENTIFIER::={ internet 1 }
mgmt OBJECT IDENTIFIER::={ internet 2 }
experimental OBJECT IDENTIFIER::={ internet 3 }
private OBJECT IDENTIFIER::={ internet 4 }

For example, the initial Internet standard MIB would be assigned management document number 1. -> { mgmt 1 } -> 1.3.6.1.2.1

The private(4) subtree is used to identify objects defined unilaterally. Administration of the private(4) subtree is delegated by the IAB to the Internet Assigned Numbers Authority for the Internet. Initially, this subtree has at least one child:

enterprises OBJECT IDENTIFIER::={ private 1 }

The enterprises(1) subtree is used to permit parties providing networking subsystems to register models of their products.

Specific organizations have developed subtrees for private use for their products. One such tree is the ATM UNI MIB. Vendors can define private ATM UNI MIB extensions to support additional or proprietary features of their products. Objects in the MIB are defined using the subset of Abstract Syntax Notation One (ASN.1) defined by the SMI. The syntax of an object type defines the abstract data structure corresponding to that object type. The ASN.1 language is used for this purpose. The SMI purposely restricts the ASN.1 constructs which may be used. These restrictions are made for simplicity. The structure of ATM UNI ILMI MIB is illustrated in the following figure:

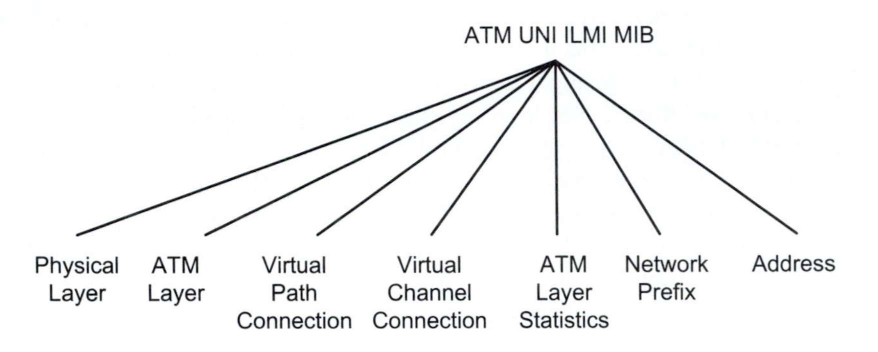

An entire tree group is either optional, conditionally required, or required. If a group is required, then every element in the group is required. If a group is conditionally required, every element in the group is required, if implemented.

Protocol Limitations

The following are some known SNMP limitations:

- ATM messages must be formatted according to SNMP version 1, not SNMP version 2.
- ALL SNMP messages will use the community name ILMI.
- In all SNMP Traps, the agent address field always has an IP Address value of 0.0.0.0.
- The supported traps are coldStart and enterpriseSpecific.
- In all SNMP traps, the timestamp field contains the value of the agent's sysUpTime MIB object at the time of trap generation. In all of the standard SNMP traps, the enterprise field in the Trap PDU contains the value of the agents sysObjectID MIB object.
- The size of messages can be up to 484 octets.

18

IP Switching Protocols

In order to fill an ATM 155 Mbps line, an IP router may need to send up to 100,000 packets per second. This is beyond the capability of most routers, mainly because each IP datagram needs to be routed separately (IP is not connection oriented).

Ipsilon developed the IP switch, to solve this problem by implementing Cut Through Routing, thus allowing IP routing to be 5 times faster then other IP routers on the market. This is done by detecting several classes of IP flows during the routing process. A **flow** is a sequence of packets, having the same source and destination addresses, as well as common higher-level protocol types (UDP, TCP), type of service and other characteristics (as indicated by the information in the IP packet header).

Once a flow is detected and classified, the IP switch signals the upstream node (where the data comes from) to use a new VC for that flow. The same is done by the downstream node, making the switch send flow packets using a new VC.

When the flow is received and transmitted through dedicated VCs, it can be switched using the ATM switching hardware, with no routing involved. In addition, a layer 2 label is attached to the header of each flow packet, enabling faster lookup in cache routing tables.

Signalling between IP switches is done using two protocols: Ipsilon Flow Management Protocol (IFMP) and General Switch Management Protocol (GSMP).

IFMP

RFC 1953 1996-05 http://www.cis.ohio-state.edu/htbin/rfc/rfc1953.html

The Ipsilon Flow Management Protocol (IFMP), is a protocol for instructing an adjacent node to attach a layer 2 label to a specified IP flow to route it through an IP switch. The label allows more efficient access to cached routing information for that flow and allows the flow to be switched rather than routed in certain cases.

IFMP is composed of two sub-protocols: the Adjacency Protocol and the Redirection Protocol. IFMP messages are encapsulated within an IPv4 packet. They are sent to the IP limited broadcast address (255.255.255.255). The protocol field in the IP header contains the value 101 (decimal) to indicate an IFMP message.

The structure of the IFMP header is shown in the following illustration:

Version (1)	Op Code (1)	Checksum (2)

IFMP header structure

Version

The ICMP protocol Version number. The current version is 1.

OP Code

The function of the message.
Four Op Codes are defined for the IFMP Adjacency Protocol.
0 SYN.
1 SYNACK.
2 RSTACK.
3 ACK.

Five Codes are defined for the IFMP Redirection Protocol.
4 REDIRECT.
5 RECLAIM.
6 RECLAIM ACK.
7 LABEL RANGE.
8 ERROR.

Checksum

The CRC value.

GSMP

RFC 1987 1996-08 http://www.cis.ohio-state.edu/htbin/rfc/rfc1987.html

The General Switch Management Protocol (GSMP), is a general purpose protocol to control an ATM switch. GSMP allows a controller to establish and release connections across the switch, add and delete leaves on a point-to-multipoint connection, manage switch ports, request configuration information and request statistics.

GSMP packets are variable length and are encapsulated directly into AAL5 with an LLC/SNAP header 0x00-00-00-88-0C to indicate GSMP messages.

The structure of the GSMP header is shown in the following illustration:

Version (1)	Message type (1)	Result (1)	Code (1)

GSMP header structure

Version
The GSMP protocol version number, currently version 1.

Message type
The GSMP message type. There are five classes, each of which have a number of different message types. The classes are: Connection Management, Port Management, Statistics, Configuration, and Events.

Result
Used in request messages. Indicates whether a response is required if the outcome is successful.

Code
Further information concerning the result in a response message. Mostly used to pass an error code in a failure response. In a request message, the code field is not used and is set to 0.

19

ISDN

ITU SR-NWT-001953 1991-06, ETS 300 102-1 1990-12,
AT&T 801-802-100 1989-05

ISDN (Integrated Services Digital Network) is an all digital communications line that allows for the transmission of voice, data, video and graphics, at very high speeds, over standard communication lines. ISDN provides a single, common interface with which to access digital communications services that are required by varying devices, while remaining transparent to the user. Due to the large amounts of information that ISDN lines can carry, ISDN applications are revolutionizing the way businesses communicate.

ISDN is not restricted to public telephone networks alone; it may be transmitted via packet switched networks, telex, CATV networks, etc.

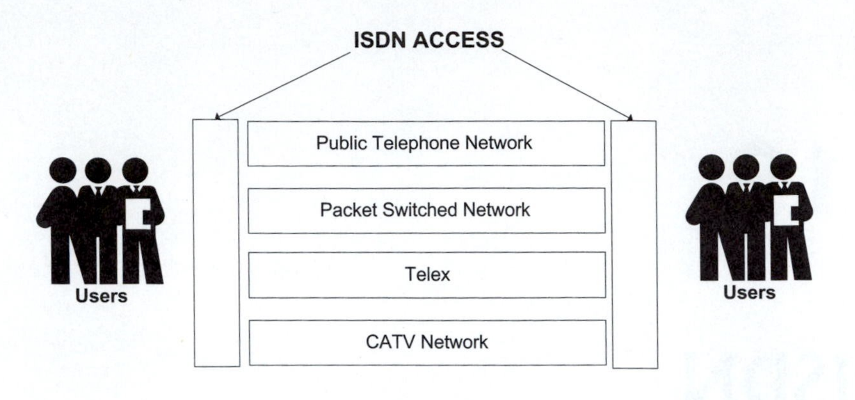

ISDN applications

The following protocols are described in this chapter:

- LAPD: Link Access Protocol - Channel D.
- ISDN: Integrated Services Digital Network.

LAPD

ITU Q.921 (Blue Book)

The LAPD (Link Access Protocol - Channel D) is a layer 2 protocol which is defined in CCITT Q.920/921. LAPD works in the Asynchronous Balanced Mode (ABM). This mode is totally balanced (i.e., no master/slave relationship). Each station may initialize, supervise, recover from errors, and send frames at any time. The protocol treats the DTE and DCE as equals.

The format of a standard LAPD frame is as follows:

Flag	Address field	Control field	Information	FCS	Flag

LAPD frame structure

Flag

The value of the flag is always (0x7E). In order to ensure that the bit pattern of the frame delimiter flag does not appear in the data field of the frame (and therefore cause frame misalignment), a technique known as Bit Stuffing is used by both the transmitter and the receiver.

Address field

The first two bytes of the frame after the header flag is known as the address field. The format of the address field is as follows:

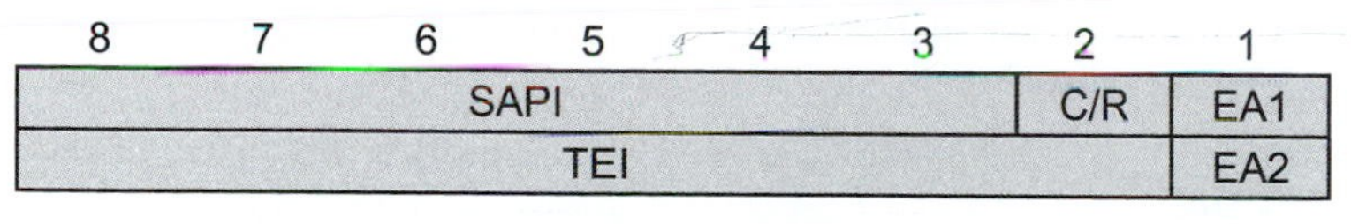

LAPD address field

EA1 First Address Extension bit which is always set to 0.

C/R Command/Response bit. Frames from the user with this bit set to 0 are command frames, as are frames from the network with this bit set to 1. Other values indicate a response frame.

SAPI Service Access Point Identifier. Valid values are as follows:

0	Call-Control procedures.
1	Packet-mode communications using I.451 call-control procedures.
16	Packet communication conforming to X.25 level 3.
63	Layer 2 management procedures.

EA2 Second Address Extension bit which is always set to 1.

TEI Terminal Endpoint Identifier. Valid values are as follows:

0-63	Used by non-automatic TEI assignment user equipment.
64-126	Used by automatic TEI assignment equipment.
127	Used for a broadcast connection meant for all Terminal Endpoints.

Control field

The field following the Address Field is called the Control Field and serves to identify the type of the frame. In addition, it includes sequence numbers, control features and error tracking according to the frame type.

FCS

The Frame Check Sequence (FCS) enables a high level of physical error control by allowing the integrity of the transmitted frame data to be checked. The sequence is first calculated by the transmitter using an algorithm based on the values of all the bits in the frame. The receiver then performs the same calculation on the received frame and compares its value to the CRC.

Window size

LAPD supports an extended window size (modulo 128) where the number of possible outstanding frames for acknowledgement is raised from 8 to 128. This extension is generally used for satellite transmissions where the acknowledgement delay is significantly greater than the frame transmission times. The type of the link initialization frame determines the modulo of the session and an "E" is added to the basic frame type name (e.g., SABM becomes SABME).

Frame types

The following are the Supervisory Frame Types in LAPD:

RR	Information frame acknowledgement and indication to receive more.
REJ	Request for retransmission of all frames after a given sequence number.
RNR	Indicates a state of temporary occupation of station (e.g., window full).

The following are the Unnumbered Frame Types in LAPD:

DISC	Request disconnection.
UA	Acknowledgement frame.
DM	Response to DISC indicating disconnected mode.
FRMR	Frame reject.
SABM	Initiator for asynchronous balanced mode. No master/slave relationship.
SABME	SABM in extended mode.
UI	Unnumbered Information.
XID	Exchange Information.

There is one Information Frame Type in LAPD:

Info	Information transfer frame.

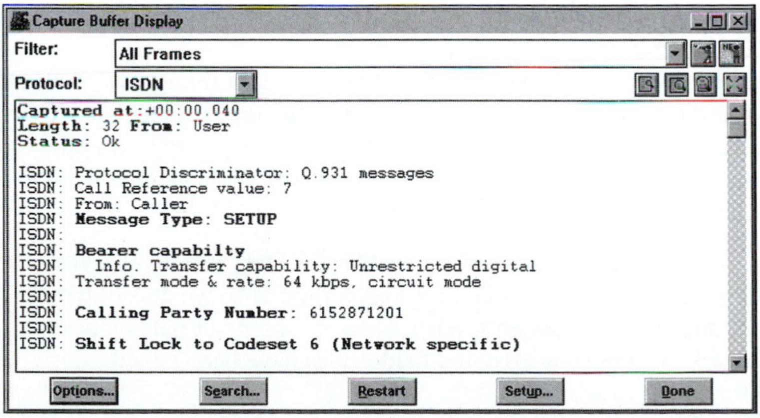

```
Capture Buffer Display
Filter:    All Frames
Protocol:  ISDN
Captured at:+00:00.040
Length: 32 From: User
Status: Ok

ISDN: Protocol Discriminator: Q.931 messages
ISDN: Call Reference value: 7
ISDN: From: Caller
ISDN: Message Type: SETUP
ISDN:
ISDN: Bearer capabilty
ISDN:   Info. Transfer capability: Unrestricted digital
ISDN: Transfer mode & rate: 64 kbps, circuit mode
ISDN:
ISDN: Calling Party Number: 6152871201
ISDN:
ISDN: Shift Lock to Codeset 6 (Network specific)
Options...  Search...  Restart  Setup...  Done
```

ISDN decode

International Variants of ISDN

The organization primarily responsible for producing the ISDN standards is the CCITT. The CCITT study group responsible for ISDN first published a set of ISDN recommendations in 1984 (Red Books). Prior to this publication, various geographical areas had developed different versions of ISDN. This resulted in the CCITT recommendation of a common ISDN standard for all countries, in addition to allocated variants definable for each country.

The use of nation-specific information elements is enabled by using the Codeset mechanism which allows different areas to use their own information elements within the data frames.

The following is a description of most ISDN variants:

National ISDN1 (Bellcore)

SR-NWT-001953 1991-06
This variant is used in the USA by Bellcore. It has four network-specific message types. It does not have any single octet information elements. In addition to Codeset 0 elements it has four Codeset 5 and five Codeset 6 information elements.

National ISDN-2 (Bellcore)

SR-NWT-002361 1992-12
The main difference between National ISDN-1 and ISDN-2 is parameter downloading via components (a component being a sub-element of the Extended Facility information element). These components are used to communicate parameter information between ISDN user equipment, such as an ISDN telephone, and the ISDN switch.

Other changes are the addition of the SEGMENT, FACILITY and REGISTER message types and the Segmented Message and Extended Facility information elements. Also, some meanings of field values have changed and some new accepted field values have been added.

5ESS (AT&T)

AT&T 801-802-100 1989-05
This variant is used in the USA by AT&T. It is the most widely used of the ISDN protocols and contains 19 network-specific message types. It has no

Codeset 5, but does have 18 Codeset 6 elements and an extensive information management element.

Euro ISDN (ETSI)

ETS 300 102-1 1990-12

This variant is to be adopted by all of the European countries. Presently, it contains single octet message types and has five single octet information elements. Within the framework of the protocol there are no Codeset 5 and Codeset 6 elements, however each country is permitted to define its own individual elements.

VN3, VN4 (France)

DGPT: CSE P 22-30 A 1994-08

These variants are prevalent in France. The VN3 decoding and some of its error messages are translated into French. It is a sub-set of the CCITT document and only has single octet message types. The more recent VN4 is not fully backward compatible but closely follows the CCITT recommendations. As with VN3, some translation has taken place. It has only single octet message types, five single octet information elements, and two Codeset 6 elements.

1TR6 (Germany)

1 TR 6 1990-08

This variant is prevalent in Germany. It is a sub-set of the CCITT version, with minor amendments. The protocol is part English and part German.

ISDN 30 [DASS-2] (England)

BTNR 190 1992-07

This variant is used by British Telecom in addition to ETSI (see above). At layers 2 and 3 this standard does not conform to CCITT structure. Frames are headed by one octet and optionally followed by information. However most of the information is IA5 coded, and therefore ASCII decoded

Australia

AP IX-123-E

This protocol is being superseded by a new Australian protocol. (The name of the protocol has not been released). It is a subset of the CCITT standard and has only single octet message types and information elements; it only has Codeset 5 elements.

TS014 Australia

TS014 (Austel) 1995

This is the new Australian ISDN PRI standard issued by Austel. This standard is very similar to ETSI.

NTT-Japan

INS-NET Interface and Services 1993-03

The Japanese ISDN service provided by NTT is known as INS-Net and its main features are as follows:

- Provides a user-network interface that conforms to the CCITT Recommendation Blue Book.
- Provides both basic and primary rate interfaces.
- Provides a packet-mode using Case B.
- Supported by Signalling System No. 7 ISDN User Part with the network.
- Offered as a public network service.

ARINC 746

ARINC Characteristic 746-4 1996-04

In passenger airplanes today there are phones in front of each passenger. These telephones are connected in a T1 network and the conversation is transferred via a satellite. The signalling protocol used is based on Q.931, but with a few modifications and is known as ARINC 746. The leading companies in this area are GTE and AT&T. In order to analyze ARINC, the LAPD variant should also be specified as **ARINC**.

ARINC 746 Attachment 11

ARINC Characteristic 746-4 1996-04

ARINC (Aeronautical Radio, INC.) Attachment 11 describes the Network Layer (layer 3) message transfer necessary for equipment control and circuit switched call control procedures between the Cabin Telecommunications Unit (CTU) and SATCOM system, North American Telephone System (NATS), and Terrestrial Flight Telephone System (TFTS). The interface described in this attachment is derived from the CCITT recommendations Q.930, Q.931 and Q.932 for call control and the ISO/OSI standards DIS 9595 and DIS 9596 for equipment control. These Network Layer messages should be transported in the information field of the Data Link Layer frame.

ARINC 746 Attachment 17

ARINC Characteristic 746-4 1996-04

ARINC (Aeronautical Radio, INC.) Attachment 17 represents a system which provides passenger and cabin crew access to services provided by the

CTU and intelligent cabin equipment. The distribution portion of the CDS transports the signalling and voice channels from headend units to the individual seat units. Each zone within the aircraft has a zone unit that controls and services seat units within that zone.

Northern Telecom - DMS 100

NIS S208-6 Issue 1.1 1992-08

This variant represents Northern Telecom's implementation of National ISDN-1. It provides ISDN BRI user-network interfaces between the Northern Telecom ISDN DMS-100 switch and terminals designed for the BRI DSL. It is based on CCITT ISDN-1 and Q Series Recommendations and the ISDN Basic Interface Call Control Switching and Signalling Requirements and supplementary service Technical References published by Bellcore.

DPNSS1

BTNR 188 1995-01

DPNSS1 (Digital Private Network Signalling System No. 1) is a common-channel signalling system used in Great Britain. It extends facilities normally only available between extensions on a single PBX to all extensions on PBXs that are connected together in a private network. It is primarily intended for use between PBXs in private networks via time-slot 16 of a 2048 kbit/s digital transmission system. Similarly it may be used in time-slot 24 of a 1.544 kbit/s digital transmission system. Note that the LAPD variant should also be selected to be DPNSS1.

Swiss Telecom

PTT 840.73.2 1995-06

The ISDN variant operated by the Swiss Telecom PTT is called SwissNet. The DSS1 protocol for SwissNet is fully based on ETS. Amendments to this standard for SwissNet fall into the category of definitions of various options in the standard and of missing requirements. They also address SwissNet-specific conditions, e.g., assuring compatibility between user equipment and SwissNet exchanges of different evolution steps.

QSIG

ISO/IEC 11572 1995

QSIG is a modern, powerful and intelligent inter-private PABX signalling system. QSIG standards specify a signalling system at the Q reference point which is primarily intended for use on a common channel; e.g. a G.703 interface. However, QSIG will work on any suitable method of connecting the PINX equipment. The QSIG protocol stack is identical in structure to the DSSI protocol stack. Both follow the ISO reference model. Both can have an identical layer 1 and layer 2 (LAPD), however, at layer 3 QSIG and DSS1 differ.

ISDN Frame Structure

Shown below is the general structure of the ISDN frame.

8	7	6	5	4	3	2	1
Protocol discriminator							
0	0	0	0	Length of reference call value			
Flag	Call reference value						
0	Message type						
Other information elements as required							

ISDN frame structure

Protocol discriminator

The protocol used to encode the remainder of the Layer.

Length of call reference value

Defines the length of the next field. The Call reference may be one or two octets long depending on the size of the value being encoded.

Flag

Set to zero for messages sent by the party that allocated the call reference value; otherwise set to one.

Call reference value

An arbitrary value that is allocated for the duration of the specific session, which identifies the call between the device maintaining the call and the ISDN switch.

Message type

Defines the primary purpose of the frame. The message type may be one octet or two octets (for network specific messages). When there is more than one octet, the first octet is coded as eight zeros. A complete list of message types is given in *ISDN Message Types* below.

ISDN Information Elements

There are two types of information elements: single octet and variable length.

Single octet information elements

The single octet information element appears as follows:

8	7 6 5	4 3 2 1
1	Information element identifier	Information element

Single octet information element

The following are the available single octet information elements:

1 000 ----	Reserved
1 001 ----	Shift
1 010 0000	More data
1 010 0001	Sending Complete
1 011 ----	Congestion Level
1 101 ----	Repeat indicator

Variable length information elements

The following is the format of the variable length information element:

8	7 6 5 4 3 2 1
0	Information element identifier
Length of information elements	
Information elements (multiple bytes)	

Variable length information element

The information element identifier identifies the chosen element and is unique only within the given Codeset. The length of the information element informs the receiver as to the amount of the following octets belonging to each information element. The following are possible variable length information elements:

0 0000000	Segmented Message
0 0000100	Bearer Capability
0 0001000	Cause
0 0010100	Call identify
0 0010100	Call state
0 0011000	Channel identification
0 0011100	Facility

0 0011110	Progress indicator
0 0100000	Network-specific facilities
0 0100111	Notification indicator
0 0101000	Display
0 0101001	Date/time
0 0101100	Keypad facility
0 0110100	Signal
0 0110110	Switchhook
0 0111000	Feature activation
0 0111001	Feature indication
0 1000000	Information rate
0 1000010	End-to-end transit delay
0 1000011	Transit delay selection and indication
0 1000100	Packet layer binary parameters
0 1000101	Packet layer window size
0 1000110	Packet size
0 1101100	Calling party number
0 1101101	Calling party subaddress
0 1110000	Called party number
0 1110001	Called Party subaddress
0 1110100	Redirecting number
0 1111000	Transit network selection
0 1111001	Restart indicator
0 1111100	Low layer compatibility
0 1111101	High layer compatibility
0 1111110	User-user
0 1111111	Escape for extension
Other values	Reserved

ISDN Message Types

The following are possible ISDN message types:

Call Establishment

000 00001	Alerting
000 00010	Call Proceeding
000 00011	Progress
000 00101	Setup
000 00111	Connect
000 01101	Setup Acknowledge
000 01111	Connect Acknowledge

Call Information Phase

001 00000	User Information
001 00001	Suspend Reject
001 00010	Resume Reject
001 00100	Hold
001 00101	Suspend
001 00110	Resume
001 01000	Hold Acknowledge
001 01101	Suspend Acknowledge
001 01110	Resume Acknowledge
001 10000	Hold Reject
001 10001	Retrieve
001 10011	Retrieve Acknowledge
001 10111	Retrieve Reject

Call Clearing

010 00101	Disconnect
010 00110	Restart
010 01101	Release
010 01110	Restart Acknowledge
010 11010	Release Complete

Miscellaneous

011 00000	Segment
011 00010	Facility
011 00100	Register
011 01110	Notify
011 10101	Status inquiry
011 11001	Congestion Control
011 11011	Information
011 11101	Status

ISDN Terminology

BRI

The Basic Rate Interface is one of the two services provided by ISDN. BRI is comprised of two B-channels and one D-channel (2B+D). The B-channels each operate at 64 Kbps and the D-channel operates at 16 Kbps. It is used by single line business customers for typical desk-top type applications.

C/R

C/R refers to Command or Response. The C/R bit in the address field defines the frame as either a command frame or a response frame to the previous command.

Codeset

Three main Codesets are defined. In each Codeset, a section of the information elements are defined by the associated variant of the protocol:

Codeset 0	The default code, referring to the CCITT set of information elements.
Codeset 5	The national specific Codeset.
Codeset 6	The network specific Codeset.

The same value may have different meanings in various Codesets. Most elements usually appear only once in each frame.

In order to change codesets two methods are defined:

Shift	This method enables a temporary change to another Codeset. Also termed as non-locking shift, the shift only applies to the next information element.
Shift Lock	This method implements a permanent change until indicated otherwise. Shift-Lock may only change to a higher Codeset.

CPE

Customer Premises Equipment - refers to all ISDN compatible equipment connected at the user sight. Examples of devices are telephone, PC, Telex, Facsimile, etc. The exception is the FCC definition of NT1. The FCC views the NT1 as a CPE because it is on the customer sight, but the CCITT views NT1 as part of the network. Consequently the network reference point of the network boundary is dependent on the variant in use.

ISDN channels B, D and H

The three logical digital communication channels of ISDN perform the following functions:

B-Channel	Carries user service information including: digital data, video, and voice.
D-Channel	Carries signals and data packets between the user and the network.
H-Channel	Performs the same function as B-Channels, but operates at rates exceeding DS-0 (64 Kbps).

ISDN devices

Devices connecting a CPE and a network. In addition to facsimile, telex, PC, telephone, ISDN devices may include the following:

TA — Terminal Adapters - devices that are used to portray non-ISDN equipment as ISDN compatible.

LE — Local Exchange - ISDN central office (CO). The LE implements the ISDN protocol and is part of the network.

LT — Local Termination - used to express the LE responsible for the functions associated with the end of the Local Loop.

ET — Exchange Termination - used to express the LE responsible for the switching functions.

NT — Network Termination equipment exists in two forms and is referred to accordingly. The two forms are each responsible for different operations and functions.

- NT1 - Is the termination of the connection between the user sight and the LE. NT1 is responsible for performance, monitoring, power transfer, and multiplexing of the channels.
- NT2 - May be any device that is responsible for providing user sight switching, multiplexing, and concentration: LANs, mainframe computers, terminal controllers, etc. In ISDN residential environments there is no NT2.

TE — Terminal Equipment - any user device e.g.: telephone or facsimile. There are two forms of terminal equipment:

- TE1 - Equipment is ISDN compatible.
- TE2 - Equipment is not ISDN compatible.

ISDN reference points

Reference points define the communication points between different devices and suggest that different protocols may be used at each side of the point. The main points are as follows:

R A communication reference point between a non-ISDN compatible TE and a TA.

S A communication reference link between the TE or TA and the NT equipment.

T A communication reference point between user switching equipment and a Local Loop Terminator.

U A communication reference point between the NT equipment and the LE. This reference point may be referred to as the network boundary when the FCC definition of the Network terminal is used.

The following diagram illustrates the ISDN Functional Devices and Reference Points:

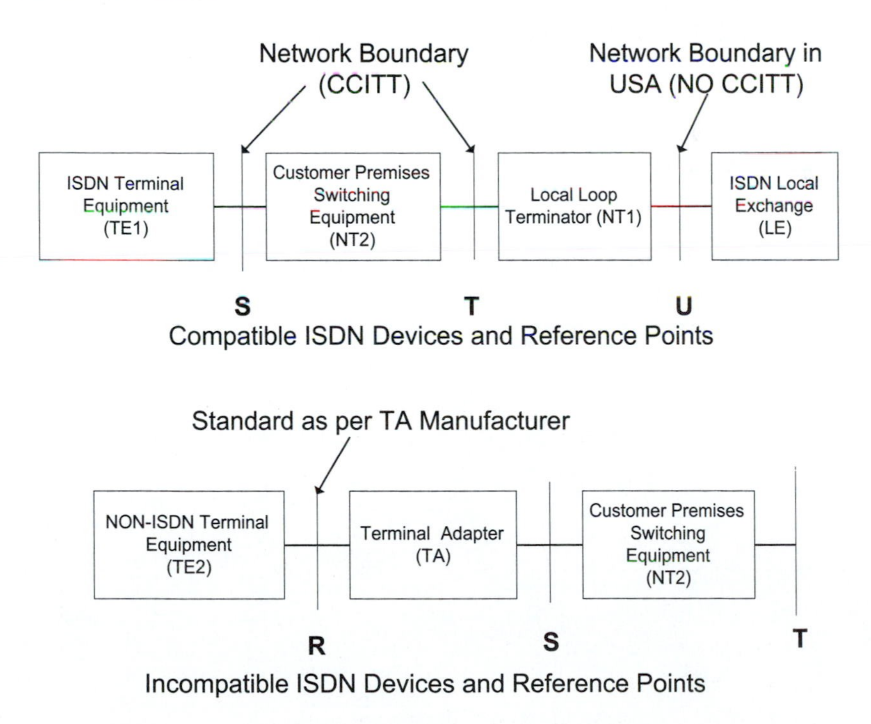

Compatible ISDN Devices and Reference Points

Incompatible ISDN Devices and Reference Points

LAPD

The Link Access Protocol on the D-channel. LAPD is a bit orientated protocol on the data link layer of the OSI reference model. Its prime function is ensuring the error free transmission of bits on the physical layer (layer 1).

PRI

The Primary Rate Interface is one of the two services provided by ISDN. PRI is standard dependent and thus varies according to country. In North America, PRI has twenty-three B-channels and one D-channel (23B+D). In Europe, PRI has thirty B-channels and one D-channel (30B+D).

The American B- and D-channels operate at an equal rate of 64 Kbps. Consequently, the D-channel is sometimes not activated on certain interfaces, thus allowing the time slot to be used as another B-channel. The 23B+D PRI operates at the CCITT designated rate of 1544 Kbps.

The European PRI is comprised of thirty B-channels and one D-channel (30B+D). As in the American PRI all the channels operate at 64 Kbps. However, the 30B+D PRI operates at the CCITT designated rate of 2048 Kbps.

SAPI

Service Access Point Identifier, the first part of the address of each frame.

TEI

Terminal End Point Identifier, the second part of the address of each frame.

20

ISO Protocols

The Institute of Electrical and Electronic Engineers (IEEE) defines the International Standards Organization (ISO) protocols. The ISO protocol suite is a complete, seven-layer protocol conforming to the Open System Interconnection (OSI) networking model. The ISO protocol suite includes the following protocols:

- IS-IS: Intermediate System to Intermediate System.
- ES-IS: End-System to Intermediate System.
- ISO-IP: Internetworking Protocol.
- ISO-TP: Transport Protocol.
- ISO-SP: Session Protocol.
- ISO-PP: Presentation Protocol.
- ACSE: Application Control Service Element
- CCITT X.400: Consultative Committee Protocol.

The following diagram illustrates the ISO protocol suite in relation to the OSI model:

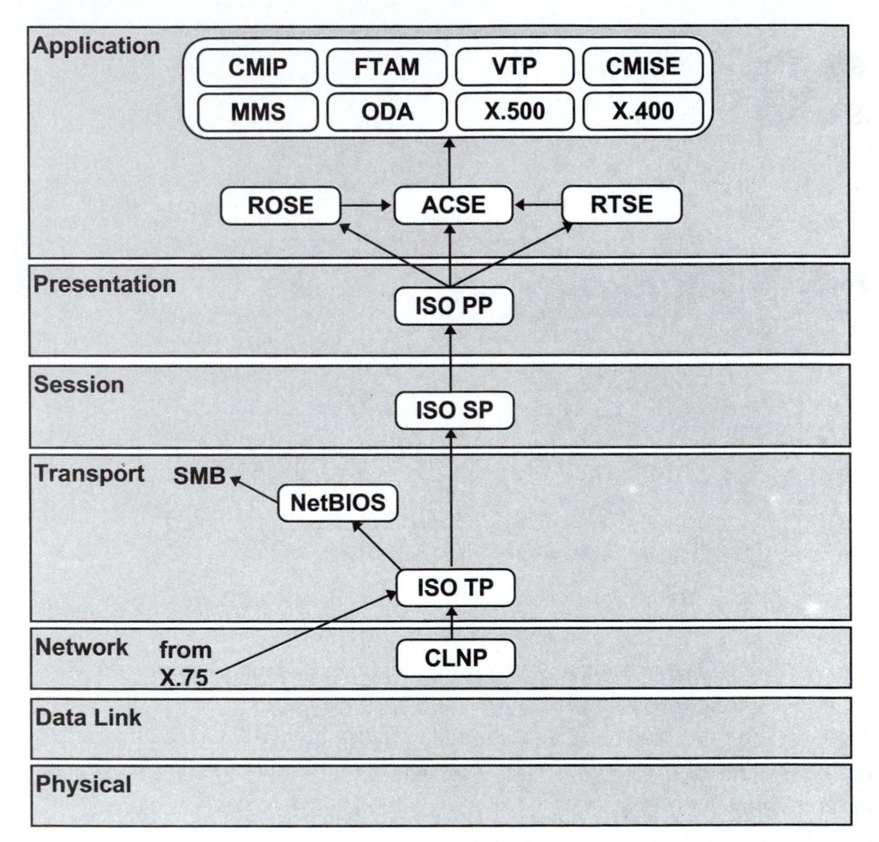

ISO protocol suite in relation to the OSI model

IS-IS

ISO 10589

IS-IS (Intermediate System to Intermediate System) is a protocol of the network layer. It permits intermediate systems within a routing domain to exchange configuration and routing information to facilitate the operation of the routing and relaying functions of the network layer. IS-IS is designed to operate in close conjunction with ES-IS (ISO 9542) and CLNS (ISO 8473).

The format of the IS-IS header which is common to all PDUs is as follows:

8	7	6	5	4	3	2	1	Octet
Intradomain routing protocol discriminator								1
Length indicator								2
Version/protocol ID extension								3
ID length								4
R	R	R	PDU type					5
Version								6
Reserved								7
Maximum area addresses								8

IS-IS header structure

Intradomain routing protocol discriminator
Network layer protocol identifier assigned to this protocol (= 83 decimal).

Length indicator
Length of the fixed header in octets.

Version/protocol ID extension
Equal to 1.

ID length
Length of the ID field of NSAP addresses and NETs used in this routing domain.

R
Reserved bits.

PDU type

Type of PDU. Bits 6, 7 and 8 are reserved.

Version

Equal to 1.

Maximum area addresses

Number of area addresses permitted for this intermediate system's area.

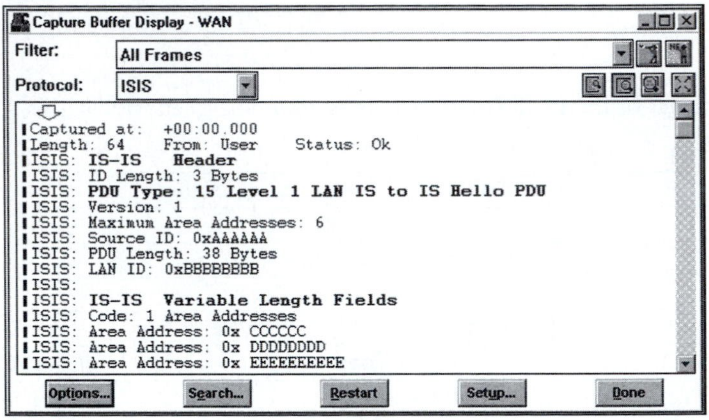

```
Capture Buffer Display - WAN
Filter:    All Frames
Protocol:  ISIS

Captured at:  +00:00.000
Length: 64    From: User    Status: Ok
ISIS: IS-IS   Header
ISIS: ID Length: 3 Bytes
ISIS: PDU Type: 15 Level 1 LAN IS to IS Hello PDU
ISIS: Version: 1
ISIS: Maximum Area Addresses: 6
ISIS: Source ID: 0xAAAAAA
ISIS: PDU Length: 38 Bytes
ISIS: LAN ID: 0xBBBBBBBB
ISIS:
ISIS: IS-IS  Variable Length Fields
ISIS: Code: 1 Area Addresses
ISIS: Area Address: 0x CCCCCC
ISIS: Area Address: 0x DDDDDDDD
ISIS: Area Address: 0x EEEEEEEEEE

Options...   Search...   Restart   Setup...   Done
```

IS-IS decode

ES-IS

ISO 9542

The End System to Intermediate System (ES-IS) protocol distributes routing information among ISO hosts.

ISO-IP

IETF RFC 1069 http://www.cis.ohio-state.edu/htbin/rfc/rfc1069.html

The ISO documents IS 8473 and IS 8348 define the ISO Internetworking Protocol (ISO-IP) or CLNP which includes built-in error signaling to aid in routing management. ISO-IP is intended to facilitate the interconnection of open systems. It is used in the network layer and provides connectionless-mode network service.

Each PDU contains the following, according to this order:

1. Fixed part
2. Address part
3. Segmentation part, if present
4. Options part, if present
5. Data part, if present

The ISO-IP PDU has the following format:

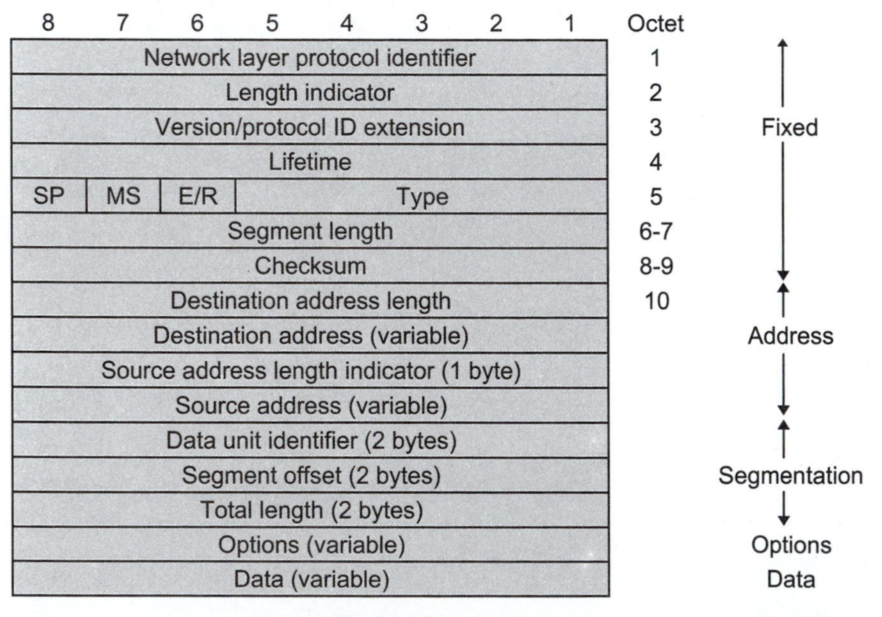

ISO-IP PDU - fixed part

Fixed Part

The ISO-IP fixed part contains the following fields:

Network layer protocol identifier
Set to binary 1000 0001 to identify this network layer as CLNP. A value of 0000 0000 identifies the inactive network layer protocol subset.

Length indicator
Length of the header in octets.

Version/protocol ID extension
Value of binary 0000 0001 identifies the ISO 8473 standard.

Lifetime
Remaining lifetime of the PDU in units of 500 milliseconds.

SP
Segmentation permitted flag. Value of 1 indicates that segmentation is permitted. The value is determined by the originator of the PDU and cannot be changed by any other network entity for the lifetime of the initial PDU and any derived PDUs.

MS
More segments flag. Value of 1 indicates that segmentation has occurred and the last octet of the NSDU is not contained in this PDU.

E/R
Error report flag. A value of 1 indicates to generate an error report PDU according to the standard.

Type
Type code identifies the type of the protocol data unit: DT or ER.

Segment length
Length in octets of the PDU, including header and data.

Checksum
Checksum value computed on the entire PDU header. A value of 0 indicates that the checksum is to be ignored.

Address Part

The address part has the following parameters:

Destination
Destination network service access point address.

Source
Source network service access point address.

Segmentation Part

When the segmentation permitted flag (SP) is set, ISO-IP frames also include the following fields:

Data unit identifier
The sequence number used to identify the order of segments when fragmentation is enabled.

Segment offset
The offset in the original data unit where the segment is located.

Total length
The total length of the initial data unit before fragmentation.

Frame Options

The following options can be present in ISO-IP frames:

Source routing
Specifies the network path by providing a set of addresses that the frame must travel. The following parameters are present in frames with the source route option:

Type of routing	Represented as complete or partial.
NextNET	Next network entity title in the route list that is to be processed.

Record route

Causes each node encountered by the frame to record its network entity title in the frame. The following parameters are present in frames with the record route option:

Type of routing	Represented as complete or partial.
#NETs	Number of network entity titles currently in the route listing.

Priority

Requested priority ranging from 0 to 14 with priority 14 being highest.

Padding

Number of pad bytes used to produce the desired frame alignment.

Security

A security format code and parameters which are displayed as (Code) Parameters. The parameters that follow the code indicate the security level.

1	Source address specific.
2	Destination address specific.
3	Globally unique.

QoS

Quality of Service requested for the connection as (Code) parameters. The parameters that follow the code indicate the Quality of Service.

1	Source address specific.
2	Destination address specific.
3	Globally unique.

Frame Error Messages

The following are possible error messages for ISO-IP frames:

Error Message	*Description*
{not specified}	Unknown error.
{protocol error}	Protocol procedure error.
{bad checksum}	Checksum is invalid.
{too congested}	Frame discarded due to congestion.
{bad PDU header}	PDU header syntax error.
{fragment needed}	Segmentation needed, but not permitted.
{incomplete PDU}	Incomplete PDU received.
{duplicate option}	Option already implemented.
{dest unreachable}	Destination IP address unreachable.

Error Message	*Description*
{destinat unknown}	Destination IP address unknown.
{unknown SR error}	Unknown source routing error.
{SR syntax error}	Syntax error in source routing field.
{bad SR address}	Unknown address in source routing field.
{bad SR path}	Source route path not acceptable.
{TTL expired}	Lifetime expired while in transit.
{reasmbly expired}	Lifetime expired during reassembly.
{bad option}	Specified option not supported.
{bad protocol ver}	Specified protocol version not supported.
{bad security opt}	Specified security option not supported.
{bad SR option}	Source routing option not supported.
{bad RR option}	Record routing option not supported.
{reassmbly failed}	Reassembly failed due to interference.

ISO-TP

ISO 8073

ISO-TP describes the TP (Transport Protocol). This protocol is intended to be simple but general enough to cater to the entire range of Network Service qualities possible, without restricting future extensions. It is structured to give rise to classes of protocols which are designed to minimize possible incompatibilities and implementation costs.

The format of the ISO-TP header is as follows:

LI	Fixed part	Variable part

ISO-TP header structure

LI is the length indicator field providing the length of the header in bytes. The format of the fixed part of non-data PDUs is as follows:

PDU type	DST-REF	SRC-REF	Variable
1 byte	2 bytes	2 bytes	1 byte

ISO-TP non-data PDU: fixed part

The PDU types are listed above. The meanings of DST-REF, SRC-REF and the last byte are related to the type of PDU. Refer to the ISO-TP standard for the exact structure for each PDU type.

ISO-SP

ISO/IEC 8327-1 09-1996; ITU-T X.225

The ISO-SP protocol specifies procedures for a single protocol for the transfer of data and control information from one session entity to a peer session entity.

The Session protocol data units are transferred using the Transport Data Transfer Service. The TSDU (Transport Service Data Unit) is comprised of a number of SPDUs (Session Protocol Data Units). There can be up to 4 SPDUs depending on the concatenation method that is being used (basic or extended) and on the SPDU type.

Each SPDU contains one or more octets. The SPDU structure is as follows:

SI	LI	Parameter field	User information field

SPDU structure

SI

SPDU indicator. This field indicates the type of SPDU.

LI

The length indicator signifies the length of the associated parameter field.

Parameter field

In the SPDU the parameter field contains the PGI or PI units defined for the SPDU.

The structure of a PGI unit is shown in the following illustration:

PGI	LI	Parameter field

Parameter field: PGI structure

The structure of the PI units is shown in the following illustration:

PI	LI	Parameter field

Parameter field: PI structure

PGI Parameter group identifier. Identifies the parameter group.

PI	Parameter identifier. Indicates the type of information contained in the parameter field.
Parameter field	In the PGI, the parameter field contains a single parameter value, or one or more PI units. In the PI, the parameter field contains the parameter value.

User information field

This contains segments of a segmented SSDU.

ISO-PP

ISO document IS 8823 defines the ISO Presentation Protocol (PP) which performs context negotiation and management between open systems.

Frames

PP frames may be one of the following commands:

[Connect Presentation]	Requests a presentation layer connection.
[Connect Presentation Accept]	Acknowledges the presentation connection.

Parameters

PP frames can contain the following parameter:

X.410 Mode {1984 X.400}

This frame is based on the CCITT Recommendation X.410. This usually means that the application is a CCITT 1984 X.400 Message Handling System.

ACSE

ITU-T Recommendation X.227

The Application Control Service Element (ACSE) protocol provides services for establishing and releasing application-associations. The ACSE protocol also includes two optional functional units. One functional unit supports the exchange of information in support of authentication during association establishment. The second functional unit supports the negotiation of application context during association establishment. The ACSE services apply to a wide range of application-process communications requirements.

CCITT X.400 Message Handling System

The International Telegraph and Telephone Consultative Committee (known as CCITT, after its French name) produced this standard for exchange of electronic mail between computers. The CCITT recommendations X.400 through X.430 define an application layer protocol and a minimal presentation layer protocol. CCITT X.400 uses the ISO Session Layer services and protocol, documented in ISO documents IS 8326 and IS 8327, respectively.

The CCITT X.400 standard describes message transfer agents (MTAs) that are responsible for delivery of electronic mail between computers. The MTAs use a protocol known as P1 to carry out transfer of message protocol data units (MPDUs). MTAs exchange two kinds of MPDUs: User and Service. User MPDUs contain messages, while Service MPDUs supply information about message transfers. Two kinds of Service MPDUs exist: Delivery Report and Probe.

X.400 frames can be one of the following types:

[User MPDU Message]	Normal mail handling system (MHS) message.
[DeliveryReport MPDU]	Sent to discover status of prior message.
[Probe MPDU]	Sent to discover if the message was delivered.

Each CCITT X.400 frame contains the following parameter:

Protocol	P1.

21

LAN Data Link Layer Protocols

The Data Link Layer defines how data is formatted for transmission and how access to the network is controlled. This layer has been divided by the IEEE 802 standards committee into two sublayers: media access control (MAC) and logical link control (LLC).

The following data link layer protocols are described:

- Ethernet.
- Token Ring.
- FDDI: Fiber Distributed Data Interface.
- LLC: Logical Link Control.
- SNAP: SubNetwork Access Protocol.
- CIF: Cells in Frames.
- GARP: Generic Attribute Registration Protocol.
- GMRP: GARP Multicast Registration Protocol.

- GVRP: GARP VLAN Registration Protocol.
- SRP MAC Protocol: Spatial Reuse Protocol.
- VLAN.

FDDI, Token Ring and Ethernet may be physical interfaces or may act as logical protocols encapsulated over a WAN protocol or ATM.

The following illustration represents the LAN protocols in relation to the OSI model:

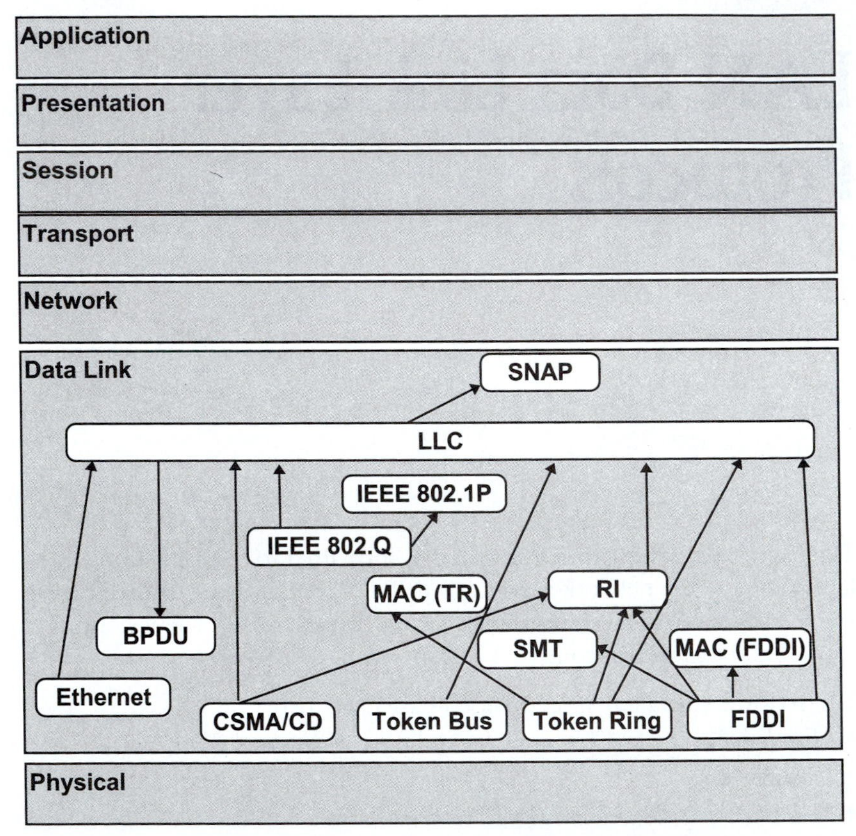

LAN protocols in relation to the OSI model

Ethernet

ANSI/IEEE 802.3 1933-00

Ethernet is a widely used data communications network standard developed by DEC, Intel, and Xerox. It uses a bus topology and CMSA/CD access method. The terms Ethernet and the IEEE 802.3 standard are often used interchangeably.

The Ethernet header structure is shown in the illustration below.

Destination	Source	Len	Data unit + pad	FCS
(6 bytes)	(6 bytes)	(2)	(46-1500 bytes)	(4 bytes)

Ethernet header structure

Destination address

The address structure is as follows:

I/G	U/L	Address bits

Ethernet destination address structure

I/G Individual/group address may be:
- 0 Individual address.
- 1 Group address.

U/L Universal/local address may be:
- 0 Universally administered.
- 1 Locally administered.

Source address

The address structure is as follows:

0	U/L	Address bits

Ethernet source address structure

0 The first bit is always 0.

U/L Universal/local address may be:
- 0 Universally administered.
- 1 Locally administered.

Length/type

In the Ethernet protocol, the value (≥0x0600 Hex) of this field is Ethernet List, indicating the protocol inside.

In the 802.3 protocol, the value (46-1500 Dec) is the length of the inner protocol, which is the LLC encapsulated inner protocol. (The LLC header indicates the inner protocol type.)

Data unit + pad

LLC protocol.

FCS

Frame check sequence.

Token Ring

IEEE 802.5 1995-00

Token Ring is a LAN protocol where all stations are connected in a ring and each station can directly hear transmissions only from its immediate neighbor. Permission to transmit is granted by a message (token) that circulates around the ring.

The Token Ring header structure is shown in the illustration below:

SDEL 1 byte
Access control 1 byte
Frame control 1 byte
Destination address 6 bytes
Source address 6 bytes
Route information 0-30 bytes
Information (LLC or MAC) variable
FCS 4 bytes
EDEL 1 byte
Frame status 1 byte

Token Ring header structure

SDEL / EDEL

Starting Delimiter / Ending Delimiter. Both the SDEL and EDEL have intentional Manchester code violations in certain bit positions so that the start and end of a frame can never be accidentally recognized in the middle of other data.

Access control

The format is as follows:

Token Ring access control format

PPP Priority bits:
- 000 Lowest priority.
- 111 Highest priority.

T Token bit:
- 0 Token.
- 1 Frame.

M Monitor count:
- 0 Initial Value.
- 1 Modified to active monitor.

RRR Reservation bits:
- 000 Lowest priority reservation.
- 111 Highest priority reservation.

Frame control

The format is as follows:

2	1	1	4 bits
Frame type	0	0	Attention

Token Ring frame control format

Frame type may have the following values:

00	MAC frame.
01	LLC frame.
11 or 10	Undefined.

The second 2 bits are always zero.

Attention indicates those frames for which the adapter does special buffering and processing.

0001	Express buffer.
0010	Beacon.
0011	Claim token.
0100	Ring purge.
0101	Active monitor present.
0110	Standby monitor present.

Destination address

The address structure is as follows:

I/G	U/L	Address bits

Token Ring destination address structure

I/G Individual/group address may be:
- 0 Individual address.
- 1 Group address.

U/L Universal/local address may be:
- 0 Universally administered.
- 1 Locally administered.

Source address

The address structure is as follows:

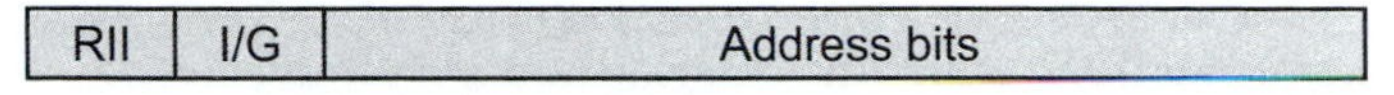

Token Ring source address structure

RII Routing information indicator:
- 0 RI absent.
- 1 RI present.

I/G Individual/group address:
- 0 Group address.
- 1 Individual address.

Route information

The structure is as follows:

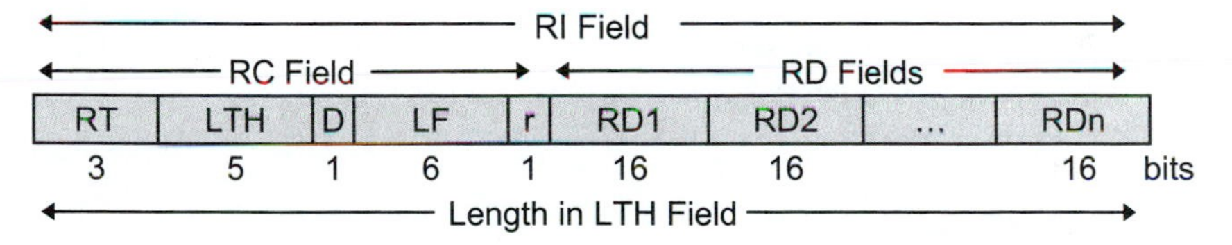

Token Ring route information structure

RC	Routing control.
RDn	Route descriptor.
RT	Routing type.
LTH	Length.
D	Direction bit.
LF	Largest frame.
r	Reserved.

Information

The Information field may be LLC or MAC. The MAC information structure is as follows:

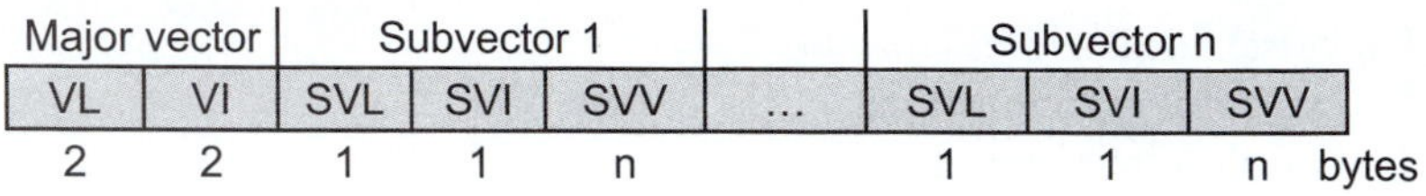

Major vector		Subvector 1				Subvector n			
VL	VI	SVL	SVI	SVV	...	SVL	SVI	SVV	
2	2	1	1	n		1	1	n	bytes

Token Ring MAC information structure

VL

Major vector length. Specifies the length of the vector in octets.

VI

Major vector identifier. A code point that identifies the vector. The VI format is as follows:

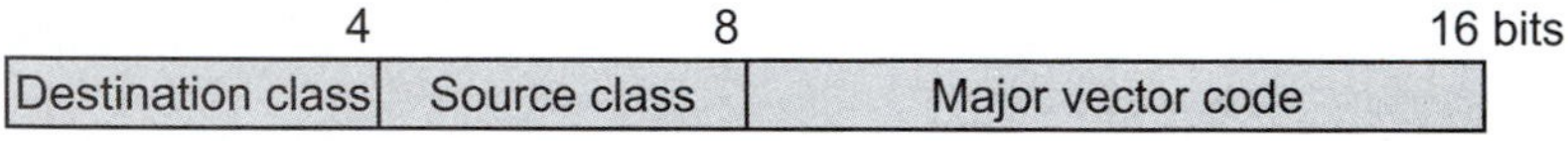

Destination class	Source class	Major vector code
4	8	16 bits

Token Ring major vector identifier

Destination class / source class

Class fields assure proper routing within a ring station:

0 Ring station.
4 Configuration report server.
5 Ring parameter server.
6 Ring error monitor.

Major vector code

The vector code uniquely defines the vector:

0x00 Response.
0x02 Beacon.
0x03 Claim token.
0x04 Ring purge.
0x05 Active monitor present.
0x06 Standby monitor present.
0x07 Duplicate address test.
0x08 Lobe media test.
0x09 Transmit forward.
0x0B Remove ring station.
0x0C Change parameters.

0x0D Initialize ring station.
0x0E Request station addresses.
0x0F Request station state.
0x10 Request station attachment.
0x20 Request initialization.
0x22 Report station addresses.
0x23 Report station state.
0x24 Report station attachment.
0x25 Report new active monitor.
0x26 Report SUA change.
0x27 Report neighbor notification incomplete.
0x28 Report active monitor error.
0x29 Report error.

SVL

Sub-vector length. Specifies the length of the sub-vector in octets.

SVI

Sub-vector identifier. A code point that identifies the sub-vector:
0x01 Beacon type.
0x02 Upstream neighbor addresses next.
0x03 Local ring number.
0x04 Assign physical drop number next.
0x05 Error timer value.
0x06 Authorized function classes next.
0x07 Authorized access priority.
0x08 Authorized environment.
0x09 Correlation.
0x0A SA of last AMP or SMP.
0x0B Physical drop number.
0x20 Response code.
0x21 Reserved.
0x22 Product instance ID.
0x23 Ring station version number.
0x26 Wrap data.
0x27 Frame forward.
0x28 Station identifier.
0x29 Ring station status.
0x2A Transmit status code.
0x2B Group address(es).
0x2C Functional address(es).

0x2D Isolating error count.
0x2E Non-isolating error count.
0x2F Function request ID.
0x30 Error code.

SVV

Sub-vector value, variable length sub-vector information.

FCS

Frame check sequence.

Frame status

Contains bits that may be set on by the recipient of the frame to signal recognition of the address and whether the frame was successfully copied.

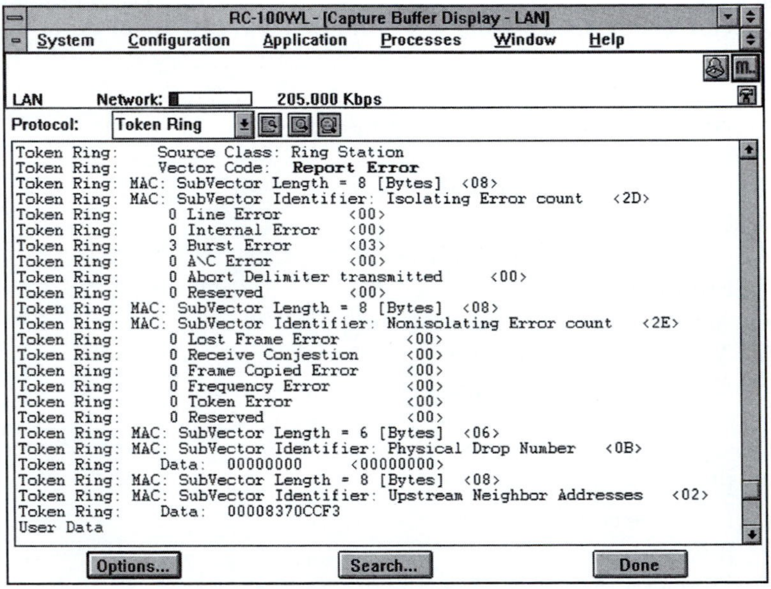

```
RC-100WL - [Capture Buffer Display - LAN]
System   Configuration   Application   Processes   Window   Help
LAN   Network:   205.000 Kbps
Protocol:  Token Ring
Token Ring:     Source Class: Ring Station
Token Ring:     Vector Code:  Report Error
Token Ring: MAC: SubVector Length = 8 [Bytes]  <08>
Token Ring: MAC: SubVector Identifier: Isolating Error count    <2D>
Token Ring:      0 Line Error          <00>
Token Ring:      0 Internal Error      <00>
Token Ring:      3 Burst Error         <03>
Token Ring:      0 A\C Error           <00>
Token Ring:      0 Abort Delimiter transmitted        <00>
Token Ring:      0 Reserved            <00>
Token Ring: MAC: SubVector Length = 8 [Bytes]  <08>
Token Ring: MAC: SubVector Identifier: Nonisolating Error count    <2E>
Token Ring:      0 Lost Frame Error          <00>
Token Ring:      0 Receive Conjestion        <00>
Token Ring:      0 Frame Copied Error        <00>
Token Ring:      0 Frequency Error           <00>
Token Ring:      0 Token Error               <00>
Token Ring:      0 Reserved                  <00>
Token Ring: MAC: SubVector Length = 6 [Bytes]  <06>
Token Ring: MAC: SubVector Identifier: Physical Drop Number    <0B>
Token Ring:     Data:  00000000       <00000000>
Token Ring: MAC: SubVector Length = 8 [Bytes]  <08>
Token Ring: MAC: SubVector Identifier: Upstream Neighbor Addresses    <02>
Token Ring:     Data:  00008370CCF3
User Data
Options...   Search...   Done
```

Token Ring decode

FDDI

The Fiber Distributed Data Interface (FDDI) is a 100 Mbyte fiber-optic media using a timed token over a dual ring of trees. FDDI is standardized by the American National Standards Institute (ANSI).

The FDDI header structure is shown in the illustration below.

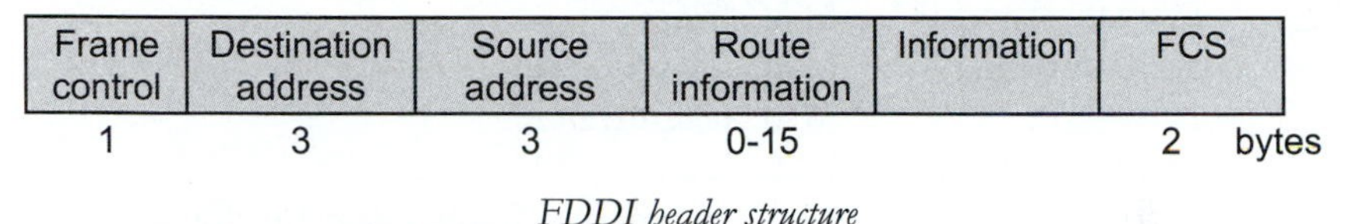

FDDI header structure

Frame control

The frame control structure is as follows:

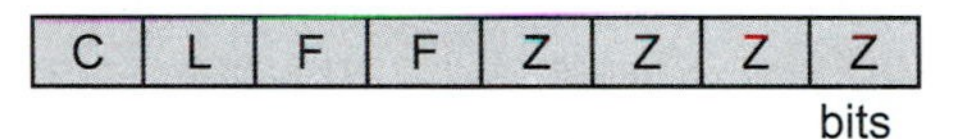

FDDI frame control structure

C Class bit:
- 0 Asynchronous frame.
- 1 Synchronous frame.

L Address length bit:
- 0 16 bits (never).
- 1 48 bits (always).

FF Format bits.

ZZZZ Control bits.

The following is a description of the various Frame Control field values (CLFF ZZZZ to ZZZZ):

0x00 0000	Void frame.
1000 0000	Non-restricted token.
1100 0000	Restricted token.
0L00 0001 to 1111	Station management frame.
1L00 1111	SMT next station addressing frame.
1L00 0001 to 1111	MAC frame.
1L00 0010	MAC beacon frame.
1L00 0011	MAC claim frame.
CL01 r000 to r111	LLC frame.
0L01 rPPP	LLC information frame (asynchronous, PPP=frame priority).
0L01 rrrr	LLC information frame (synchronous, r=reserved).
CL10 r000 to r111	Reserved for implementer.
CL11 rrrr	Reserved for future standardization.

Destination address

The address structure is as follows:

I/G	U/L	Address bits

FDDI destination address structure

I/G Individual/group address may be:
- 0 Individual address.
- 1 Group address.

U/L Universal/local address may be:
- 0 Universally administered.
- 1 Locally administered.

Source address

The address structure is as follows:

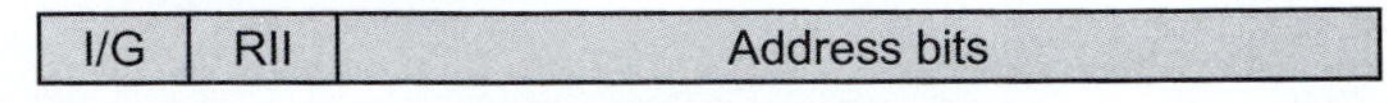

FDDI source address structure

I/G Individual/group address:
- 0 Group address.
- 1 Individual address.

RII Routing information indicator:
- 0 RI absent.
- 1 RI present.

Route Information

The structure of the route information is as follows:

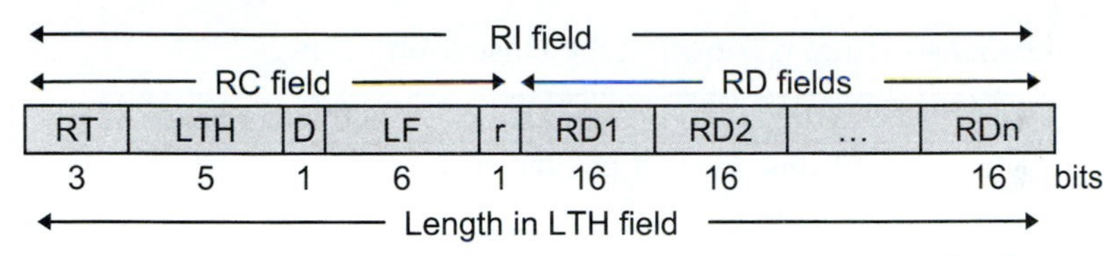

FDDI route information structure

RC Routing control (16 bits).
RDn Route descriptor.
RT Routing type.
LTH Length.
D Direction bit.
LF Largest frame.
r reserved.

Information

The Information field may be LLC, MAC or SMT protocol.

FCS

Frame check sequence.

LLC

ISO 8802-2 1989-12
RFC 2364 http://www.cis.ohio-state.edu/htbin/rfc/rfc2364.html

The IEEE 802.2 Logical Link Control (LLC) protocol provides a link mechanism for upper layer protocols. LLC type I service provides a data link connectionless mode service, while LLC type II provides a connection-oriented service at the data link layer.

The LLC header structure is shown in the illustration below.

DSAP	SSAP	Control	LLC information
1 byte	1 bytes	1 or 2 bytes	

LLC header structure

DSAP

The destination service access point structure is as follows:

I/G	Address bits

LLC DSAP structure

I/G Individual/group address may be:
- 0 Individual DSAP.
- 1 Group DSAP.

SSAP

The source service access point structure is as follows:

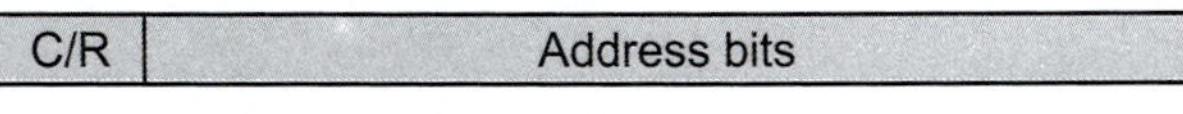

LLC SSAP structure

C/R Command/response:
- 0 Command.
- 1 Response.

Control

The structure of the control field is as follows:

	1				8	9	16 bits
Information	0	N(S)				P/F	N(R)
Supervisory	1	0	SS	XXXX		P/F	N(R)
Unnumbered	1	1	MM	P/F	MMM		

LLC control field structure

N(S) Transmitter send sequence number.

N(R) Transmitter receive sequence number.

P/F Poll/final bit. Command LLC PDU transmission/ response LLC PDU transmission.
S Supervisory function bits:
00 RR (receive ready).
01 REJ (reject).
10 RNR (receive not ready).

X Reserved and set to zero.

M Modifier function bits.

LLC information

LLC data or higher layer protocols.

SNAP

RFC 1042 http://www.cis.ohio-state.edu/htbin/rfc/rfc1042.html

The SubNetwork Access Protocol (SNAP) is used for encapsulating IP datagrams and ARP requests and replies on IEEE 802 networks. IP datagrams are sent on IEEE 802 networks encapsulated within the 802.2 LLC and SNAP data link layers and the 802.3, 802.4 or 802.5 physical network layers. The SNAP header follows the LLC header and contains an organization code indicating that the following 16 bits specify the EtherType code.

The SNAP header is as shown in the following illustration:

DSAP	SSAP	Control
1 byte	1 byte	1 byte

LCC header structure

Organization code	EtherType
3 bytes	2 bytes

SNAP header structure

When SNAP is present the DSAP and SSAP fields within the LLC header contain the value 170 (decimal) each and the Control field is set to 3 (unnumbered information).

Organization code

Set to 0.

EtherType

Specifies which protocol is encapsulated within the IEEE 802 network: IP = 2048, ARP = 2054.

CIF

ATM Forum Cells in Frames Version 1.0 21.10.1996 1.0

CIF (Cells In Frames) describes the mechanism by which ATM traffic is carried across a media segment and a network interface card conforming to the specification for Ethernet Version 2, IEEE 802.5 Token Ring, or IEEE 802.3. ATM cells can be carried over many different physical media, from optical fiber to spread spectrum radio. ATM is not coupled to any particular physical layer. CIF defines a new pseudo-physical layer over which ATM traffic can be carried. It is not simply a mechanism for translation between frames and cells; neither is it simple encapsulation. CIF carries ATM cells in legacy LAN frames. This defines a protocol between CIF end system software and CIF attachment devices (CIF-AD) which makes it possible to support ATM services, including multiple classes of service, over an existing LAN NIC just as if an ATM NIC were in use. CIF specifies how the ATM layer protocols can be made to work over the existing LAN framing protocols in such a way that the operation is transparent to an application written to an ATM compliant API. Over Ethernet, CIF frames have an Ethernet header and trailer. CIF frames are encapsulated in Token Ring and LLC by use of a SNAP header.

The format of the header is shown in the following illustration:

1 8 9 11 16 bits

P	CIF format	P	F F	Format flags
P	Format flags	GFC	VPI
VPI	VCI		
VCI	PT	C	HEC

CIF header format

P

Even Parity bit for an octet.

CIF format

CIF format identifier. Only three format types are defined. Formats 0 and 1 are used for CIF signalling. Format 2 is the default format for carrying user traffic. Formats 112-127 are reserved for use in experimentation and for pre-standard CIF implementations.

FF
CIF format independent flags. These bits contain flags that are independent of any CIF format type. These CIF format independent flags are reserved. They are set to 0 when sent and are ignored when received.

Format flags
CIF format dependent flags. The CIF format dependent flags differ depending on the CIF format type.

GFC
Generic Flow Control. The structure and semantics of octets 3-7 in the CIF header are the same as those of an ATM UNI cell header. These octets are collectively known as the "CIF cell header template".

VPI
Virtual Path Identifier.

VCI
Virtual Channel Identifier.

PT
Payload Type.

C
Cell Loss Priority.

HEC
Header Error Check. The sender of a LAN frame always calculates and fills in the HEC field. The receiver may either rely on the LAN CRC to detect errors in the frame (i.e., not validate the received HECs), or it may check the correctness of the HEC.

GARP

IEEE 802.1P http://standards.ieee.org/catalog/IEEE802.1.html

The Generic Attribute Registration Protocol (GARP) provides a generic attribute dissemination capability that is used by participants in GARP applications to register and de-register attribute values with other GARP participants within a Bridged LAN. A GARP participant in a bridge or an end station consists of a GARP application component and a GARP Information Declaration (GID) component associated with each port of the bridge. The propagation of information between GARP participants for the same application in a bridge is carried out by the GARP Information Propagation (GIP) component. Protocol exchanges take place between GARP participants by means of LLC Type 1 services, using the group MAC address and PDU format defined for the GARP application concerned.

The format of the GARP PDU is shown in the following illustration:

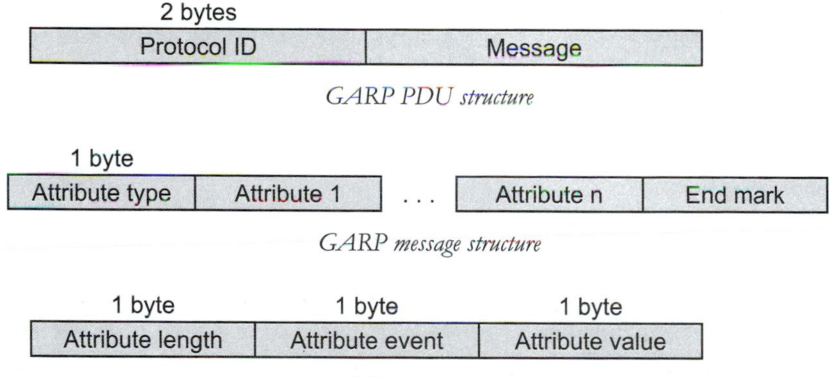

GARP PDU structure

GARP message structure

GARP attribute structure

Protocol ID

Identifies the GARP protocol.

Identifier

Decimal value which aids in matching requests and replies.

Attribute type

Defines the attribute. Values may be:

1 Group attribute.

2` Service Requirement attribute.

Attribute length

Length of the Attribute.

Attribute event

The values of the attribute event can be:

0 Leave_all

1 Join_Empty operator

2 Join_In operator

3 Leave_Empty operator

4 Leave_In operator

5 Empty operator

The default is reserved.

Attribute value

This is encoded in accordance with the specification for the Attribute Type.

End mark

Coded as 0.

GMRP

IEEE 802.1P http://standards.ieee.org/catalog/IEEE802.1.html

The GARP Multicast Registration Protocol (GMRP) provides a mechanism that allows bridges and end stations to dynamically register group membership information with the MAC bridges attached to the same LAN segment and for that information to be disseminated across all bridges in the Bridged LAN that supports extended filtering services. The operation of GMRP relies upon the services provided by the GARP.

The format of the GMRP packet is that of the GARP. However, the attribute type is specific to GMRP: it can be as follows:

1 Group Attribute Type.

2 Service Requirement Attribute Type.

GVRP

IEEE 802.1P http://standards.ieee.org/catalog/IEEE802.1.html

The GARP VLAN Registration Protocol (GVRP) defines a GARP application that provides the VLAN registration service. It makes use of GID and GIP, which provide the common state machine descriptions and the common information propagation mechanisms defined for use in GARP-based applications.

The format of the GVRP packet is that of the GARP. However, the attribute type is specific to GVRP:

1 VID Group Attribute Type.

SRP MAC Protocol

Internet Draft: http://search.ietf.org/internet-drafts/draft-tsiang-srp-00.txt (1999) and PentaCom Ltd.

Packets used by (SRP Spatial Reuse Protocol) can be sent over any point-to-point link layer (e.g., SONET/SDH, ATM, point to point Ethernet connections). The maximum transfer unit (MTU) is 9216 octets; the minimum MTY for data packets is 55 octets.
The overall SRP packet format is as shown in the following illustration:

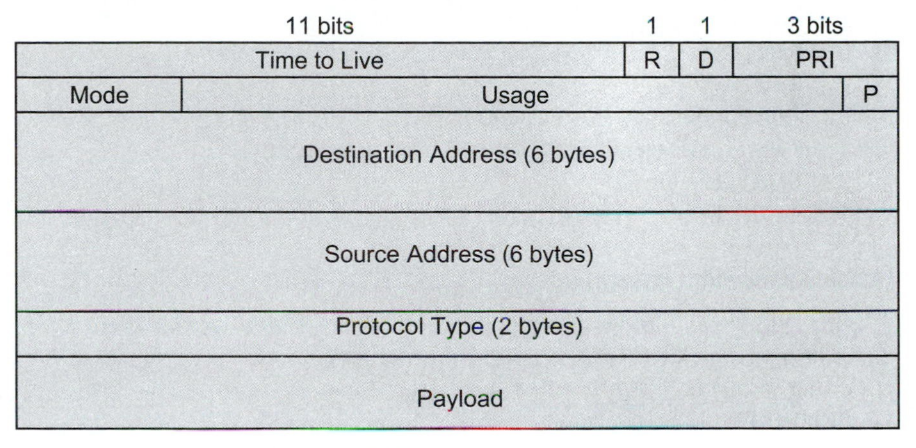

SRP packet structure

Time to live
11 bit field representing the hop-count. It is decremented every time a node forwards a packet. If the TTL reaches zero it is stripped off the ring, thus allowing for a total node space of 2048 nodes on a ring. However, due to certain failure conditions (e.g. when the ring is wrapped) the total number of nodes that are supported by SRP is 1024. When a packet is first sent onto the ring the TTL should be set to at least twice the total number of nodes on the ring.

Ring identifier (R)
1-bit field which identifies to which ring this packet is designated, as follows:

Inner ring	0
Outer ring	1

Designation strip (D)

1-bit field used to indicate whether or not the destination node should strip the packet. This bit is almost always set to 1.

Priority field (PRI)

3-bit field that indicates the priority level of the SRP packet (0-7). Higher values indicate higher priority.

Mode

3-bit field used to identify the mode of the packet as follows:

000	Reserved
001	Reserved
010	Reserved
011	Reserved
100	Control message (pass to host)
101	Control message (locally buffered for host)
110	MAC keep alive
111	Packet data

Usable bandwidth (Usage)

12-bit field which indicates the current usable bandwidth this node sees according to the SRP Fairness Algorithm (SRP-fa).

Parity bit (P)

Indicates the parity value over the last 31 bits of the SRP header to provide additional data integrity over the header. Odd parity is used (i.e., the number of ones including the parity bit is an odd number).

Destination address

Globally unique 48-bit address assigned by IEEE.

Source address

Globally unique 48-bit address assigned by the IEEE.

Protocol type

2-octet field similar to EtherType field. Possible values are as follows:

0x2007	SRP Control
0x0800	IP version 4
0x0806	ARP

VLAN

IEEE 802.1Q http://standards.ieee.org/catalog/IEEE802.1.html

A VLAN is a logical group of LAN segments, independent of physical location, with a common set of requirements. VLAN tagged frames carry an explicit identification of the VLAN to which it belongs. The value of the VID in the Tag header signifies the particular VLAN it belongs to. This additional tag field appears in the Ethernet and SNAP protocols.

The format of the Ethernet encoded Tag header is shown in the following illustration:

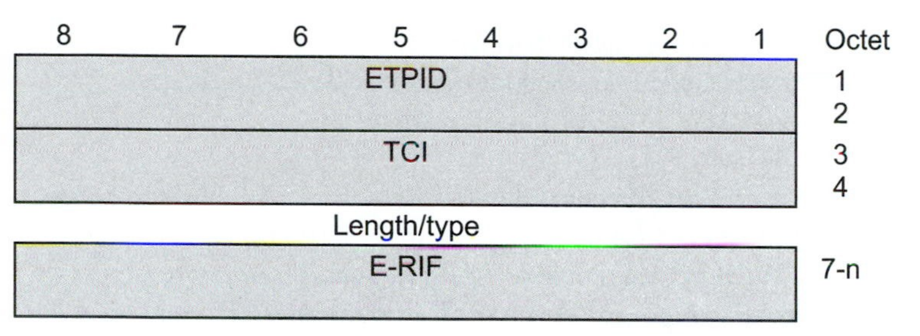

Ethernet-encoded tag header

ETPID

Ethernet-coded Tag Protocol Identifier. Value is 81-00.

TCI

Tag Control Information. The structure of the TCI field is as follows:

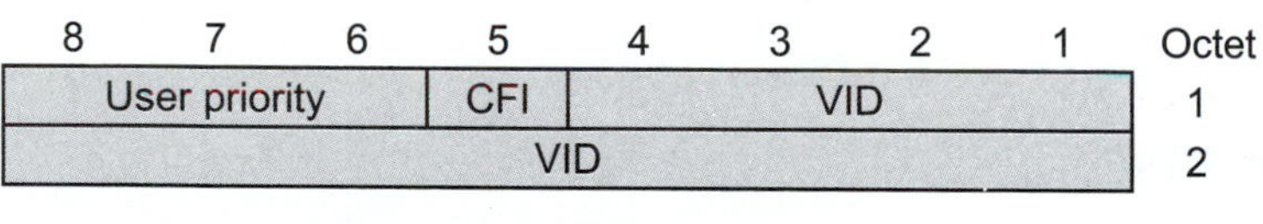

TCI structure

User priority

3-bit binary number representing 8 priority levels, 0-7.

CFI

Canonical Format Indicator. When set, the E-RIF field is present and the NCFI bit determines whether MAC address information carried by the frame is in canonical or non-canonical format. When reset, indicates that the E-RIF field is not present and that all MAC information carried by the frame is in canonical format.

VID

VLAN Identifier. Uniquely identifies the VLAN to which the frame belongs.

0 Null VLAN ID. Indicates that the tag header contains only user priority information, no VLAN ID.

1 Default PVID value used for classifying frames on ingress through a bridge port.

FFF Reserved for implementation use.

All other values are available for general use as VLAN identifiers.

E-RIF

Embedded RIF format. Present only if CFI is set in TCI. When present, immediately follows the Length/Type field. E-RIF consists of two components: 2-octet Route Control (RC) field and 0 or more octets of Route Descriptors, up to a maximum of 28 octets. E-RIF may be 2-30 octets.

The format of the RC is as follows:

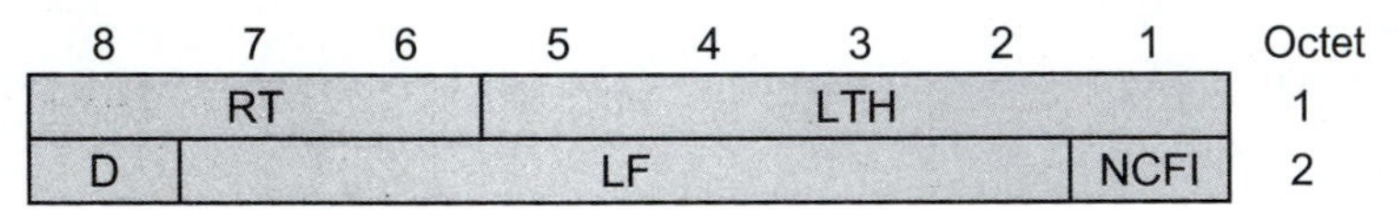

RC structure

RT

Routing type.

LTH

Length field.

D

Direction bit.

LF

Largest frame.

NCFI

Non-canonical format indicator. When reset, all MAC address information in the frame is in non-canonical format. When set, all MAC address information in the frame is in canonical format.

The format of the SNAP-encoded Tag header on 802.5 is shown in the following illustration:

8 7 6 5 4 3 2 1	Octet
SNAP header (AA-AA-03)	1-3
SNAP PID (00-00-00)	4-6
Tag protocol type (81-00)	7-8
TCI	9-10

SNAP-encoded tag header on 802.5

Values of the SNAP header, SNAP PID and Tag protocol type are given in the illustration. TCI value is as described for Ethernet-encoded tag header.

The format of the SNAP-encoded Tag header on FIDDI is shown in the following illustration:

8 7 6 5 4 3 2 1	Octet
SNAP header (AA-AA-03)	1-3
SNAP PID (00-00-00)	4-6
Tag protocol type (81-00)	7-8
TCI	9-10
E-RIF	11, 12

SNAP-encoded tag header on FDDI

Values of the SNAP header, SNAP PID and Tag protocol type are given in the illustration. The TCI and E-RIF fields are as described for Ethernet-encoded tag header. The E-RIF field is present only if CFI is set in TCI and RII is reset. It may be between 2 and 30 octets long.

22

LAN Emulation

In order to make it possible to continue using existing LAN application software, while taking advantage of the increased bandwidth of ATM transmission, standards have been developed to allow the running of LAN layer protocols over ATM. LAN Emulation (LANE) is one such method, enabling the replacement of 10 Mbps Ethernet or 4/16 Mbps Token Ring LANs with dedicated ATM links. It also allows the integration of ATM networks with legacy LAN networks. This software protocol running over ATM equipment offers two major features:

- The ability to run all existing LAN applications over ATM without change. The immediate benefit is that it is not necessary to reinvest in software applications.
- The ability to interconnect ATM equipment and networks to existing LANs, and also the ability to link logically separate LANs via one ATM backbone. The benefit is that ATM equipment may be introduced only where it is needed.

The function of LANE is to emulate a LAN (either IEEE 802.3 Ethernet or 802.5 Token Ring) on top of an ATM network. Basically, the LANE protocol defines a service interface for higher layer protocols which is identical to that of existing LANs. Data is sent across the ATM network encapsulated in the appropriate LAN MAC packet format. Thus, the LANE protocols make an ATM network look and act like a LAN, only much faster. (ATM Forum Standards version 1.0 for LAN emulation)

```
Capture Buffer Display
Filter: All Frames
S  Protocol...
Captured at:   +00:00.280
Length: 78     From: User     Status: Ok
ATM/SAR: BASize - 31582
ATM/SAR: Length - 9011
LE 802.3: LEC ID: 33121
Ethernet: Destination Address 00C07B5E8161
Ethernet: Source Address 00A024D17E73
Ethernet: 802.3, Length = 96 [Bytes]
LLC: DSAP=AA (SNAP ), Individual SAP
LLC: SSAP=AA (SNAP ), Command
LLC: Unnumbered   UI  , Poll
LLC: Protocol:SNAP
SNAP: OUI: Routed non ISO protocol 0x000000
SNAP: Ethernet V.2, Type DOD IP
IP: Version = 4
IP: Total Length = 79
IP: Identifiers = 17408
IP: Flags & Fragment Offset: 0x0000
IP:    .0............. May Fragment
IP:    ..0............ Last Fragment
IP:    Fragment Offset = 0 [Bytes]
IP: Time to Live = 32 [Seconds/Hops]
IP: Protocol: 88 IGRP
IP: Header Checksum = 0x1D6E
Options...   Search...   Restart   Setup...   Done
```

LAN protocols transmitted via LAN Emulation

LAN Emulation Components

There are several participants in the LAN emulation (LE) protocol operation: the LAN Emulation Client (LEC), the LAN Emulation Server (LES), the LAN Emulation Configuration Server (LECS), and the Broadcast and Unknown Server (BUS). Each of these is described below.

LAN Emulation Client (LEC)

The LEC is the user requiring LAN emulation services. Typically, it is the workstation running the application or the ATM bridge which connects the ATM network with the legacy LAN. There can be many LAN Emulation Clients in an emulated LAN.

LAN Emulation Server (LES)

The LES implements address registration (allowing stations to register their MAC and ATM addresses) and provides address resolution (answers ARP, Address Resolution Protocol, requests by converting between MAC and ATM addresses). Each emulated LAN can have only one LES. However, a physical LAN can serve several emulated LANs, each with its own LES.

LAN Emulation Configuration Server (LECS)

The LECS provides configuration information, including the address of the LES, the type of emulated LAN and the maximum frame size. Each network can only have one LECS.

Broadcast/Unknown Server (BUS)

The BUS performs all broadcasts and multicasts. Frames are sent through the BUS in two instances:

- When the information is to be transferred (broadcast) to all stations.
- When a source LEC has sent an ARP to the LES, and does not wish to wait for a response before starting the data transfer to the destination LEC. In this case, the source LEC transmits the information to the BUS, which in turn floods the entire network.

Each emulated LAN can have only one BUS. However, a physical LAN can serve several emulated LANs, each with its own BUS.

Location of LAN Emulation Service Components

While the ATM Forum specifies that there are three separate *logical* components to the LAN Emulation service (the LES, LECS and BUS), it deliberately does not specify whether they are *physically* separate or united. This decision is left to the vendors.

Many vendors merge the LES, LECS and BUS into a single physical unit. There have been two popular choices where to place this unit:

1. Adding the LE service functionality into switches.
2. Providing an external station which connects up to any switch and provides LE services.

Data Transfer

There are several stages pertaining to connection establishment over an emulated LAN. They are described in this section from the LEC point of view.

LAN emulation uses the signalling protocol to establish transient connections. Each LEC has a unique ATM address (in one of the address formats supported by ATM Forum's UNI 3.0 signalling). In the SETUP message, the Broadband Low-Layer Information (B-LLI) element is used to identify LAN emulation connections.

Initially, the protocol ID (PID) in the B-LLI element is ISO/IEC TR/9577. There are several types of supported PIDs:

- Control connections.
- 802.3 data connections.
- 802.5 data connections.
- 802.3 multicast forward connections.
- 802.5 multicast forward connections.

Initialization

In the initialization phase, the LEC must identify the type of emulated LAN that it is joining and determine the addresses of the LECS and LES.

To determine the ATM address of the LECS, the LEC performs the following:

1. Attempts to extract this address from the switch with the use of ILMI. If successful, the LEC attempts to connect to that address.
2. If unsuccessful, uses a well-known ATM address to try and establish the SVC.
3. If unsuccessful, uses a PVC at VPI=0, VCI=17 to establish the connection.
4. If still unsuccessful, tries to contact the LES.

Configuration

Once a connection to the LECS has been established, the following information is exchanged between the LEC and the LECS:

The LEC sends its ATM address, its MAC address and the requested LAN types and frame sizes.

The LECS returns the LES address and the LAN type and frame size to use.

Joining

In this phase, the LEC attempts to join the emulated LAN. To do this it:

- Creates a control direct bi-directional VCC with the LES.
- Transmits a Join Request (ATM address, LAN information, Proxy indication, optional MAC address).
- Possibly accepts a Control Distribute VCC before a Join Request is received.

This operation may time out or fail. An example of joining an emulated LAN is given below.

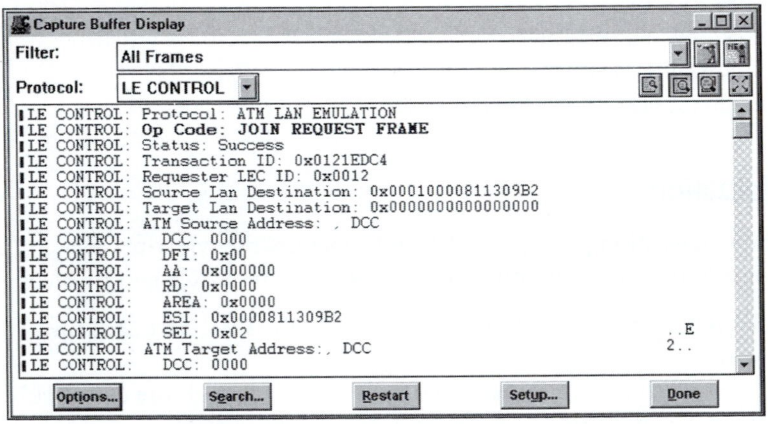

```
Capture Buffer Display
Filter: All Frames
Protocol: LE CONTROL
LE CONTROL: Protocol: ATM LAN EMULATION
LE CONTROL: Op Code: JOIN REQUEST FRAME
LE CONTROL: Status: Success
LE CONTROL: Transaction ID: 0x0121EDC4
LE CONTROL: Requester LEC ID: 0x0012
LE CONTROL: Source Lan Destination: 0x00010000811309B2
LE CONTROL: Target Lan Destination: 0x0000000000000000
LE CONTROL: ATM Source Address: , DCC
LE CONTROL:    DCC: 0000
LE CONTROL:    DFI: 0x00
LE CONTROL:    AA: 0x000000
LE CONTROL:    RD: 0x0000
LE CONTROL:    AREA: 0x0000
LE CONTROL:    ESI: 0x0000811309B2
LE CONTROL:    SEL: 0x02                                   ..E
LE CONTROL: ATM Target Address:, DCC                       2..
LE CONTROL:    DCC: 0000
Options...   Search...   Restart   Setup...   Done
```

Decode of emulated LAN

Registration and BUS Initialization

The BUS takes care of processing broadcast requests from a LEC to other LAN emulation clients. To do this, it must be aware of all ATM stations on the line. Thus, when each LEC comes up, it registers at the BUS. The LEC must:

- Register any MAC addresses.
- Resolve the 0xFFFFFFFFFFFF MAC address (broadcast address) to get the ATM address of the BUS.

- Create a unidirectional multicast send VCC to the BUS. This VCC will be used when the LEC desires to perform a broadcast.
- Accept a unidirectional multicast forward VCC from the BUS. This is the VCC that the BUS will use when performing broadcasts to the LEC.

Data Movement

When data movement is required, the top-level application sends the driver the information with the desired MAC address. The LAN emulation driver can then proceed with the following steps:

- Verify that the internal cache contains the association between the MAC address and the ATM address.
- If not, inquire of the LES.
- While waiting for a response, the LEC may transmit frames using the BUS.
- Once a response has been received, a direct connection is established using the signalling protocol. The association of the ATM address and the MAC address are added to the cache.
- Connections are deleted based on inactivity.

LAN Emulation Protocol Stack

LAN emulation is implemented in all equipment participating in the emulated LAN, namely workstations, switches, network interface cards, bridges, etc. The following diagram illustrates the LAN emulation protocol stack.

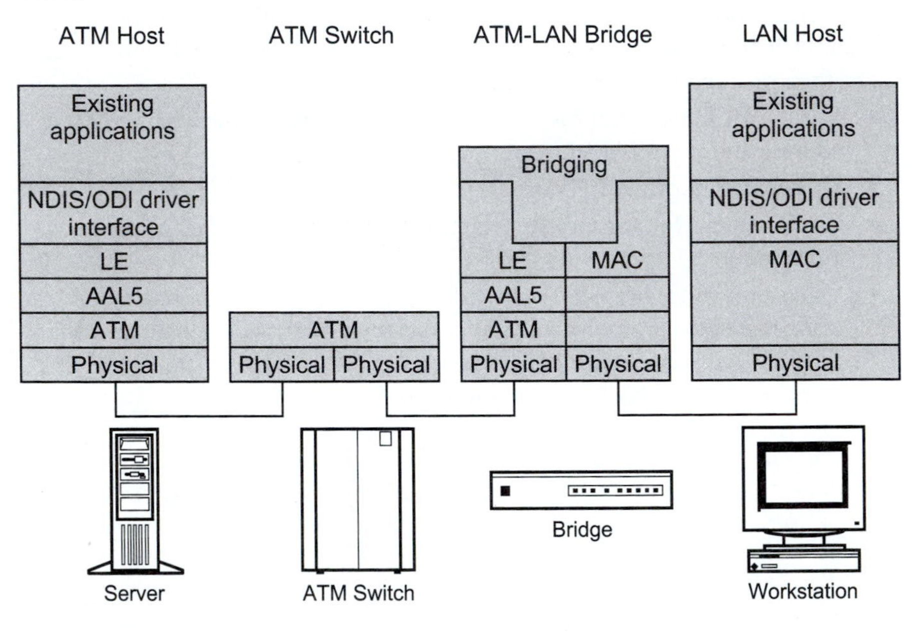

LAN Emulation protocol stack

The end-points of the connection (the server and the workstation) run the same applications over an NDIS/ODI driver (in this example). It is the underlying layers that are different: the server runs LAN Emulation over AAL5 and the workstation runs Ethernet. The upper-layer application remains unchanged; it does not have to realize that ATM is running underneath. The bridge has both technologies: a LAN port connects to the legacy LAN and an ATM port connects to the newer ATM network. The switch continues to run its normal stack, switches cells and creates connections.

LAN Emulation Packet Formats

Data Packets

LAN emulation provides for two possible data packet formats: Ethernet and Token Ring. The LAN emulation data frames preserve all the information contained in the original 802.3 or 802.5 frames, but add a 2-byte LEC ID (the source ID), which is unique to each LEC. The first format, based on Ethernet IEEE 802.3, is shown below.

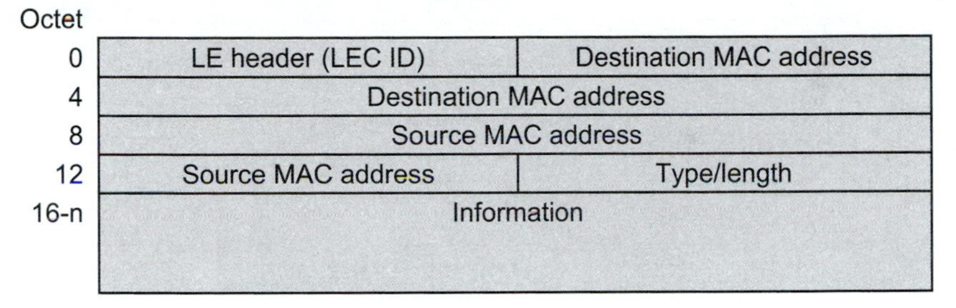

Ethernet IEEE 802.3 packet format

The second packet format, based on IEEE 802.5 Token Ring, is shown below.

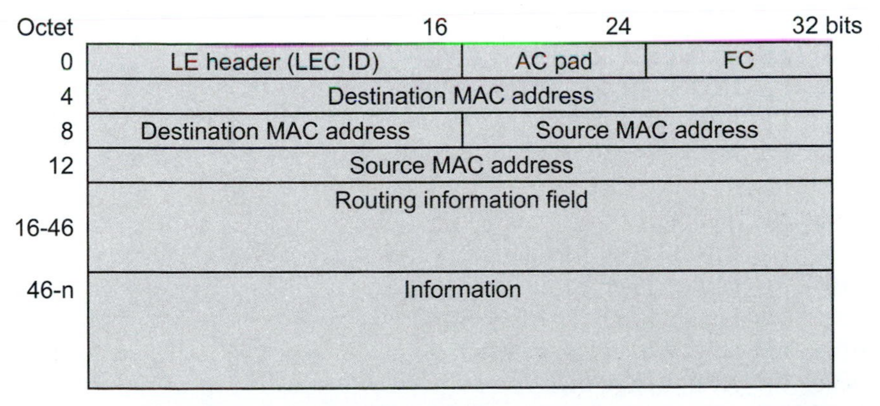

Token Ring IEEE 802.5 packet format

The original 802.3 or 802.5 frame is maintained since it may be needed at some nodes. For example, an ATM-to-Ethernet bridge will receive LAN emulation Ethernet frames from the ATM side, strip off the first two bytes, and send the Ethernet frame on to the Ethernet side.

Control Frames

The format of all LAN emulation control frames, except for READY_IND and READY_QUERY, is shown below.

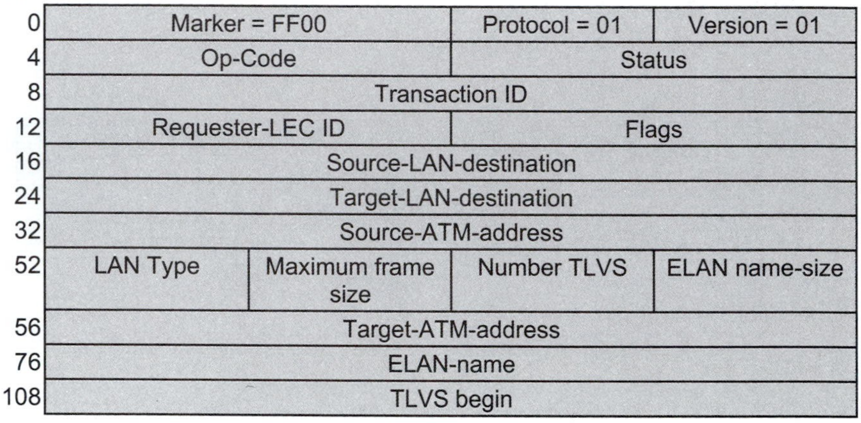

LAN Emulation control frame format

Op-Code

Control frame type. Values may be:

0001 LE_CONFIGURE_REQUEST.
0101 LE_CONFIGURE_RESPONSE.
0002 LE_JOIN_REQUEST.
0102 LE_JOIN_RESPONSE.
0003 READY_QUERY.
0103 READY_IND.
0004 LE_REGISTER_REQUEST.
0104 LE_REGISTER_RESPONSE.
0005 LE_UNREGISTER_REQUEST.
0105 LE_UNREGISTER_RESPONSES.
0006 LE_ARP_REQUEST.
0106 LE_ARP_RESPONSE.
0007 LE_FLUSH_REQUEST.
0107 LE_FLUSH_RESPONSE.
0008 LE_NARP_REQUEST.

0108 Undefined.
0009 LE_TOPOLOGY_REQUEST.
0109 Undefined.

Transaction ID

Arbitrary value supplied by the requester and returned by the responder to allow the receiver to discriminate between different responses.

Requester LEC ID

LEC ID of LAN emulation client sending the request (0000 if unknown).

Flags

Bit flags:
0001 Remote Address used with LE_ARP_RESPONSE.
0080 Proxy Flag used with LE_JOIN_REQUEST.
0100 Topology change used with LE_TOPOLOGY_REQUEST.

The meaning of the remaining fields depends on the Op-Code value.

LUNI 2.0

LAN Emulation LLC-multiplexed data and control frames have the following additional fields which appear before the fields in the data and control frames described above:

byte:				
0	LLC-X"AA"	LLC-X"AA"	LLC-X"03"	OUI-X"00"
4	OUI-X"A0"	OUI-X"3E"	Frame Type	
8	ELAN ID			

Additional fields for LLC-multiplexed data and control

The LUNI 2.0 standard defines the following additional Flags for Control frames:
0002 V2 capable used with LE_CONFIG_REQUEST and LE_JOIN_REQUEST.
0004 Selective multicast used with LE_JOIN_REQUEST.
0008 V2 required used with LE_JOIN_RESPONSE.

The LUNI 2.0 standard defines the following additional Op-Code values for Control frames:
000A LE_VERIFY_REQUEST.
010A LE_VERIFY_RESPONSE.

23

MPLS

Multiprotocol Label Switching (MPLS) is an Internet Engineering Task Force (IETF)-specified framework that provides for the designation, routing, forwarding and switching of traffic flows through the network.

It performs the following :

- Specifies mechanisms to manage traffic flows of various granularities, such as flows between different hardware, machines, or even flows between different applications
- Remains independent of the layer-2 and layer-3 protocols
- Provides a means to map IP addresses to simple, fixed-length labels used by different packet-forwarding and packet-switching technologies
- Interfaces to existing routing protocols, such as Resource ReSerVation Protocol (RSVP) and Open Shortest PathFirst (OSPF)
- Supports IP, ATM, and Frame Relay layer-2 protocols

In MPLS, data transmission occurs on Label-Switched Paths (LSPs). LSPs are a sequence of labels at each and every node along the path from the source to the destination. LSPs are established either prior to data transmission (control-driven) or upon detection of a certain flow of data

(data-driven). The labels, are underlying protocol-specific identifiers, There are several label distribution protocols used today, such as Label Distribution Protocol (LDP) or RSVP or piggybacked on routing protocols like border gateway protocol (BGP) and OSPF. Each data packet encapsulates and carries the labels during their journey from source to destination. High-speed switching of data is possible because the fixed-length labels are inserted at the very beginning of the packet or cell and can be used by hardware to switch packets quickly between links.

MPLS is a versatile solution to address the problems faced by present-day networks-speed, scalability, quality-of-service (QoS) management, and traffic engineering. MPLS has emerged as an elegant solution to meet the bandwidth-management and service requirements for next-generation IP-based backbone networks.

This chapter describes the MPLS protocol family. Protocols described in this chapter include:

- MPLS: Multi Protocol Label Switching
  MPLS Signalling Protocols
- LDP: Label Distribution Protocol.
- CR-LDP
- RSVP-TE

MPLS

IETF draft-rosen-tag-stack-02.txt

Multi Protocol Label Switching (MPLS) is a set of procedures for augmenting network layer packets with "label stacks", thereby turning them into labeled packets. It defines the encoding used by a label switching router to transmit such packets over PPP and LAN links. It is an Ethernet Tag Switching protocol. This protocol attaches labels to IP and IPv6 protocols in the network layer, after the data link layer headers, but before the network layer headers. It inserts a 4 or 8 byte label.

The format of the MPLS label stack is shown in the following illustration:

1	2	3	4	5	6	7	8 bits
Label (20 bits)							
				CoS			S
TTL							

MPLS label stack

Label

The field contains the actual value for the label. This gives information on the protocol in the network layer and further information needed to forward the packet.

CoS

Class of Service. The setting of this field affects the scheduling and/or discard algorithms which are applied to the packet as it is transmitted through the network.

S

Bottom of the Stack, 1-bit field set to one for the last entry in the label stack and zero for all other label stack entries.

TTL

Time to Live, 8-bit field used to encode a time to live value.

LDP

Internet Draft: draft-ietf-mpls-ldp-07.txt

In the MPLS (Multi Protocol Label Switching) 2 label switching routers (LSR) must agree on the meaning of the labels used to forward traffic between and through them. LDP (Label Distribution Protocol) is a new protocol that defines a set of procedures and messages by which one LSR (Label Switched Router) informs another of the label bindings it has made.

The LSR uses this protocol to establish label switched paths through a network by mapping network layer routing information directly to data-link layer switched paths. These LSPs may have an endpoint at a directly attached neighbor (like IP hop-by-hop forwarding), or may have an endpoint at a network egress node, enabling switching via all intermediary nodes. A FEC (Forwarding Equivalence Class) is associated with each LSP created. This FEC specifies which packets are mapped to that LSP.

Two LSRs (Label Switched Routers) which use LDP to exchange label mapping information are known as LDP peers and they have an LDP session between them. In a single session, each peer is able to learn about the others label mappings, in other words, the protocol is bi-directional.

There are 4 sorts of LDP messages:
1. Discovery messages
2. Session messages
3. Advertisement messages
4. Notification messages.

Using discovery messages, the LSRs announce their presence in the network by sending Hello messages periodically. This hello message is transmitted as a UDP packet. When a new session must be established, the hello message is sent over TCP. Apart from the Discovery message; all other messages are sent over TCP.

The notification messages signal errors and other events of interest.

There are 2 kinds of notification messages:

1. Error notifications; these signal fatal errors and cause termination of the session

2. Advisory notifications; these are used to pass on LSR information about the LDP session or the status of some previous message received from the peer.

All LDP messages have a common structure that uses a Type-Length-Value (TLV) encoding scheme. This TLV encoding is used to encode much of the information carried in LDP messages. The Value part of a TLV-encoded object (TLV), may itself contain one or more TLVs.

Messages are sent as LDP PDUs. Each PDU can contain more than one LDP message. Each LDP PDU is an LDP header followed by one or more LDP message:

The structure of the LDP header is shown below:

2 bytes	2 bytes
Version	PDU Length
LDP Identifier 6 bytes	

LDP header structure

Version
The protocol version number. The present number is 1.

PDU Length
The total length of the PDU excluding the version and the PDU length field.

LDP identifier
This field uniquely identifies the label space of the sending LSR for which this PDU applies. The first 4 octets encode the IP address assigned to the LSR. The lst 2 indicate a label space within the LSR.

LDP messages

All LDP messages have the following format.

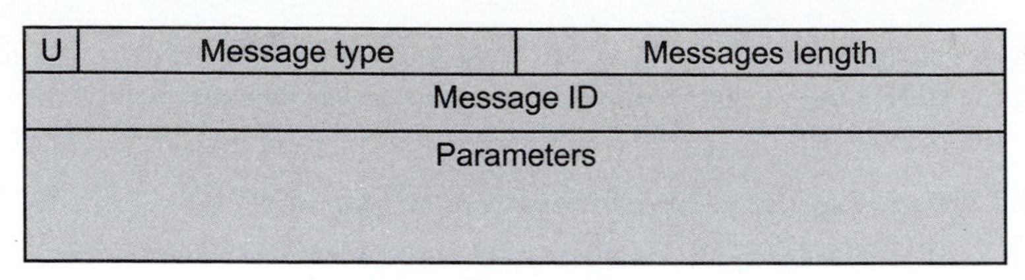

LDP Message format

U

The U bit is an unknown message bit.

Message type

The type of message. The following message types exist:

0x001	Notification
0x100	Hello
0x200	Initialization
0x201	Keep Alive
0x300	Address
0x301	Address Withdraw
0x400	Label Mapping
0x401	Label Request
0x404	Label Abort request
0x402	Label Withdraw
0x403	Label Release
default	Unknown Message Name

Message length

The length in octets of the message ID, mandatory parameters and optional parameters:

Message ID

32-bit value used to identify the message.

Parameters:

The parameters contain the TLVs. There are both mandatory and optional parameters. Some messages have no mandatory parameters, and some have no optional parameters.

TLV format

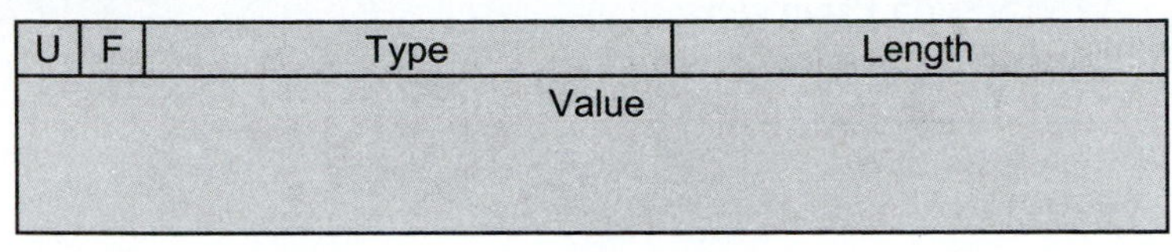

TLV format

U

The U bit is an unknown TLV bit.

F

Forward unknown TLV bit.

Type

Encodes how the Value field is to be interpreted.

Length

Specifies the length of the Value field in octets.

Value

Octet string of Length octets that encodes information to be interpreted as specified by the Type field.

The following are optional TLV parameters:

Optional TLV Parameters

100	Fec
101	Address List
103	Hop Count
104	Path Vector
200	Generic Label
201	ATM Label
202	Frame Relay Label
300	Status
301	Extended Status
302	Returned PDU
303	Returned Message
400	Common Hello Parameters.
401	Transport Address
402	Configuration Sequence Number

500 Common Session Parameters
501 ATM Session Parameters
502 Frame Relay Session Parameters
600 Label Request Message ID

CR-LDP

Draft-ietf-mpls-ldp-06;
Draft-ietf-mpls-cr-ldp-03;
Draft-fan-mpls-lambada-signalling-00

CR-LDP (constraint-based LDP) contains extensions for LDP to extend its capabilities. This allows extending the information used to setup paths beyond what is available for the routing protocol

CR-LDP is the same as LDP but has the following additional TLV parameters.

Value	Parameter
821	LSPID
822	ResCls
503	Optical Session Parameters
800	Explicit Route
801-804	ER-Hop TLVS
810	Traffic Parameters
820	Preemption
823	Route Pinning
910	Optical Interface Type
920	Optical Trail Desc
930	Optical Label
940	Lambada Set

RSVP-TE

draft-ietf-rsvp-spec-13.txt;
draft-ietf-rsvp-md5-02.txt;
draft-ietf-mpls-rsvp-lsp-tunnel-05.txt;
draft-fan-mpls-lambda-signaling-00.txt

The RSVP-TE (traffic extension) protocol is an addition to the RSVP protocol (see *TCP*) with special extensions to allows it to set up optical paths in an agile optical network.

The RSVP protocol defines a *session* as a data flow with a particular destination and transport-layer protocol. However, when RSVP and MPLS are combined, a flow or session can be defined with greater flexibility and generality. The ingress node of an LSP (Label Switched Path) uses a number of methods to determine which packets are assigned a particular label. Once a label is assigned to a set of packets, the label effectively defines the *flow* through the LSP. We refer to such an LSP as an *LSP tunnel* because the traffic through it is opaque to intermediate nodes along the label switched path. New RSVP Session, Sender and Filter Spec objects, called LSP Tunnel IPv4 and LSP Tunnel IPv6 have been defined to support the LSP tunnel feature. The semantics of these objects, from the perspective of a node along the label switched path, is that traffic belonging to the LSP tunnel is identified solely on the basis of packets arriving from the "previous hop" (PHOP) with the particular label value(s) assigned by this node to upstream senders to the session. In fact, the IPv4(v6) that appears in the object name only denotes that the destination address is an IPv4(v6) address. When referring to these objects generically, the qualifier LSP Tunnel is used.

In some applications it is useful to associate sets of LSP tunnels. This can be useful during reroute operations or in spreading a traffic trunk over multiple paths. In the traffic engineering application, such sets are called traffic engineered tunnels (TE tunnels). To enable the identification and association of such LSP tunnels, two identifiers are carried. A tunnel ID is part of the Session object. The Session object uniquely defines a traffic engineered tunnel. The Sender and Filter Spec objects carry an LSP ID. The Sender (or Filter Spec) object, together with the Session object, uniquely identify an LSP tunnel.

Apart from the existing message types listed in RSVP an additional message type is available:

Value	Message type
14	Hello

In addition, the following additional Protocol Object Types exist:

Value	Object type
16	Label
19	Optical
20	Explicit Route
21	Record Route
22	Hello
207	Attribute Session

24

Novell Protocols

The Novell NetWare protocol suite was greatly influenced by the design and implementation of the Xerox Network System (XNS) protocol architecture. It provides comprehensive support for every major desktop operating system, including DOS, Windows, Macintosh, OS/2, and UNIX. In addition, Novell provides extensive support for local area networks and asynchronous wide area communications. The Novell suite includes the following protocols:

- IPX: Internetwork Packet Exchange.
- BCAST: Broadcast.
- BMP: Burst Mode Protocol.
- DIAG: Diagnostic Responder.
- NCP: NetWare Core Protocol.
- NDS: NetWare Directory Services.
- NLSP: NetWare Link Services Protocol.
- NovelNetBIOS.
- RIPX: Routing Information Protocol.
- SAP: Service Advertising Protocol.

- SER: Serialization.
- SPX: Sequenced Packet Exchange.
- WDOG: Watchdog.

The following diagram represents the Novell protocol suite in relation to the OSI model :

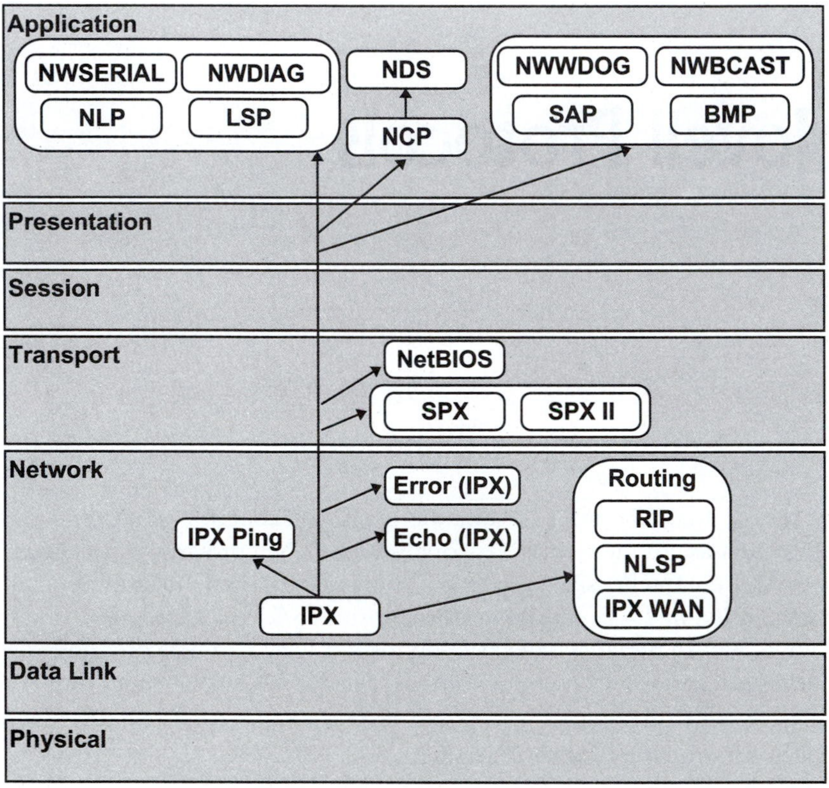

Novell protocol suite in relation to the OSI model

IPX

"Novell's Guide to NetWare LAN Analysis" by Laura A. Chappell and Dan E. Hakes, Novell Press, 1994

Internet Protocol Exchange (IPX) is Novell's implementation of the Xerox Internet Datagram Protocol (IDP). IPX is a connectionless datagram protocol that delivers packets across the Internet and provides NetWare workstations and file servers with addressing and internetworking routing services.

The structure of the IPX packet is shown in the following illustration:

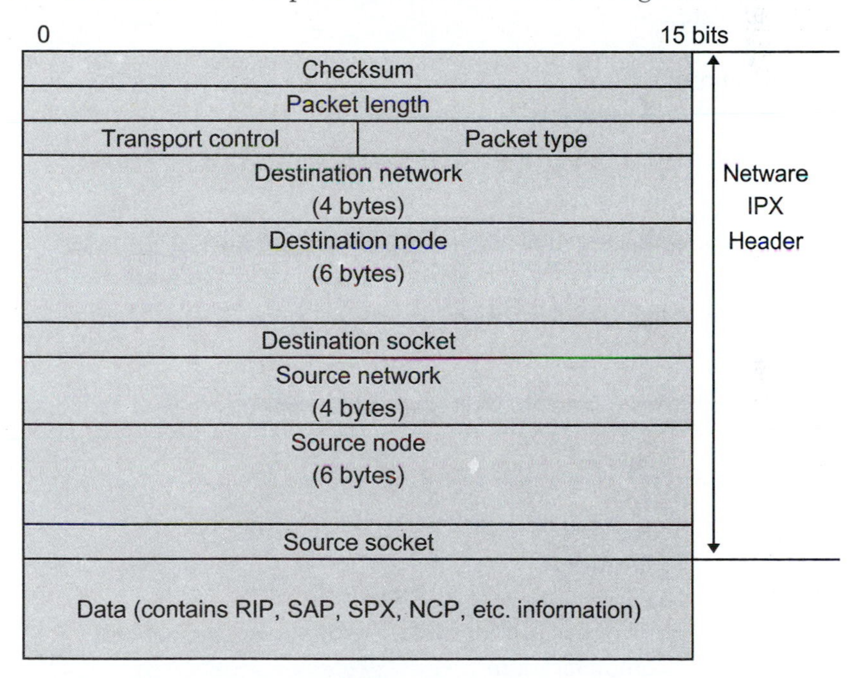

IPX packet structure

Checksum
Set to FFFFH.

Packet length
Length of the IPX datagram in octets.

Transport control

Used by NetWare routers. Set to zero by IPX before packet transmission.

Packet type

Specifies the packet information:

0	Hello or SAP.
1	Routing Information Protocol.
2	Echo Packet.
3	Error Packet.
4	NetWare 386 or SAP.
5	Sequenced Packet Protocol.
16-31	Experimental protocols.
17	NetWare 286.

Network number

A 32-bit number assigned by the network administrator; set to 0 on the local network.

Node number

A 48-bit number that identifies the LAN hardware address. If the node number is FFFF FFFF FFFF, it means broadcast, and if the node number is 0000 0000 0001, it means it is the server (on NetWare 3.x and 4.x only).

Socket number

A 16-bit number that identifies the higher-layer packet:

0451H	NCP.
0452H	SAP.
0453H	RIP.
0455H	NetBIOS.
0456H	Diagnostics.
0x457	Serialization packet (SER).
4000-6000H	Ephemeral sockets, used for file server and network communication.

BCAST

"Novell's Guide to NetWare LAN Analysis" by Laura A. Chappell and Dan E. Hakes, Novell Press, 1994

The Broadcast (BCAST) protocol deals with announcements from the network informing the user that he has received a message.

The format of the BCAST packet is shown in the following illustration:

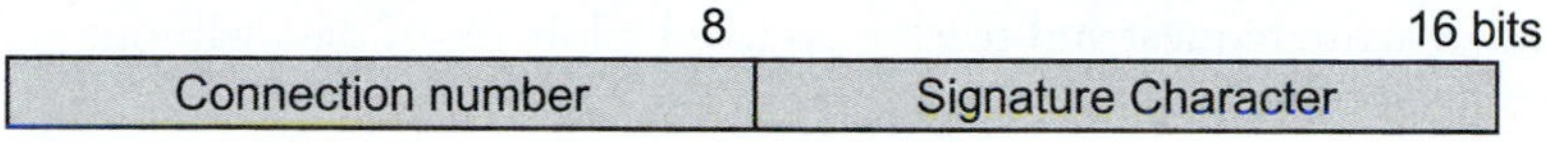

8	16 bits
Connection number	Signature Character

BCAST packet structure

Connection number

Given to the station during the login process.

Signature character

The value is 0x21 (ASCII character!) which means *Broadcast Message Waiting*.

BMP (Burst)

"Novell's Guide to NetWare LAN Analysis" by Laura A. Chappell and Dan E. Hakes, Novell Press, 1994

The Burst Mode Protocol (BMP) is actually a type of NCP packet (Request type = 7777H). BMP was designed to allow multiple responses to a single request for file reads and writes. Burst Mode increases the efficiency of client/server communications by allowing workstations to submit a single file read or write request and receive up to 64 kilobytes of data without submitting another request.

The format of the BMP packet is shown in the following illustration:

16	24	32 bits
Request type	Stream type flags	Stream type
Source connection ID		
Destination connection ID		
Packet sequence number		
Send delay time		
Burst sequence number	ACK sequence number	
Total burst length		
Total burst offset		
Packet length	Number of list entries	
Missing fragment list		
Function code		
File handle		
Starting offset		
Bytes to write		

BMP packet structure

Request type

Identical to Request Type in NCP and always will be set to 7777H (Burst mode packet).

Stream type flags

Available flags.

Stream type
Burst mode control bits.

Source connection ID
Connection ID number assigned to source workstation.

Destination connection ID
Connection ID number assigned to destination workstation.

Packet sequence number
Used by workstation and file server to identify packets sent and received.

Send delay time
Time delay between packets.

Burst sequence number
Burst number being transmitted.

ACK sequence number
Next accepted burst sequence number.

Total burst length
Length of transmitted burst (in octets).

Burst offset
Location of packet data within the burst.

Packet length
Length of packet burst data (in octets).

Number of list entries
Number of elements in the missing fragment list.

Missing fragment list
Data fragments not yet received.

If "Burst Offset = 0", four additional fields follow.

Function code
Write/read function.

Starting offset

Offset to start writing(/reading).

Bytes to write

No. of bytes to write(/read).

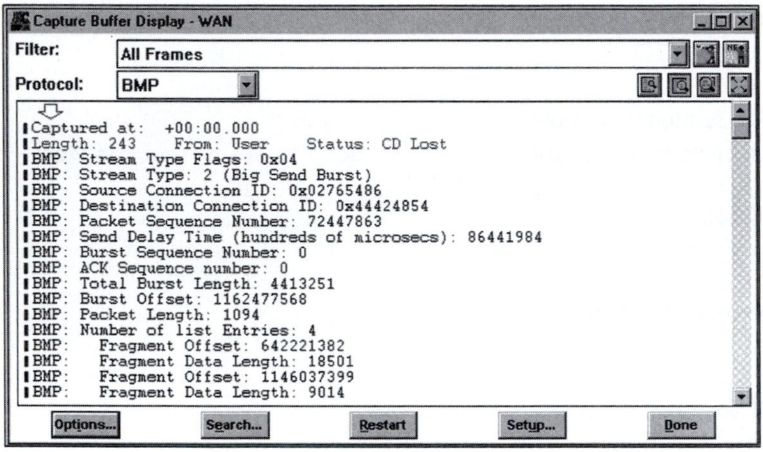

```
Capture Buffer Display - WAN
Filter:   All Frames
Protocol: BMP

Captured at:  +00:00.000
Length: 243     From: User     Status: CD Lost
BMP: Stream Type Flags: 0x04
BMP: Stream Type: 2 (Big Send Burst)
BMP: Source Connection ID: 0x02765486
BMP: Destination Connection ID: 0x44424854
BMP: Packet Sequence Number: 72447863
BMP: Send Delay Time (hundreds of microsecs): 86441984
BMP: Burst Sequence Number: 0
BMP: ACK Sequence number: 0
BMP: Total Burst Length: 4413251
BMP: Burst Offset: 1162477568
BMP: Packet Length: 1094
BMP: Number of list Entries: 4
BMP:   Fragment Offset: 642221382
BMP:   Fragment Data Length: 18501
BMP:   Fragment Offset: 1146037399
BMP:   Fragment Data Length: 9014

Options...   Search...   Restart   Setup...   Done
```

BMP decode

DIAG

"Novell's Guide to NetWare LAN Analysis" by Laura A. Chappell and Dan E. Hakes, Novell Press, 1994

The Diagnostic Responder (DIAG) protocol is useful in analyzing NetWare LANs. DIAG can be used for connectivity testing, configuration and information gathering.

The DIAG request packet structure is shown in the following illustration:

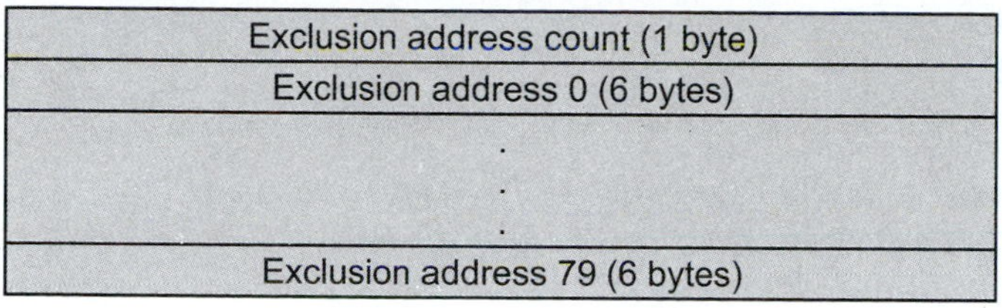

Exclusion address count (1 byte)
Exclusion address 0 (6 bytes)
. . .
Exclusion address 79 (6 bytes)

DIAG request packet structure

Exclusion address count

The number of stations that will be requested not to respond. A value of 0 in this field indicates that all stations should respond. The maximum value for this field is 80 (exclusion address 0-79).

The DIAG response packet structure is shown in the following illustration:

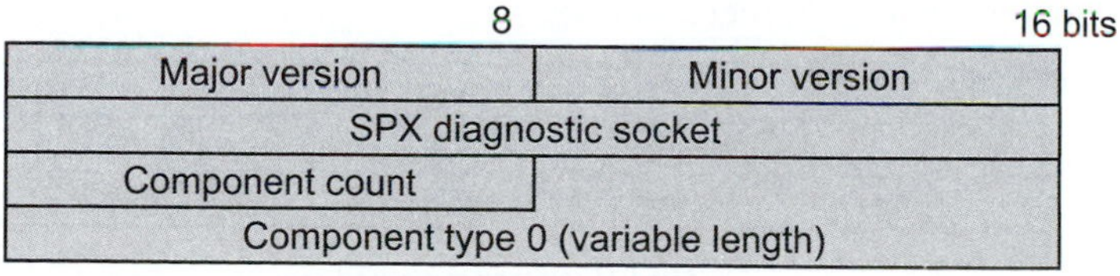

8	16 bits
Major version	Minor version
SPX diagnostic socket	
Component count	
Component type 0 (variable length)	

DIAG response packet structure

Major/minor version

Version of the diagnostic responder that is installed in the responding station.

SPX diagnostic socket

The socket No. to which all SPX diagnostic responses can be addressed.

Component count
Number of components found within this response packet.

Component type
Contains information about one of the components or active process at the responding node.

Simple:
0 = IPX/SPX.
1 = Router drivers.
2 = LAN drivers.
3 = Shells.
4 = VAPs.

Extended:
5 = Router.
6 = File Server/Router.
7 = Nondedicated IPX/SPX.

Each extended type will be followed by additional fields.

Number of local networks (1 byte)

DIAG additional field

Number of local networks
Number of local networks with which this component can communicate.

For each local network, there will be:

Local network type (1 byte)
Network address (4 bytes)
Node address (6 bytes)

Format for local networks

Local network type
Contains a number indicating the type of network with which the component communicates.

Network address
Contains the 4-byte network address assigned to the network listed in the Local Network Type field.

Node address

Contains the 6-byte node address that accompanies the network address listed above. These will depend on the number of local networks.

```
Capture Buffer Display - WAN
Filter: All Frames
Protocol...
IPX: Destination Socket: 0x4456
IPX: Source Network: 0x5637FF01
IPX: Source Node: NetWare Server
IPX: Source Socket: Diagnostic
DIAG: Packet Type: Response
DIAG: Major Version: 2
DIAG: Minor Version: 2
DIAG: SPX Diagnostic Socket: 0x0F02
DIAG: Component Count: 4
DIAG: Component ID: 7 (Nondedicated IPX\SPX)
DIAG:   Number of Local Networks: 2
DIAG: Component ID: 1 (Bridge Driver)
DIAG: Component ID: 2 (LAN Driver)
DIAG: Component ID: 3 (Shell)
User Data
OFFST DATA                                              ASCII
003E: FF FF 98 55 44 44 33 DE ED AA A2 A3 D3 DE 45 09   ...UDD3.......E.
004E: 88 70 91 22 33 99 BB FF BB CC AE 24 EE 44 56 17   .p."3......$.DV.
(....)
Options...   Search...   Restart   Setup...   Done
```

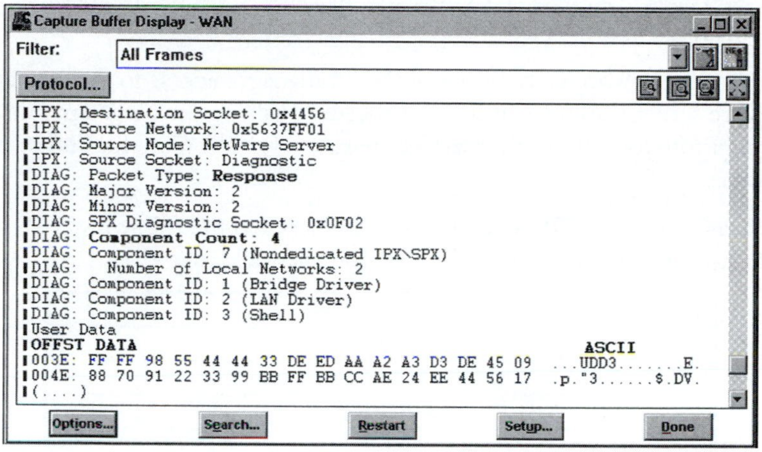

DIAG decode

NCP

"Novell's Guide to NetWare LAN Analysis" by Laura A. Chappell and Dan E. Hakes, Novell Press, 1994

The Novell NetWare Core Protocol (NCP) manages access to the primary NetWare server resources. It makes procedure calls to the NetWare File Sharing Protocol (NFSP) that services requests for NetWare file and print resources.

The format of the NCP Request header is shown in the following illustration. The request type is 2 bytes; all other fields are 1 byte.

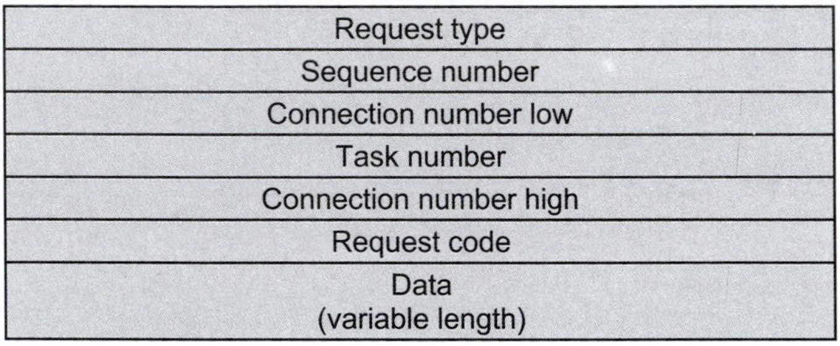

Request type
Sequence number
Connection number low
Task number
Connection number high
Request code
Data (variable length)

NCP request header

Request type

Identifies the packet type:

1111H Allocate slot request.
2222H File server request.
3333H File server reply.
5555H Deallocate slot request.
7777H Burst mode packet (BMP).
9999H Positive acknowledge.
H signifies hexadecimal notation.

Sequence number

Number used by the workstation and file server to identify packets which are sent and received.

Connection number low

Low connection ID number assigned to the workstation.

Task number

Identifies the operating system e.g., DOS, task.

Connection number high

High Connection ID number assigned to the workstation. Used only on the 1000-user version of NetWare, on all other versions will be set to 0.

Request code

Identifies the specific request function code.

The structure of the NCP Reply header is the same as the Request header, but the last 2 bytes differ after Connection Number High. This is shown in the following illustration:

Completion code
Connection status

NCP reply header: last 2 bytes

Completion code

The completion code indicates whether or not the Client's request was successful. A value of 0 in the Completion Code field indicates that the request was successful. Any other value indicates an error.

Connection status

The fourth bit in this byte will be set to 1 if DOWN is typed at the console prompt, to bring the server down.

NDS

"Novell's Guide to NetWare LAN Analysis" by Laura A. Chappell and Dan E. Hakes, Novell Press, 1994

NetWare Directory Services (NDS) is a globally distributed network database that replaces the bindery used in previous versions of NetWare. In an NDS-based network, one logs into the entire network providing access to all network services.

The format of the NDS packet is shown in the following illustration:

8	16	24	32 bits
Fragger handle			
Maximum fragment size			
Message size			
Fragment flag			
Internal verb			

NDS packet structure

Fragger handle
Fragmented request/reply handle.

Maximum fragment size
Maximum number of data bytes that can be sent as a reply.

Message size
Actual size of the message being sent.

Fragment flag
Flag field, always set to 0.

Internal verb
Number of the NDS verb that should be executed.

NLSP

Novell publication NetWare Link Services Protocol Specification rev 1.0

The NetWare Link Services Protocol (NLSP™) provides link state routing for Internetwork Packet Exchange networks. It is a protocol for information exchange among routers geared to the needs of large IPX networks. IPX is the network layer protocol used by the Novell NetWare operating system.

The general format of the header is shown in the following illustration:

8	16	24	32 bits
Protocol ID	Length	Minor ver	Reserved
NR | R | Pkt type	Major ver	Reserved	
Packet length			

NLSP header structure

Protocol ID
Identifies the NLSP routing layer.

Length indicator
The number of bytes in the fixed portion of the header.

Minor version
The value of this field is one.

NR
Multi-homed non-routing server, a 1-bit field. When the value of this field is one, the system has more than one network interface, but does not forward traffic from one network segment to another.

R
Reserved, 2-bit field.

Packet type
The packet type.

Major version
The value of this field is one.

Packet length
The entire length of the packet in bytes, including the fixed portion of the NLSP header.

NovelNetBIOS

This is a proprietary protocol developed by Novell based on NetBIOS.

The data stream type field is a 1-byte fixed field. All of the other fields are variable. Possible values for the data stream type field are:

1	Find Name.
2	Name Recognized.
3	Check Name.
4	Name in Use.
5	De-Register Name.
6	Session Data.
7	Session End.
8	Session End Ack.
9	Status Query.
10	Status Response.
11	Directed Datagram.

RIPX

"Novell's Guide to NetWare LAN Analysis" by Laura A. Chappell and Dan E. Hakes, Novell Press, 1994

The Routing Information Protocol (RIP), is used to collect, maintain and exchange correct routing information among gateways within the Internet. This protocol should not be confused with RIP of the TCP/IP suite of protocols.

The format of the RIPX packet is shown in the following illustration.

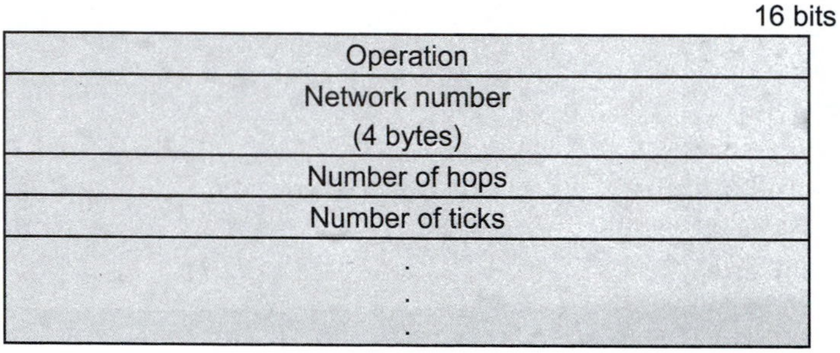

16 bits
Operation
Network number (4 bytes)
Number of hops
Number of ticks
. . .

Operation

Specifies the packet operation:

1 RIP Request.

2 RIP Response.

Network number

The 32-bit address of the specified network.

Number of hops

The number of routers that must be passed to reach the specified network. Routers broadcast "going down", containing the value 16 in this field, which means the route is no longer available.

Number of ticks

A measure of time needed to reach the specified network (18.21 ticks/second).

SER

"Novell's Guide to NetWare LAN Analysis" by Laura A. Chappell and Dan E. Hakes, Novell Press, 1994

To ensure that a single version of NetWare is not being loaded on multiple servers, the operating system broadcasts copy-protection packets, called Serialization packets, to determine whether there are multiple copies of the same operating system on the network.

Serialization packets contain only one field, the 6-byte Serialization Data field.

SAP

"Novell's Guide to NetWare LAN Analysis" by Laura A. Chappell and Dan E. Hakes, Novell Press, 1994

Before a client can communicate with a server it must know what servers are available on the network. This information is made available through Novell's Service Advertising Protocol (SAP). SAP services provide information on all the known servers throughout the entire internetwork. These servers can include file servers, print servers, NetWare access servers, remote console servers and so on.

The format of the SAP response packet is shown in the following illustration:

Operation (2 bytes)
Service type (2 bytes)
Server name (48 bytes)
Network address (4 bytes)
Node address (6 bytes)
Socket address (2 bytes)
Hops (2 bytes)
. . .

SAP response packet structure

The SAP packet may have up to 7 sets of server information.

Operation

Specifies the operation that the packet will perform:

1 General service request.
2 General service response.
3 Nearest service request.
4 Nearest service response.

Service type

Specifies the service performed. Examples include:

01H User.
04H File service.
07H Print server.

21H NAS SNA gateway.
23H NACS.
27H TCP/IP gateway.
98H NetWare access server.
107H NetWare 386 STOREXP Spec.
137H NetWare 386 print queue.
H signifies hexadecimal notation.

Server name

48-byte field containing the server's name in quotation marks.

Network address

32-bit network number of server.

Node address

48-bit node number of server.

Socket address

16-bit socket number of server.

Hops

Number of routers that must be passed through to reach the specified network. If the value in this field is 16 the service is not available.

The structure of the Request header is shown in the following illustration:

Operation (2 bytes)
Service type (2 bytes)

SAP request header

SPX

"Novell's Guide to NetWare LAN Analysis" by Laura A. Chappell and Dan E. Hakes, Novell Press, 1994

The Sequential Packet Exchange (SPX), is Novell's version of the Xerox Sequenced Packet Protocol (SPP). It is a transport layer protocol providing a packet delivery service for third party applications.

In July 1991, Novell established an SPX development team to create an improved version of SPX called SPX II. The primary improvements provided by SPX II include utilization of larger packets sizes and implementation of a windowing protocol.

The structure of the SPX packet is shown in the following illustration:

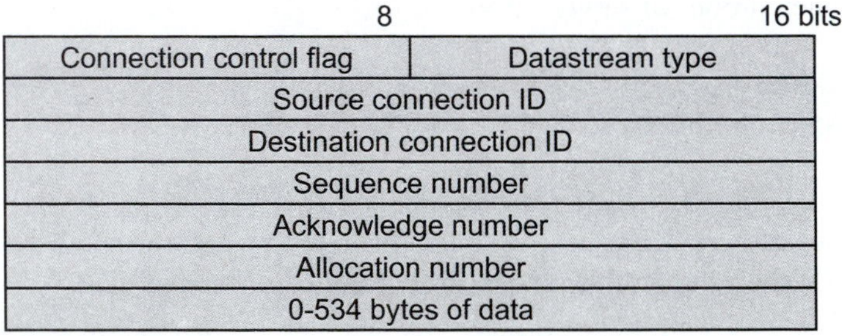

8	16 bits
Connection control flag	Datastream type
Source connection ID	
Destination connection ID	
Sequence number	
Acknowledge number	
Allocation number	
0-534 bytes of data	

SPX packet structure

Connection control flag

Four flags which control the bi-directional flow of data across an SPX connection. These flags have a value of 1 when set and 0 if not set.

Bit 4 Eom: End of message.
Bit 5 Att: Attention bit, not used by SPX.
Bit 6 Ack: Acknowledge required.
Bit 7 Sys: Transport control.

Datastream type

Specifies the data within the packet:

0-253 Ignored by SPX.
254 End of connection.
255 End of connection acknowledgment.

Source connection ID
A 16-bit number assigned by SPX to identify the connection.

Destination connection ID
The reference number used to identify the target end of the transport connection.

Sequence number
A 16-bit number, managed by SPX, which indicates the number of packets transmitted.

Acknowledge number
A 16-bit number, indicating the next expected packet.

Allocation number
A 16-bit number, indicating the number of packets sent but not yet acknowledged.

The SPX II header is the same as the SPX header described above, except for the following differences:

Connection control flag
Bit 2 - Size negotiation.
Bit 3 - SPX II type.

Datastream type
252 - Orderly release request.
253 - Orderly release acknowledgment.

There is also an additional 2-byte Extended Acknowledgement field at the end.

WDOG

"Novell's Guide to NetWare LAN Analysis" by Laura A. Chappell and Dan E. Hakes, Novell Press, 1994

The Watchdog (WDOG) protocol provides constant validation of active workstation connections and notifies the NetWare operating system when a connection may be terminated as a result of lengthy periods without communication.

The format of the WDOG packet is shown in the following illustration:

Connection number (bits 1–8)	Signature character (bits 9–16)

WDOG packet structure

Connection number
Given to the station during the login process.

Signature character
Contains 0x3F (ASCII character ?) or 0x59 (ASCII character Y).

25

PPP Suite

The Point-to-Point Protocol (PPP) suite includes the following protocols, in addition to PPP:

- MLP: Multilink PPP.
- PPP-BPDU: PPP Bridge Protocol Data Unit.
- PPPoE: PPP over Ethernet.
- BAP: Bandwidth Allocation Protocol.
- BSD.
- CHAP: Challenge Handshake Authentication Protocol.
- DESE: Data Encryption Standard Encryption.
- EAP: Extensible Authentication Protocol.
- IPHC: IP Header Compression
- LCP: Link Control Protocol.
- LEX: LAN Extension Interface Protocol.
- LQR: Link Quality Report.
- MPPC: Microsoft Point-to-Point Compression Protocol.
- PAP: Password Authentication Protocol.

PPP control protocols

- ATCP: AppleTalk Control Protocol.
- BACP: Bandwidth Allocation Control Protocol.
- BCP: Bridging Control Protocol.
- BVCP: PPP Banyan Vines Control Protocol.
- CCP: Compression Control Protocol.
- DNCP: PPP DECnet Phase IV Control Protocol.
- ECP: Encryption Control Protocol.
- IPCP: IP Control Protocol.
- IPv6CP: IPv6 Control Protocol.
- IPXCP: IPX Control Protocol.
- LEXCP: LAN Extension Interface Control Protocol.
- NBFCP: PPP NetBios Frames Control Protocol.
- OSINLCP: OSI Network Layer Control Protocol.
- SDCP: Serial Data Control Protocol.
- SNACP: SNA PPP Control Protocol.

The following diagram shows the PPP suite in relation to the OSI model :

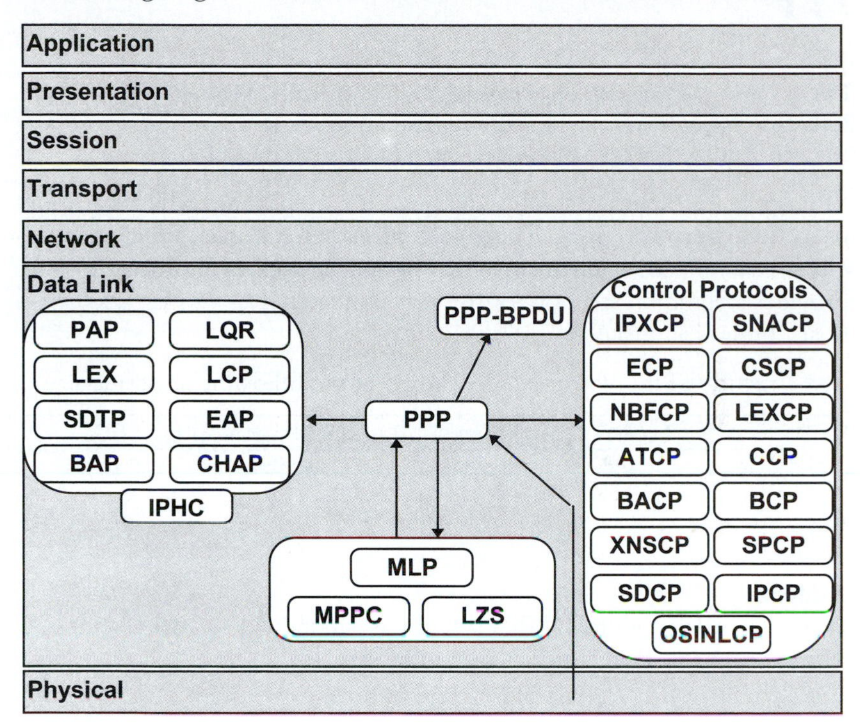

PPP protocol suite in relation to the OSI model

PPP

RFC 1548 http://www.cis.ohio-state.edu/htbin/rfc/rfc1548.html
RFC 1661 http://www.cis.ohio-state.edu/htbin/rfc/rfc1661.html
RFC 1662 http://www.cis.ohio-state.edu/htbin/rfc/rfc1662.html

PPP (Point-to-Point Protocol) is designed for simple links which transport packets between two peers. These links provide full-duplex simultaneous bi-directional operation and are assumed to deliver packets in order. PPP provides a common solution for the easy connection of a wide variety of hosts, bridges and routers.

The structure of the PPP header is shown in the following illustration:

Address	Control	Protocol	Information	FCS
1 byte	1 byte	2 bytes	variable	2 bytes

PPP header structure

Address
HDLC broadcast address. PPP does not assign individual station addresses. The value of this field is always set to FF Hex.

Control
HDLC command for Unnumbered Information (UI) with the Poll/Final bit set to zero. The value of this field is always set to 03 Hex. Frames containing any other value in this field are discarded.

Protocol
Identifies the encapsulated protocol within the Information field of the frame.

Information
Higher-level protocol data.

FCS
Value of the frame checksum calculation. PPP verifies the contents of the FCS field upon receipt of the packet.

MLP (PPP Multilink)

RFC 1717 http://www.cis.ohio-state.edu/htbin/rfc/rfc1717.html
RFC 1990 http://www.cis.ohio-state.edu/htbin/rfc/rfc1990.html

Multilink is based on a PCP option negotiation that permits a system to indicate to its peer that it is capable of combining multiple physical links into a "bundle". Only under exceptional conditions would a given pair of systems require the operation of more than one bundle connecting them.

Multilink is negotiated during the initial LCP option negotiation. A system indicates to its peer that it is willing to do multilink by sending the multilink option as part of the initial LCP option negotiation. This negotiation indicates the following:

1. The system offering the option is capable of combining multiple physical links into one logical link.
2. The system is capable of receiving upper layer PDUs fragmented using the multilink header and reassembling the fragments back into the original PDU for processing.
3. The system is capable of receiving PDUs of size N octets where N is specified as part of the option even if N is larger than the maximum receive unit (MRU) for a single physical link.

Using the PPP Multilink protocol, network protocol packets are first encapsulated (but not framed) according to normal PPP procedures; large packets are broken up into multiple segments sized appropriately for the multiple physical links. A new PPP header consisting of the Multilink protocol identifier, and the Multilink header is inserted before each section. Thus, the first fragment of a multilink packet in PPP will have two headers: one for the fragment, followed by the header for the packet itself.

PPP Multilink fragments are encapsulated using the protocol identifier 0x00-0x3d. Following the protocol identifier is a 4-byte header containing a sequence number, and two 1-byte fields indicating whether the fragment begins or terminates a packet. After negotiation of an additional PPP LCP option, the 4-byte header may be optionally replaced by a 2-byte header with a 12-bit sequence space. Address/Control and Protocol ID compression are assumed to be in effect.

The following is the format for the long sequence number fragment:

PPP Header	Address 0xff								Control 0x03
	PID (H) 0x00								PID (L) 0x3d
MP Header	B	E	0	0	0	0	0	0	Sequence number
	Sequence number (L)								
	Fragment data . . .								
PPP FCS	FCS								

PPP Multilink long sequence number fragment

The following is the format for the short sequence number fragment:

PPP Header	Address 0xff				Control 0x03
	PID (H) 0x00				PID (L) 0x3d
MP Header	B	E	0	0	Sequence Number
	Fragment Data . . .				
PPP FCS	FCS				

PPP Multilink short sequence number fragment

PID

Protocol ID compression.

B

Beginning fragment bit. A 1-bit field which is set to 1 on the first fragment derived from a PPP packet and set to 0 for all other fragments from the same PPP packet.

E

Ending fragment bit. A 1-bit field which is set to 1 on the last fragment and set to 0 for all other fragments. A fragment may have both the beginning and ending fragment bits set to 1.

Sequence number

24-bit or 12-bit number that is incremented for every fragment transmitted. By default, the sequence field is 24 bits long, but can be negotiated to be only 12 bits with an LCP configuration option.

0

A reserved field between the ending fragment bit and the sequence number. This field is not used at present and must be set to zero. It is 2 bits long when short sequence numbers are used; otherwise it is 6 bits in length.

FCS

Frame check sequence. This value is inherited from the normal framing mechanism from the member link on which the packet is transmitted.

PPP-BPDU

RFC 1638 http://www.cis.ohio-state.edu/htbin/rfc/rfc1638.html

There exist two basic types of bridges: those that interconnect LANs directly, called Local Bridges, and those that interconnect LANs via an intermediate WAN medium such as a leased line, called Remote Bridges. The PPP-BPDU (Bridge Protocol Data Unit) is used to connect Remote Bridges.

The format of the PPP-BPDU packet is shown in the following illustration:

4	8	16		32 bits
F I Z 0	Pads	MAC type	LAN ID high word (optional)	
LAN ID low word (optional)		Pad byte	Frame control	
Destination MAC address				
Destination MAC address		Source MAC address		
Source MAC address				
LLC data				

PPP-BPDU packet structure

F
F flag, set if the LAN FCS field is present.

I
I flag, set if the LAN ID field is present.

Z
Z flag, set if IEEE 802.3 pad must be zero filled to minimum size.

0
0 flag reserved, must be zero.

Pads
Any PPP frame may have padding inserted in the Optional Data Link Layer Padding field. This number tells the receiving system how many pad octets to strip off.

MAC type

Values of the MAC Type field.

1 Bridge-Identification.
2 Line-Identification.
3 MAC-Support.
4 Tinygram-Compression.
5 LAN-Identification.
6 MAC-Address.
7 Spanning-Tree-Protocol.

LAN ID

Optional 32-bit field that identifies the Community of LANs which may be interested in receiving this frame. If the LAN ID flag is not set, then this field is not present and the PDU is four octets shorter.

Frame control

On 802.4, 802.5 and FDDI LANs, there are a few octets preceding the Destination MAC Address, one of which is protected by the FCS. The MAC Type of the frame determines the contents of the Frame Control field. A pad octet is present to provide a 32-bit packet alignment.

Destination MAC address

As defined by the IEEE. The MAC Type field defines the bit ordering.

Source MAC address

As defined by the IEEE. The MAC Type field defines the bit ordering.

LLC data

This is the remainder of the MAC frame which is (or would be were it present) protected by the LAN FCS.

PPPoE

RFC 2516 http://www.cis.ohio-state.edu/htbin/rfc/rfc2516.html

PPPoE is a method for transmitting PPP over Ethernet. It provides the ability to connect a network of hosts over a simple bridging access device to a remote access concentrator. With this model, each host utilizes its own PPP stack and the user is presented with a familiar user interface. Access control, billing and type of service can be done on a per-user, rather than a per-site, basis.

To provide a point-to-point connection over Ethernet, each PPP session must learn the Ethernet address of the remote peer, as well as establish a unique session identifier. PPPoE includes a discovery protocol that performs this.

PPPoE has two distinct stages. There is a discovery stage and a PPP session stage. When a host wishes to initiate a PPPoE session, it must first perform discovery to identify the Ethernet MAC address of the peer and establish a PPPoE SESSION_ID. While PPP defines a peer-to-peer relationship, discovery is inherently a client-server relationship. In the discovery process, a host (the client) discovers an access concentrator (the server). Based on the network topology, there may be more than one access concentrator that the host can communicate with. The discovery stage allows the host to discover all access concentrators and then select one. When discovery completes successfully, both the host and the selected access concentrator have the information they then use to build their point-to-point connection over Ethernet.

The discovery stage remains stateless until a PPP session is established. Once a PPP session is established, both the host and the access concentrator must allocate the resources for a PPP virtual interface.

The EtherType field in the Ethernet frame is set to either 0x8863 for the discovery stage or 0x8864 for the PPP session stage.

The Ethernet payload for PPPoE is as shown in the following illustration:

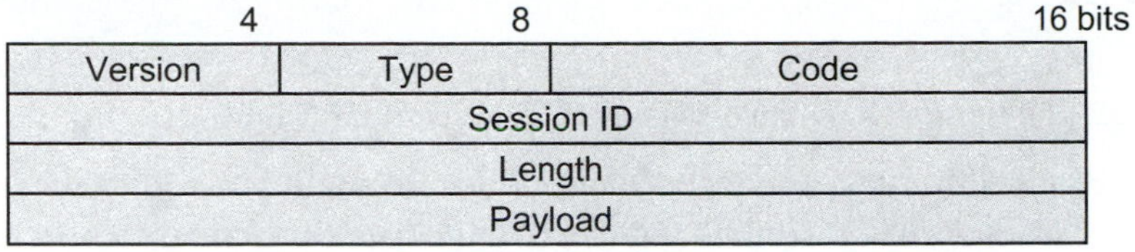

4	8	16 bits
Version	Type	Code
Session ID		
Length		
Payload		

Ethernet payload for PPPoE

Version

Specifies the version number: 0x1 for the current version of PPPoE (RFC 2516).

Type

Set to 0x1 for the current version of PPPoE (RFC 2516).

Code

Value of the code depends on the packet sent. Values may be as follows:

Packet	*Code*
Discovery stage:	
Active Discovery Initiation (PADI)	0x09
Active Discovery Offer (PADO)	0x07
Active Discovery Request (PADR)	0x19
Active Discovery Session-confirmation (PADS)	0x65
Active Discovery Terminate (PADT)	0xa7
PPP Session Stage	0x00

Session ID

Unsigned value in network byte order which defines a PPP session along with the Ethernet source and destination addresses. 0xffff is reserved for future use.

Length

Length of the PPPoE payload, not including the Ethernet or PPPoE headers.

BAP

RFC 2125 http://www.cis.ohio-state.edu/htbin/rfc/rfc2125.html

The Bandwidth Allocation Protocol (BAP) can be used to manage the number of links in a multi-link bundle. BAP defines datagrams to coordinate adding and removing individual links in a multi-link bundle, as well as specifying which peer is responsible for various decisions regarding managing bandwidth during a multi-link connection.

The format of the BAP packet is shown in the following illustration:

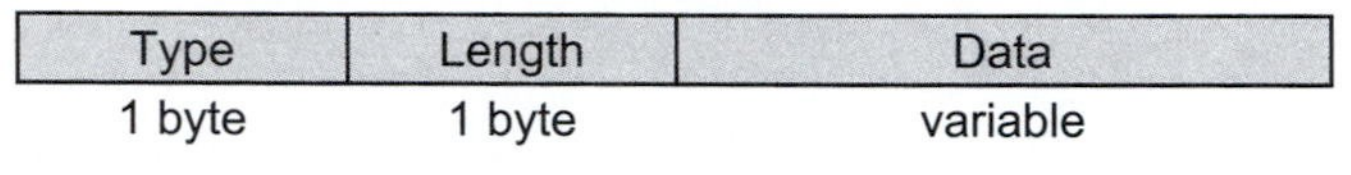

BAP packet structure

Type

One-octet field which indicates the type of the BAP Datagram Option. This field is binary coded hexadecimal.

1 Call-Request.
2 Call-Response.
3 Callback-Request.
4 Callback-Response.
5 Link-Drop-Query-Request.
6 Link-Drop-Query-Response.
7 Call-Status-Indication.
8 Call-Status-Response.

Length

One-octet field which indicates the length of this BAP option including the Type, Length and Data fields.

Data

Zero or more octets and contains information specific to the BAP option. The format and length of the Data field is determined by the Type and Length fields.

BSD

RFC 1977: http://www.cis.ohio-state.edu/htbin/rfc/rfc1977.html

BSD is the freely and widely distributed UNIX compress command source. It provides the following features:

- Dynamic table clearing when compression becomes less effective.
- Automatic turning off of compression when the overall result is not smaller than the input.
- Dynamic choice of code width within predetermined limits.
- Heavily used for many years in networks, on modem and other point-to-point links to transfer net news.
- An effective code width, requires less than 64 Kbytes of memory on both send and receive.

Before any BSD Compress packets may be communicated, PPP must reach the network layer protocol phase, and the CCP control protocol must reach the opened state. Exactly one BSD Compress datagram is encapsulated in the PPP information field, where the PPP protocol field contains 0xFD, or 0xFB. 0xFD is used above MLP. 0xFB is used below MLP to compress independently on individual links of a multilink bundle. The maximum length of the BSD Compress datagram transmitted over a PPP link is the same as the maximum length of the information field of a PPP encapsulated packet. Only packets with PPP protocol numbers in the range 0x0000 to 0x3FFF and neither 0xFD, nor 0xFB are compressed. Other PPP packets are always sent uncompressed. Control packets are infrequent and should not be compressed for robustness.

CHAP

RFC 1334 http://www.cis.ohio-state.edu/htbin/rfc/rfc1334.html

Challenge Handshake Authentication Protocol (CHAP) is used to periodically verify the identity of the peer using a 3-way handshake. This is done upon initial link establishment and may be repeated any time after the link has been established.

Exactly one CHAP packet is encapsulated in the Information field of a PPP data link layer frame where the protocol field indicates type hex c223. The structure of the CHAP packet is shown in the following illustration:

Code	Identifier	Length	Data . . .
1 byte	1 byte	2 bytes	variable

CHAP packet structure

Code

Identifies the type of CHAP packet. CHAP codes are assigned as follows:

1 Challenge.
2 Response.
3 Success.
4 Failure.

Identifier

Aids in matching challenges, responses and replies.

Length

Length of the CHAP packet including the Code, Identifier, Length and Data fields.

Data

Zero or more octets, the format of which is determined by the Code field. The format of the Challenge and Response data fields is shown in the following illustration:

Value size	Value	Name
1 byte		1 byte

CHAP Challenge and Response data structure

Value size

Indicates the length of the Value field.

Value

Challenge value is a variable stream of octets which must be changed each time a challenge is sent.
Response value is the one-way hash calculated over a stream of octets consisting of the Identifier, followed by the "secret" and the Challenge value.

Name

Identification of the system transmitting the packet.

For Success and Failure, the data field contains a variable message field which is implementation dependent.

DESE

RFC 1969: http://www.cis.ohio-state.edu/htbin/rfc/rfc1969.html

The DES (Data Encryption Standard) Encryption algorithm is a well studied, understood and widely implemented encryption algorithm which was designed for efficient implementation in hardware. DESE is an option of ECP which indicates that the issuing implementation is offering to employ the DES encryption for decrypting communications on the link and may be thought of as a request for its peer to encrypt packets in this manner.

The format of the DESE packet is shown in the following illustration:

8	16	24	32 bits
Address	Control	0 0 0 0	Protocol ID
Seq no high	Seq no low	Cyphertext (variable)	

DESE packet structure

The address and control fields are as described for PPP.

Protocol ID

Values may be 0x53 or 0x55; the latter indicates that cyphertext includes headers for MLP and requires that the Individual Link Encryption Control Protocol has reached the opened state. The leading zero may be absent if the PPP Protocol Field Compression (PFC) option has been negotiated.

Seq no high/low

Sequence numbers are 16-bit numbers which are assigned by the encryptor sequentially starting with 0 (for the first packet transmitted once ECP has reached the open state).

Cyphertext

Generation of encrypted data is described in the DESE standard.

EAP

draft-ietf-pppext-eap-auth-03.txt

The Extensible Authentication Protocol (EAP) is a PPP extension that provides support for additional authentication methods within PPP.

The format of the EAP header is shown in the following illustration:

Code	Identifier	Length	Type	Data
1 byte	1 byte	2 bytes	1 byte	variable

EAP packet structure

Code

Decimal value which indicates the type of EAP packet:

1 Request.
2 Response.
3 Success.
4 Failure.

Identifier

Aids in matching responses with requests.

Length

Indicates the length of the EAP request and response packets including the Code, Identifier, Length, Type, and Data fields. Octets in the packet outside the range of the Length field are treated as Data Link Layer padding and are ignored on reception.

Type

Indicates the EAP type. The following types are supported: KEA validate, KEA public, GTC, OTP, MD5, NQK, Notification, Identity.

IPHC

RFC 2509
IP Header Compression (IPHC) may be used for compression of both IPv4 and IPv6 datagrams or packets encapsulated with multiple IP headers. IPHC is also capable of compressing both TCP and UDP transport protocol headers. The IP/UDP/RTP header compression defined in CRTP fits within the framework defined by IPHC so that it may also be applied to both IPv4 and IPv6 packets.

In order to establish compression of IP datagrams sent over a PPP link, each end of the link must agree on a set of configuration parameters for the compression. The process of negotiating link parameters for network layer protocols is handled in PPP by a family of network control protocols (NCPs). Since there are separate NCPs for IPv4 and IPv6, this document defines configuration options to be used in both NCPs to negotiate parameters for the compression scheme.

Configuration option format

Both the network control protocol for IPv4, IPCP (RFC1332) and the IPv6 NCP, IPV6CP (RFC2023) may be used to negotiate IP Header Compression parameters for their respective protocols. The format of the configuration option is the same for both IPCP and IPV6CP.

Type	Length	IP compression protocol
TCP_SPACE		NON_TCP_SPACE
TCP_SPACE		F_MAX_TIME
MAX_HEADER		Suboptions.....
1 byte	1 byte	2 bytes

IPHC configuration option format

The following parameters are available:

Type
The value of the Type field is 2.

Length

Indicates the length of the suboption in its entirety, including the lengths of the type and length fields. The value of this field is 14.

IP compression protocol

The value of this field is 0061 (hex).

TCP_SPACE

The TCP_SPACE field indicates the maximum value of a context identifier in the space of context identifiers allocated for TCP. The suggested value of this field is: 15. TCP_SPACE must be at least 0 and at most 255 (The value 0 implies having one context).

F_MAX_PERIOD

Maximum interval between full headers. No more then F_MAX_PERIOD compressed non-TCP headers may be sent between full header headers. Suggested value: 256

F_MAX_TIME

Maximum time interval between full headers. Suggested value: 5 seconds. A value of zero implies infinity.

MAX_HEADER

The largest header size in octets that may be compressed. Suggested value: 168 octets.

The value of MAX_HEADER should be large enough so that at least the outer network layer header can be compressed. To increase compression efficiency MAX_HEADER should be set to a value large enough to cover common combinations of network and transport layer headers.

Suboptions

The Suboptions field consists of zero or more suboptions. Each suboption consists of a type field, a length field and zero or more parameter octets, as defined by the suboption type. The value of the length field indicates the length of the suboption in its entirety, including the lengths of the type and length fields.

LCP

RFC 1570 http://www.cis.ohio-state.edu/htbin/rfc/rfc1570.html
RFC 1661 http://www.cis.ohio-state.edu/htbin/rfc/rfc1661.html

In order to be sufficiently versatile to be portable to a wide variety of environments, PPP provides the Link Control Protocol (LCP) for establishing, configuring and testing the data link connection. LCP is used to automatically agree upon the encapsulation format options, handle varying limits on sizes of packets, detect a looped-back link and other common misconfiguration errors, and terminate the link. Other optional facilities provided are authentication of the identity of its peer on the link, and determination when a link is functioning properly and when it is failing.

The format of the LCP packet is shown in the following illustration:

Code	Identifier	Length	Data
1 byte	1 byte	2 bytes	variable

LCP packet structure

Code

Decimal value which indicates the type of LCP packet:

1 Configure-Request.
2 Configure-Ack.
3 Configure-Nak.
4 Configure-Reject.
5 Terminate-Request.
6 Terminate-Ack.
7 Code-Reject.
8 Protocol-Reject.
9 Echo-Request.
10 Echo-Reply.
11 Discard-Request.
12 Link-Quality Report.

Identifier

Decimal value which aids in matching requests and replies.

Length

Length of the LCP packet, including the Code, Identifier, Length and Data fields.

Data

Variable length field which may contain one or more configuration options. The format of the LCP configuration options is as follows:

Type	Length	Data

LCP configuration options

Type

One-byte indication of the type of the configuration option.

Length

Length of the configuration option including the Type, Length and Data fields.

Data

Value of the Data field.

```
Capture Buffer Display - WAN
Filter: All Frames
Protocol: LCP
Captured at:  -00:02.456
Length: 38    From: User    Status: Ok
LCP: Code = Configure-Ack
LCP: Identifier = 1
LCP: Length = 32
LCP:
LCP:
LCP:    Options Type = 1 Maximum Receive Unit
LCP:    Options Length = 4
LCP:    Max Receive Unit = 1518
LCP:
LCP:    Options Type = 4 Quality Protocol
LCP:    Options Length = 8
LCP:    Quality Protocol = Link Quality Report
LCP:     Reporting Period = 0 Hundredths Seconds
LCP:          (Peer Does Not Need To Maintain A Timer)
LCP:
LCP:    Options Type = 5 Magic Number
LCP:    Options Length = 6
LCP:    Magic Number = 1463999748
Options...  Search...  Restart  Setup...  Done
```

LCP decode

LEX

RFC 1841 http://www.cis.ohio-state.edu/htbin/rfc/rfc1841.html

A LAN extension interface unit is a hardware device installed at remote sites (such as a home office or small branch office) that connects a LAN across a WAN link to a router at a central site. To accommodate this LAN extension interface architecture, a PPP Network Control Protocol was developed: the LAN extension interface protocol, PPP-LEX. The basic functionality of LEX is to encapsulate LAN extension interface control and data packets. Consequently packets can be control packets or data packets. Control packets are described under LEXCP.

The frame format for a PPP-LEX data packet is shown in the following illustration. The MAC frame is transferred except for the FCS field. The LAN extension interface unit computes the FCS for packets transferred to the LAN and strips the FCS for packets destined for the host router.

4 8	16	24	32 bits
HDLC flag			
Address	Control	Protocol type	
F | I | Z | 0 | Pad	MAC type	Destination MAC address	
Destination MAC address			
Source MAC address			
Source MAC address		Length/type	
LLC data			
HDLC CRC		HDLC flag	

PPP-LEX data packet structure

HDLC flag
HDLC frame delimiter.

Address
Address field containing broadcast address 0xFF.

Control
Control field containing unnumbered information 0x03.

Protocol type

Contains the IETF-assigned protocol type value. In this case this field will always contain 0x0041 to indicate a data packet.

F

Set if the LAN FCS field is present. Because PPP-LEX data packets do not contain the LAN FCS field, this bit should not be set (field=0).

I

Set if the LAN ID field is present. Because PPP-LEX data packets do not contain the field, this bit should not be set (field=0).

Z

Set if IEEE 802.3 Pad must be zero filled to minimum size.

0

Reserved, must be zero.

Pad

Any PPP frame may have padding inserted in the Optional Data Link Layer Padding field. The value tells the receiving system how many pad octets to strip off. The LAN extension interface protocol does not support the Optional Data Link Layer Padding field, so the value of this field should be zero.

MAC type

This field contains the most up-to-date value of the MAC type as specified in the most recent *Assigned Numbers RFC*. The current value (according to RFC 1841) indicates IEEE 802.3/Ethernet with canonical addresses.

Destination MAC address

6-octet field containing the MAC address of the destination system as defined by IEEE. The MAC type field defines the bit ordering.

Source MAC address

6-octet field containing the MAC address of the source system as defined by IEEE. The MAC Type field defines the bit ordering.

Length / type

Ethernet protocol type. For IEEE 802.3 frames, this is a length field.

LLC data
Remainder of the MAC frame which is (or would be if it were present) protected by the LAN FCS.

HDLC CRC
16-bit cyclic redundancy check field.

LQR

RFC 1333 http://www.cis.ohio-state.edu/htbin/rfc/rfc1333.html

The Link Quality Report (LQR) protocol specifies the mechanism for link quality monitoring within PPP. Packets are sometimes dropped or corrupted due to line noise, equipment failure, buffer overruns, etc. and it is often desirable to determine when and how often the link drops data. Routers may temporarily allow another route to take precedence, or an implementation may have the option of disconnecting and switching to an alternate link. For these reasons, such a quality monitoring mechanism is necessary.

One LQR packet is encapsulated in the information field of PPP data link layer frames where the protocol field indicates type hex c025. The structure of the LQR packet is shown in the following illustration:

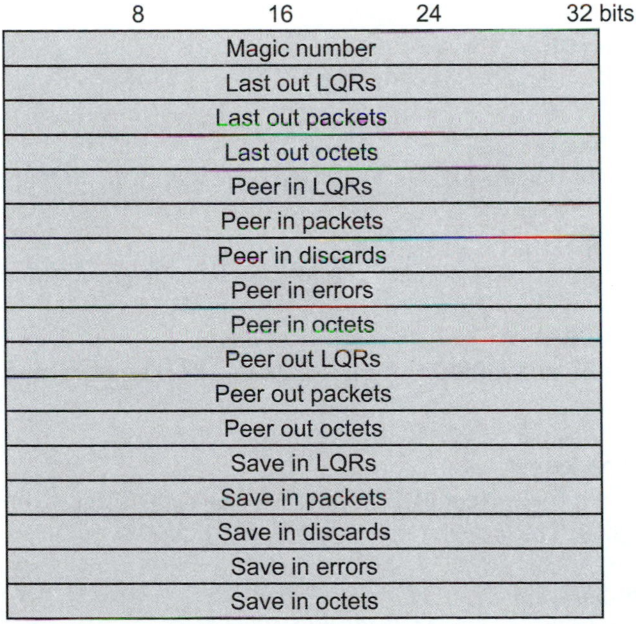

8 16 24 32 bits
Magic number
Last out LQRs
Last out packets
Last out octets
Peer in LQRs
Peer in packets
Peer in discards
Peer in errors
Peer in octets
Peer out LQRs
Peer out packets
Peer out octets
Save in LQRs
Save in packets
Save in discards
Save in errors
Save in octets

LQR packet structure

Magic number

Aids in detecting links which are in the looped-back condition.

Last out LQRs
Copied from the most recently received Peer Out LQRs on transmission.

Last out packets
Copied from the most recently received Peer Out Packets on transmission.

Last out octets
Copied from the most recently received Peer Out Octets on transmission.

Peer in LQRs
Copied from the most recently received Save In LQRs on transmission. Whenever the Peer In LQRs field is zero, the Last Out fields are indeterminate and the Peer In fields contain the initial values for the peer.

Peer in packets
Copied from the most recently received Save In Packets on transmission.

Peer in discards
Copied from the most recently received Save In Discards on transmission.

Peer in errors
Copied from the most recently received Save In Errors on transmission.

Peer in octets
Copied from the most recently received Save In Octets on transmission.

Peer out LQRs
Copied from the Out LQRs on transmission. This number must include this LQR.

Peer out packets
Copied from the current MIB ifOutUniPackets and ifOutNUniPackets on transmission. This number must include this LQR.

Peer out octets
Copied from the current MIB ifOutOctets on transmission. This number must include this LQR.

The following fields are not actually transmitted over the inbound link. Rather, they are logically appended to the packet by the Rx process.

Save in LQRs

Copied from In LQRs on reception. This number must include this LQR.

Save in packets

Copied from the current MIB ifInUniPackets and ifInNUniPackets on reception. This number must include this LQR.

Save in discards

Copied from the current MIB ifInDiscards on reception. This number must include this LQR.

Save in errors

Copied from the current MIB ifInErrors on reception. This number must include this LQR.

Save in octets

Copied from the current InGoodOctets on reception. This number must include this LQR.

MPPC

RFC 2118 http://www.cis.ohio-state.edu/htbin/rfc/rfc2118.html

The Point-to-Point Protocol provides a standard method for transporting multi-protocol datagrams over point-to-point links. The PPP Compression Control Protocol provides a method to negotiate and utilize compression protocols over PPP encapsulated links. The Microsoft Point to Point Compression protocol (also referred to as MPPC) compresses PPP encapsulated packets. The Microsoft Point to Point Compression scheme is a means of representing arbitrary Point-to-Point Protocol (PPP) packets in a compressed form. The MPPC algorithm is designed to optimize processor utilization and bandwidth utilization in order to support large number of simultaneous connections. The MPPC algorithm is also optimized to work efficiently in typical PPP scenarios (1500 byte MTU, etc.). The CCP configuration Option negotiates the use of MPPC on the link. Before any MPPC packets may be transmitted, PPP must reach the Network-Layer protocol phase and the CCP Control Protocol must reach the opened state. One MPPC datagram is encapsulated in the PPP information field. Only packets with PPP protocol numbers in the range 0021 to 00FA are compressed.

The format of the MPPC packet is shown in the following illustration:

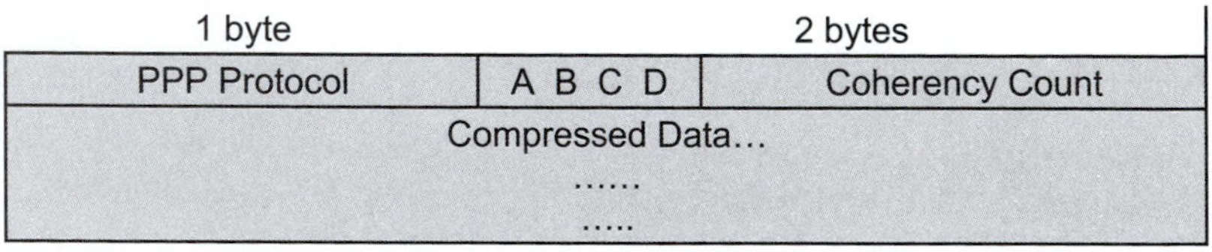

1 byte		2 bytes
PPP Protocol	A B C D	Coherency Count
Compressed Data...		
......		
.....		

MPPC packet structure

PPP Protocol

Identifies the encapsulated protocol within the information field of the frame.

Bit A

This bit indicates that the history buffer has just been initialized before the packet was generated. It is sent to inform the peer that the sender has initialized its history buffer before compressing the packet and that the receiving peer must initialize its history buffer before decompressing the packet. This bit is referred to as the flushed bit.

Bit B

This bit indicates that the packet was moved to the front end of the history buffer.

Bit C

This bit is set to indicate that the packet is compressed.

Bit D

This bit must be set to 0.

Coherency Count

This is used to assure that the packets are sent in the correct order and that no packet has been dropped.

Compressed Data

The compressed data begins with the protocol field; for example in an IP packet (0021 followed by an IIP header) the compressor will try to compress the 0021 field first and then the IP header itself.

PAP

RFC 1334 http://www.cis.ohio-state.edu/htbin/rfc/rfc1334.html

Password Authentication Protocol (PAP) provides a simple method for the peer to establish its identity using a 2-way handshake. This is done only upon initial link establishment.

The PAP packet is encapsulated in the Information field of a PPP data link layer frame where the protocol field indicates type hex c023. The structure of the PAP packet is shown in the following illustration:

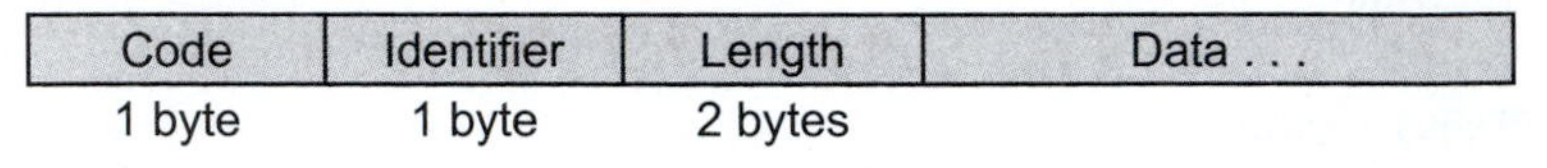

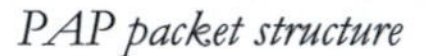
PAP packet structure

Code

One-octet field which identifies the type of PAP packet. PAP codes are assigned as follows:

1 Authenticate-Request.
2 Authenticate-Ack.
3 Authenticate-Nak.

Identifier

Aids in matching requests and replies.

Length

Indicates the length of the PAP packet including the Code, Identifier, Length and Data fields.

Data

Zero or more octets, the format of which is determined by the Code field. The format of the data field for Authenticate-Request packets is shown below:

Peer-ID length	Peer-ID	Password length	Password
1 byte	variable	1 byte	variable

Data structure for Authenticate-Request packet

Peer-ID length
Length of the Peer-ID field.

Peer-ID
Indicates the name of the peer to be authenticated.

Password length
Length of the Password field.

Password
Indicates the password to be used for authentication.

The format of the data field for Authenticate-Ack and Authenticate-Nak packets is shown below:

Message Length	Message
1 byte	variable

Data structure for Authenticate-Ack and Authenticate-Nak packets

Message length
Length of the Message field.

Message
The contents of the Message field are implementation dependent.

ATCP

RFC 1378 http://www.cis.ohio-state.edu/htbin/rfc/rfc1378.html

The AppleTalk Control Protocol (ATCP) is responsible for configuring, the AppleTalk parameters on both ends of the point-to-point link. ATCP uses the same packet exchange mechanism as the Link Control Protocol (LCP). ATCP packets may not be exchanged until PPP has reached the Network-Layer Protocol phase. ATCP packets received before this phase is reached are discarded.

The format of the ATCP packet is shown in the following illustration:

Code	Identifier	Length	Data
1 byte	1 byte	2 bytes	variable

ATCP packet structure

Code

Decimal value which indicates the type of ATCP packet.

1 Configure-Request.
2 Configure-Ack.
3 Configure-Nak.
4 Configure-Reject.
5 Terminate-Request.
6 Terminate-Ack.
7 Code-Reject.

Identifier

Decimal value which aids in matching requests and replies.

Length

Length of the ATCP packet, including the Code, Identifier, Length and Data fields.

Data

Variable length field which may contain one or more configuration options. The format of the ATCP configuration options is as follows:

Type	Length	Data

ATCP configuration options

Type

One-byte indication of the type of the configuration option.

1 AppleTalk-Address.
2 Routing-Protocol.
3 Suppress-Broadcasts.
4 AT-Compression Protocol.
6 Server-Information.
7 Zone-Information.
8 Default Router-Address.

Length

Length of the configuration option including the Type, Length and Data fields.

Data

Value of the Data field.

BACP

RFC 2125 http://www.cis.ohio-state.edu/htbin/rfc/rfc2125.html

The Bandwidth Allocation Control Protocol (BACP) is the associated control protocol for BAP. BAP can be used to manage the number of links in a multi-link bundle. BAP defines datagrams to coordinate adding and removing individual links in a multi-link bundle, as well as specifying which peer is responsible for which decisions regarding managing bandwidth during a multi-link connection. BACP defines control parameters for the BAP protocol to use.

The format of the BACP packet is shown in the following illustration:

Code	Identifier	Length	Data
1 byte	1 byte	2 bytes	variable

BACP packet structure

Code

Decimal value which indicates the type of BACP packet.

1 Configure-Request.
2 Configure-Ack.
3 Configure-Nak.
4 Configure-Reject.
5 Terminate-Request.
6 Terminate-Ack.
7 Code-Reject.

Identifier

Decimal value which aids in matching requests and replies.

Length

Length of the BACP packet, including the Code, Identifier, Length and Data fields.

Data

Variable length field which may contain one or more configuration options. The format of the BACP configuration options is as follows:

Type	Length	Data

BACP configuration options

Type

One-byte indication of the type of the configuration option.

1 Favored-Peer

Length

Length of the configuration option including the Type, Length and Data fields.

Data

Value of the Data field.

BCP

RFC 1638 http://www.cis.ohio-state.edu/htbin/rfc/rfc1638.html

The Bridging Control Protocol (BCP) is responsible for configuring the bridge protocol parameters on both ends of the point-to-point link. BCP uses the same packet exchange mechanism as the Link Control Protocol. BCP packets can not be exchanged until PPP has reached the Network-Layer Protocol phase. BCP packets received before this phase is reached are discarded.

The format of the BCP packet is shown in the following illustration:

Code	Identifier	Length	Data
1 byte	1 byte	2 bytes	variable

BCP packet structure

Code

Decimal value which indicates the type of BCP packet.

1 Configure-Request.
2 Configure-Ack.
3 Configure-Nak.
4 Configure-Reject.
5 Terminate-Request.
6 Terminate-Ack.
7 Code-Reject.

Identifier

Decimal value which aids in matching requests and replies.

Length

Length of the BCP packet, including the Code, Identifier, Length and Data fields.

Data

Variable length field which may contain one or more configuration options. The format of the BCP configuration options is as follows:

Type	Length	Data

BCP configuration options

Type

One-byte indication of the type of the configuration option.

1 Bridge-Identification.
2 Line-Identification.
3 MAC-Support.
4 Tinygram-Compression.
5 LAN-Identification.
6 MAC-Address.
7 Spanning-Tree-Protocol.

Length

Length of the configuration option including the Type, Length and Data fields.

Data

Value of the Data field.

BVCP

RFC 1763 http://www.cis.ohio-state.edu/htbin/rfc/rfc1763.html

The PPP Banyan VINES Control Protocol (BVCP) is responsible for configuring, enabling, and disabling the VINES protocol modules on both ends of the point-to-point link. BVCP uses the same packet exchange mechanism as the Link Control Protocol (LCP).

The format of the BVCP packet is shown in the following illustration:

Code	Identifier	Length	Data
1 byte	1 byte	2 byte	variable

BVCP packet structure

Code

Decimal value which indicates the type of BVCP packet.

1 Configure-Request.
2 Configure-Ack.
3 Configure-Nak.
4 Configure-Reject.
5 Terminate-Request.
6 Terminate-Ack.
7 Code-Reject.

Identifier

Decimal value which aids in matching requests and replies.

Length

Length of the BVCP packet, including the Code, Identifier, Length and Data fields.

Data

Variable length field which may contain one or more configuration options. The format of the BVCP configuration options is as follows:

Type	Length	Data

BVCP configuration options

Type

One-byte indication of the type of the configuration option.

1 BV-NS-RTP-Link-Type.
2 BV-FRP (fragmentation protocol).
3 BV-RTP.
4 BV-Suppress-Broadcast.

Length

Length of the configuration option including the type, length and data fields.

Data

Value of the data field.

CCP

RFC 1962 http://www.cis.ohio-state.edu/htbin/rfc/rfc1962.html

The Compression Control Protocol (CCP) is responsible for configuring the data compression algorithms on both ends of the point-to-point link. It is also used to signal a failure of the compression/decompression mechanism in a reliable manner. CCP uses the same packet exchange mechanism as the Link Control Protocol. CCP packets cannot be exchanged until PPP has reached the Network-Layer Protocol phase. CCP packets received before this phase is reached are discarded.

The format of the CCP packet is shown in the following illustration:

Code	Identifier	Length	Data
1 byte	1 byte	2 bytes	variable

CCP packet structure

Code

Decimal value which indicates the type of CCP packet.

1 Configure-Request.
2 Configure-Ack.
3 Configure-Nak.
4 Configure-Reject.
5 Terminate-Request.
6 Terminate-Ack.
7 Code-Reject.
14 Reset Request.
15 Reset-Ack.

Identifier

Decimal value which aids in matching requests and replies.

Length

Length of the CCP packet, including the Code, Identifier, Length and Data fields.

Data

Variable length field which may contain one or more configuration options. The format of the CCP configuration options is as follows:

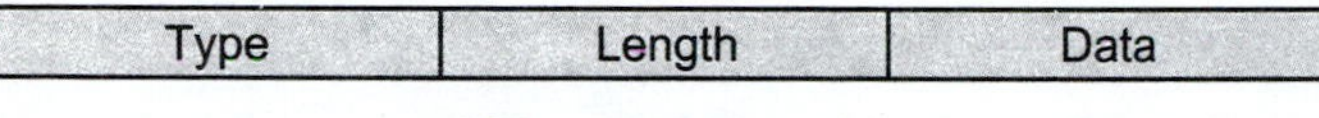

CCP configuration options

Type

One-byte indication of the type of the configuration option.

0 OUI.
1 Predictor Type 1.
2 Predictor Type 2.
3 Puddle Jumper.
16 Hewlett-Packard-PPC.
17 Stac Electronics LZS.
18 MicroSoft PPC.
19 Gandalf FZA.
20 V.42 bis Compression.
21 BSD LZW Compress.
23 LZS-DCP.

Length

Length of the configuration option including the Type, Length and Data fields.

Data

Value of the Data field.

DNCP

RFC 1376 http://www.cis.ohio-state.edu/htbin/rfc/rfc1376.html

The PPP DECnet Phase IV Control Protocol is responsible for establishing and configuring Digital's DNA Phase IV routing protocol (DECnet Phase IV) over PPP. The protocol applies only to DNA Phase IV routing messages (both data and control), and not to other DNA Phase IV protocols (MOP, LAT, etc).

The format of the DNCP packet is shown in the following illustration:

Code	Identifier	Length	Data
1 byte	1 byte	2 bytes	variable

DNCP packet structure

Code

Decimal value which indicates the type of DNCP packet.

1 Configure-Request.
2 Configure-Ack.
3 Configure-Nak.
4 Configure-Reject.
5 Terminate-Request.
6 Terminate-Ack.
7 Code-Reject.

Identifier

Decimal value which aids in matching requests and replies.

Length

Length of the DNCP packet, including the Code, Identifier, Length and Data fields.

Data

A variable length field which in similar protocols contains options, DNCP packets, however, have no options.

ECP

RFC 1968 http://www.cis.ohio-state.edu/htbin/rfc/rfc1968.html

The Encryption Control Protocol (ECP) is responsible for configuring and enabling data encryption algorithms on both ends of the point-to-point link. ECP uses the same packet exchange mechanism as the Link Control Protocol (LCP). ECP packets may not be exchanged until PPP has reached the Network-Layer Protocol phase. ECP packets received before this phase is reached are silently discarded.

The format of the header is shown in the following illustration:

Code	Identifier	Length	Data
1 byte	1 byte	2 bytes	variable

ECP header structure

Code

A one octet field identifying the type of ECP packet. When a packet is received with an unknown Code field, a Code Reject packet is transmitted.

1 Configure-Request.
2 Configure-Ack.
3 Configure-Nak.
4 Configure-Reject.
5 Terminate-Request.
6 Terminate-Ack.
7 Code-Reject.
14 Reset-Request.
15 Reset-Ack.

Identifier

Decimal value which aids in matching requests and replies.

Length

Length of the ECP packet, including the Code, Identifier, Length and Data fields.

IPv6CP

RFC 2023 http://www.cis.ohio-state.edu/htbin/rfc/rfc2023.html

The IPv6 PPP Control Protocol is responsible for configuring, enabling and disabling the IPv6 protocol modules on both ends of the point-to-point link.

The format of the IPv6CP packet is shown in the following illustration:

Code	Identifier	Length	Data
1 byte	1 byte	2 bytes	variable

IPv6CP packet structure

Code

Decimal value which indicates the type of IPv6CP packet.

1 Configure-Request.
2 Configure-Ack.
3 Configure-Nak.
4 Configure-Reject.
5 Terminate-Request.
6 Terminate-Ack.
7 Code-Reject.

Identifier

Decimal value which aids in matching requests and replies.

Length

Length of the IPv6CP packet, including the Code, Identifier, Length and Data fields.

Data

Variable length field which may contain one or more configuration options. The format of the IPv6CP configuration options is as follows:

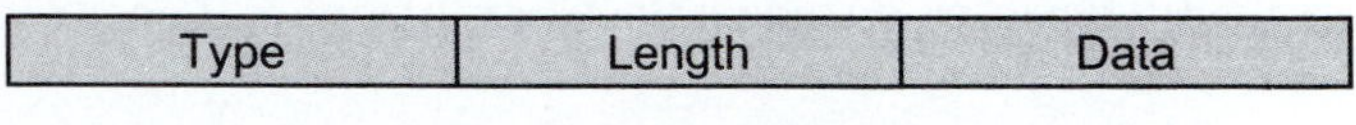

IPv6CP configuration options

Type

One-byte indication of the type of the configuration option.

1 Interface-Token.
2 IPv6-Compression-Protocol.

Length

Length of the configuration option including the type, length and data fields.

Data

Value of the Data field.

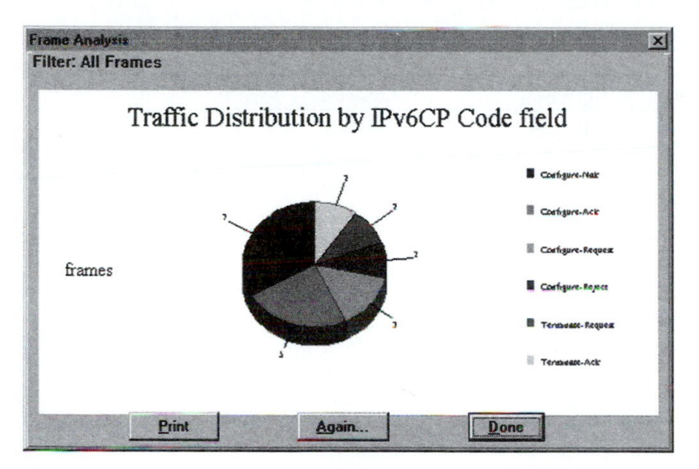

Graph of traffic distribution by IPv6CP code field

IPCP

RFC 1332 http://www.cis.ohio-state.edu/htbin/rfc/rfc1332.html

The IP Control Protocol (IPCP) is responsible for configuring the IP protocol parameters on both ends of the point-to-point link. IPCP uses the same packet exchange mechanism as the Link Control Protocol (LCP). IPCP packets may not be exchanged until PPP has reached the Network-Layer Protocol phase. Any IPCP packets received before this phase is reached are discarded.

The format of the IPCP packet is shown in the following illustration:

Code	Identifier	Length	Data
1 byte	1 byte	2 bytes	variable

IPCP packet structure

Code

Decimal value which indicates the type of IPCP packet.

1 Configure-Request.
2 Configure-Ack.
3 Configure-Nak.
4 Configure-Reject.
5 Terminate-Request.
6 Terminate-Ack.
7 Code-Reject.

Identifier

Decimal value which aids in matching requests and replies.

Length

Length of the IPCP packet, including the Code, Identifier, Length and Data fields.

Data

Variable length field which may contain one or more configuration options. The format of the IPCP configuration options is as follows:

Type	Length	Data

IPCP configuration options

Type

One-byte indication of the type of the configuration option.

1 IP Addresses - use of which is deprecated. Use of 3 is preferred.

2 IP Compression Protocol.

3 IP Address - way to negotiate the IP address to be used on the local end of the link.

Length

Length of the configuration option including the Type, Length and Data fields.

Data

Value of the Data field.

IPXCP

RFC 1552 http://www.cis.ohio-state.edu/htbin/rfc/rfc1552.html

In order to establish communication over a point-to-point link, each end of the PPP link must first send LCP packets to configure and test the data link. After the link has been established and optional facilities have been negotiated as needed by the LCP, PPP may send IPXCP packets to choose and configure the IPX network-layer protocol. Once IPXCP has reached the Opened state, IPX datagrams can be sent over the link.

The format of the IPXCP packet is shown in the following illustration:

Code	Identifier	Length	Data
1 byte	1 byte	2 bytes	variable

IPXCP packet structure

Code

Decimal value which indicates the type of IPXCP packet.

1 Configure-Request.
2 Configure-Ack.
3 Configure-Nak.
4 Configure-Reject.
5 Terminate-Request.
6 Terminate-Ack.
7 Code-Reject.

Identifier

Decimal value which aids in matching requests and replies.

Length

Length of the IPXCP packet, including the Code, Identifier, Length and Data fields.

Data

Variable length field which may contain one or more configuration options. The format of the IPXCP configuration options is as follows:

Type	Length	Data

IPXCP configuration options

Type

One-byte indication of the type of the configuration option.

1	IPX Network number.
2	IPX Node number.
3	IPX Compression protocol.
4	IPX Routing protocol.
5	IPX Router name.
6	IPX Configuration complete.

Length

Length of the configuration option including the Type, Length and Data fields.

Data

Value of the Data field.

LEXCP

RFC 1841 http://www.cis.ohio-state.edu/htbin/rfc/rfc1841.html

A LAN extension interface unit is a hardware device installed at remote sites (such as a home office or small branch office) that connects a LAN across a WAN link to a router at a central site. To accommodate this LAN extension interface architecture, a PPP Network Control Protocol was developed: the LAN extension interface protocol, PPP-LEX. The basic functionality of LEX is to encapsulate LAN extension interface control and data packets. Consequently packets can be control packets or data packets. Data packets are described under LEX.

There are two types of LEXCP packets:

- Startup options packet.
- Remote command options packets.

The startup options packet is the first LEX NCP packet that the LAN extension interface unit sends to the host router after the LCP has reached an open state. This required startup options packet configures the LAN extension interface protocol and puts the LEX NCP in an open state.

Remote command options are the LEX NCP packets that control the functioning and statistics gathering of the LAN extension interface protocol.

The format of the LEXCP packet is shown in the following illustration:

Address	Control	Protocol type
Code	Identifier	Length
Options		
1 byte	1 byte	2 bytes

PPP + LEX protocol header

Option-type	Option-length	Option-data
1 byte	1 byte	2 bytes

LEX startup options

Option-type	Option-flags	Option-length	Data
1 byte	1 byte	2 bytes	

LEX remote command options

Address
All-stations address: one-octet PPP-specified field which contains the binary sequence 11111111 (hexadecimal 0xFF). PPP does not assign individual station addresses. The all-stations address must be recognized and received by all devices.

Control
Unnumbered Information (UI) command with the P/F bit set to zero. One-octet PPP-specified field which contains the binary sequence 00000011 (hexadecimal 0x03).

Protocol type
IETF-assigned protocol type value. For control packets this field with take the value 0x8041.

Code
Identifies the type of LCP packet that the LAN extension interface packet is sending. Valid values are as follows:

Startup options:

0x01 Configure-Request.
0x02 Configure-Ack.
0x03 Configure-Nak.
0x04 Configure-Rej.

Remote command options:

0x40 LEX_RCMD_REQUEST packet.
0x41 LEX_RCMD_ACK packet.
0x42 LEX_RCMD_NAK packet.
0x43 LEX_RCMD_REJ packet.

Identifier
Randomly generated value which aids matching requests and replies. It is recommended that a non-zero value be used for the identifier. That is, zero could be used in the future for unsolicited messages from the LAN extension interface unit. Valid values are 0x01-0xFF.

Length
Length of the entire packet in octets, including the Code, Identifier, Length, and Option fields.

Option-type

Identifies the startup option or remote command option being negotiated. Valid values for startup options are as follows:

0x01 MAC Type.
0x03 MAC Address.
0x05 LAN Extension.

Valid values for remote command options are as follows:

0x01 Filter Protocol Type.
0x02 Filter MAC Address.
0x03 Set Priority.
0x04 Disable LAN Extension Ethernet Interface.
0x05 Enable LAN Extension Ethernet Interface.
0x06 Reboot LAN Extension Interface Unit.
0x07 Request Statistics.
0x08 Download Request.
0x09 Download Data.
0x0A Download Status.
0x0B Inventory Request.

Option-length

Specifies the length of the option fields, including the option-type, option-data, option-length and option-flags (for remote command) fields.

Option-data

Data relating to the value specified in the option-type field. The data for startup options is as specified here:

Option-type	*Option-data*
MAC type (0x01)	The most up-to-date value of the MAC type, currently 0x01 for IEEE 802.3/Ethernet with canonical addresses.
MAC address (0x03)	Actual MAC address in IEEE 802.3 canonical format.
LAN extension (0x05)	LAN extension interface software information. 0x01 is the current protocol version supported.

Option-flags

Specifies the remote command option, containing specific actions that must be followed. This field is only present in remote command option packets.

NBFCP

RFC 2097 http://www.cis.ohio-state.edu/htbin/rfc/rfc2097.html

The PPP NetBIOS Frames Control Protocol (NBFCP) is a network control protocol responsible for establishing and configuring the NBF protocol over PPP. The NBFCP protocol is only applicable to an end system connecting to a peer system, or the LAN to which the peer system is connected.

The format of the NBFCP packet is shown in the following illustration:

Code	Identifier	Length	Data
1 byte	1 byte	2 bytes	variable

NBFCP packet structure

Code

Decimal value which indicates the type of NBFCP packet.

1 Configure-Request.
2 Configure-Ack.
3 Configure-Nak.
4 Configure-Reject.
5 Terminate-Request.
6 Terminate-Ack.
7 Code-Reject.

Identifier

Decimal value which aids in matching requests and replies.

Length

Length of the NBFCP packet, including the Code, Identifier, Length and Data fields.

Data

Variable length field which may contain one or more configuration options. The format of the NBFCP configuration options is as follows:

Type	Length	Data

NBFCP configuration options

Type

One-byte indication of the type of the configuration option.

1 Name-Projection.
2 Peer-Information.
3 Multicast-Filtering.
4 IEEE-MAC-Address-Required.

Length

Length of the configuration option including the Type, Length and Data fields.

Data

Value of the data field.

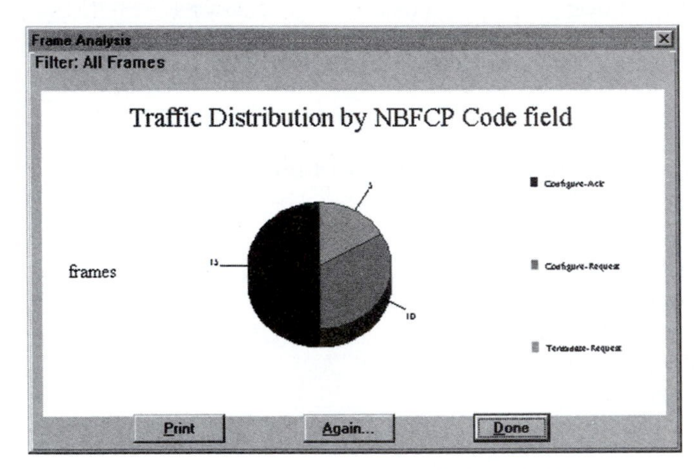

Graph of traffic distribution by NBFCP code field

OSINLCP

RFC 1377 http://www.cis.ohio-state.edu/htbin/rfc/rfc1377.html

The OSI Network Layer Control Protocol (OSINLCP) is responsible for configuring, enabling, and disabling the OSI protocol modules on both ends of the point-to-point link. OSINLCP uses the same packet exchange mechanism as the Link Control Protocol (LCP). OSINLCP packets may not be exchanged until PPP has reached the Network-Layer Protocol phase.

The format of the header is shown in the following illustration:

Code	Identifier	Length	Data
1 byte	1 byte	2 bytes	variable

OSINLCP header structure

Code

A one-octet field identifying the type of OSINLCP packet. When a packet is received with an unknown Code field, a Code Reject packet is transmitted.

1 Configure-Request.
2 Configure-Ack.
3 Configure-Nak.
4 Configure-Reject.
5 Terminate-Request.
6 Terminate-Ack.
7 Code-Reject.

Identifier

Decimal value which aids in matching requests and replies.

Length

Length of the OSINLCP packet, including the Code, Identifier, Length and Data fields.

Data

Variable length field which may contain one or more configuration options.

SDCP

RFC 1963 http://www.cis.ohio-state.edu/htbin/rfc/rfc1963.html

The PPP Serial Data Control Protocol (SDCP) is responsible for configuring, enabling and disabling the SDTP (Serial Data Transport Protocol) modules on both ends of the point-to-point link. SDCP uses the same packet exchange mechanism and state machine as the Link Control Protocol. SDCP packets may not be exchanged until PPP has reached the Network-Layer Protocol phase.

The format of the header is shown in the following illustration:

Code	Identifier	Length	Data
1 byte	1 byte	2 bytes	variable

SDCP header structure

Code

A one octet field identifying the type of SDCP packet. When a packet is received with an unknown Code field, a Code Reject packet is transmitted.

1 Configure-Request.
2 Configure-Ack.
3 Configure-Nak.
4 Configure-Reject.
5 Terminate-Request.
6 Terminate-Ack.
7 Code-Reject.

Identifier

Decimal value which aids in matching requests and replies.

Length

Length of the SDCP packet, including the Code, Identifier, Length and Data fields.

Data

Variable length field which may contain one or more configuration options. The format of the SDCP configuration options is as follows:

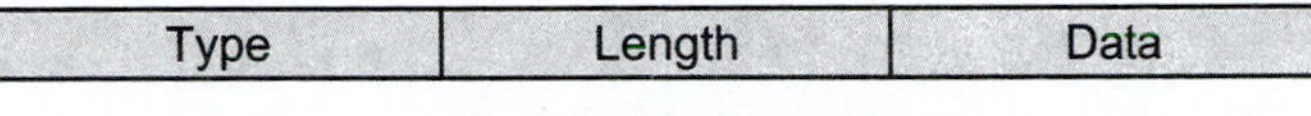

SDCP configuration options

Type

One-byte indication of the type of the configuration option.

1 Packet-Format.
2 Header-Type.
3 Length-Field-Present.
4 Multi-Port.
5 Transport-Mode.
6 Maximum-Frame-Size.
7 Allow-Odd-Frames.
8 FCS-Type.
9 Flow-Expiration-Time.

Length

Length of the configuration option including the Type, Length and Data fields.

Data

Value of the Data field.

SNACP

RFC 2043 http://www.cis.ohio-state.edu/htbin/rfc/rfc2043.html

The SNA PPP Control Protocol (SNACP) is responsible for configuring, enabling, and disabling SNA on both ends of the point-to-point link. Note that there are actually two SNA Network Control Protocols; one for SNA over LLC 802.2 and another for SNA without LLC 802.2. These SNA NCPs are negotiated separately and independently of each other. In order to establish communications over a point-to-point link, each end of the PPP link must first send LCP packets to configure and test the data link. After the link has been established and optional facilities have been negotiated as needed by the LCP, PPP must send SNACP packets to choose and configure the SNA network-layer protocol. Once SNACP has reached the opened state, SNA datagrams can be sent over the link.

The format of the SNACP packet is shown in the following illustration:

Code	Identifier	Length	Data
1 byte	1 byte	2 bytes	variable

SNACP packet structure

Code

Decimal value which indicates the type of SNACP packet.

1 Configure-Request.
2 Configure-Ack.
3 Configure-Nak.
4 Configure-Reject.
5 Terminate-Request.
6 Terminate-Ack.
7 Code-Reject.

Identifier

Decimal value which aids in matching requests and replies.

Length

Length of the SNACP packet, including the Code, Identifier, Length and Data fields.

Data

Variable length field.

26

SMDS

SMDS (Switched Multimegabit Data Service) is a broadband networking technology developed by Bellcore. It is a subset of IEEE 802.6 DQDB (Distributed Queue Dual Bus) MAN technology which was developed to be a high-speed, connectionless, public, packet-switching service. SMDS currently offers access at rates up to DS-3 or 44.736 Mbps, with plans to increase these rates to 155.520 Mbps with OC-3c. It operates by accepting high-speed customer data in increments of up to 9,188 octets, and divides it into 53-octet cells for transmission through the service provider's network. These cells are reassembled, at the receiving end, into the customer data.

SIP is a three-level protocol that controls the customer's access to the network. SIP Level 3 receives and transports frames of the upper layer protocol information. SIP Level 2, based on IEEE 802.6 DQDB standard, controls access to the physical medium. SIP Level 1 includes the PLCP and the transmission system.

The following diagram shows SMDS in relation to the OSI model:

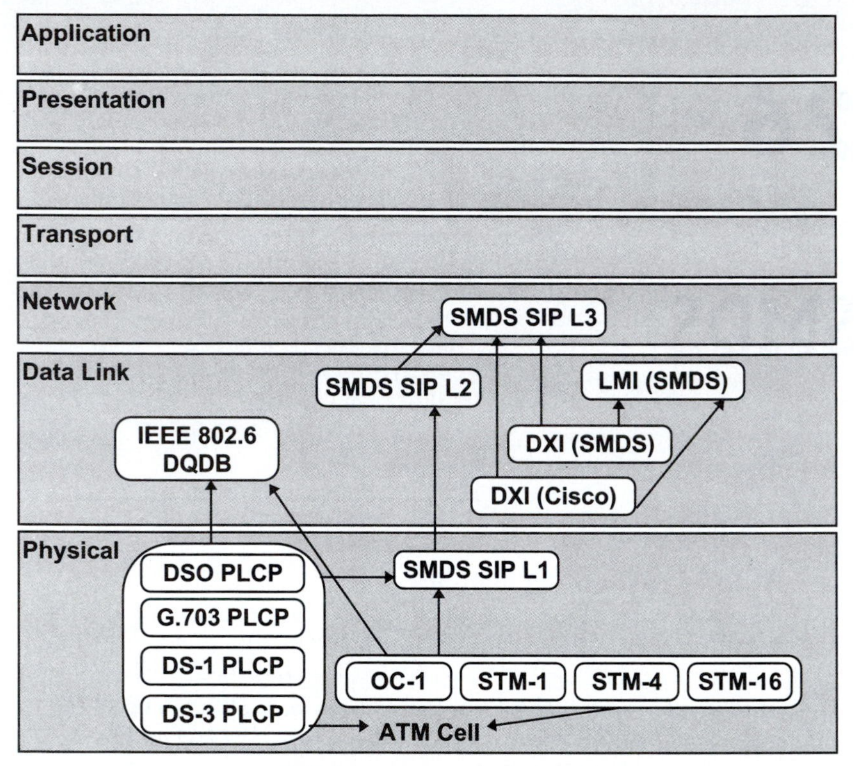

SMDS in relation to the OSI model

SIP Level 3

The SMDS SDU, which contains 9188 octets of information is passed from the upper layer protocols to the SIP Level 3 for transmission over the network. SIP Level 3 builds a L3 PDU which includes a header and trailer as shown below. The L3 PDU is then passed to SIP Level 2, where it is segmented into multiple L2 PDUs, each 53 octets in length. These PDUs are then passed to the PLCP and finally to the physical transmission medium.

The SIP Level 3 PDU is shown in the following diagram.

Header	Information	PAD	X + CRC32	Trailer	
36	≤ 9188	0-3	0,4	4	Bytes

SIP level 3 PDU

The format of the level 3 header is as follows:

6	8	12	13	16 bits
Reserved		BEtag		
BAsize				
Destination address (8 bytes)				
Source address (8 bytes)				
X+HLPI (6 bits)	PL	X+QoS	CIB	HEL
X+Bridging				
HE (12 bytes)				

SIP level 3 header structure

And the format of the level 3 trailer is as follows:

Reserved	Betag	BAsize
1 byte	1 byte	2 bytes

SIP level 3 trailer

The Level 3 PDU fields are described as follows:

Reserved

Reserved. A 1-octet field that the CPE and the SS fill with zeros.

BEtag

A 1-octet field that contains a beginning/end tag. This is a binary number with a value between 0-255 that forms an association between the first segment (containing the header) and the last segment (containing the trailer) of a Level 3 PDU.

BAsize

A 2-octet field containing the length in octets of the Level 3 PDU from the beginning of the Destination Address field and including the CRC32 field, if present.

Destination address

An 8-octet field containing the address of the intended recipient of this PDU. This field is divided into two subfields:

Address Type: 4 most significant bits indicate whether this is an Individual address (1100) or a Group address (1110).

Address: remaining 60 bits is the actual SMDS address.

Source address

An 8-octet field containing the address of the sender of this PDU. This field contains Address Type and Address subfields as described for Destination Address.

HLPI

Higher Layer Protocol Identifier. A 6-bit field that aligns the SIP and DQDB protocol formats.

PL

PAD Length. A 2-bit field that indicates the number of octets in the PAD field, which aligns the Level 3 PDU on a 32-bit boundary.

QoS

Quality of Service. A 4-bit field that aligns the SIP and DQDB protocol formats.

CIB

CRC32 Indication Bit. A 1-bit field that indicates the presence (1) or absence (0) of the CRC32 field.

HEL

Header Extension Length. A 3-bit field that indicates the number of 32-bit words in the Header Extension field.

Bridging

A 2-octet field that aligns the SIP and DQDB Bridging protocol formats.

HE

Header Extension. A 12-octet field that contains the version and carrier-selection information presented in a variable number of subfields:

Element Length: 1-octet subfield containing the combined length of Element Length, Element Type and Element Value fields, in octets.

Element Type: 1-octet subfield containing a binary value indicating the type of information found in the Element Value field.

Element Value: variable-length field with a value that depends on the Element Type and its function.

HE PAD: variable-length field, 0-9 octets in length, that assures the length of the HE field is 12 octets.

Information field

Variable-length field, up to 9,188 octets in length, that contains user information.

PAD

Variable-length field, 1-3 octets in length, filled with zeros aligning the entire PDU on a 32-bit boundary.

CRC32

2-octet field that performs error detection on the PDU, beginning with the DA field, up to and including the CRC32 field.

SIP Level 2

When the Level 3 PDU processing is complete, it is passed to the SIP Level 2 to create one or more Level 2 PDUs. The SIP Level 2 generates the 53-octet cells which are transmitted over the PLCP and physical transmission medium. The SIP Level 2 PDU contains a 5-octet header, a 44-octet Segmentation Unit (payload) and a 2-octet trailer as shown below.

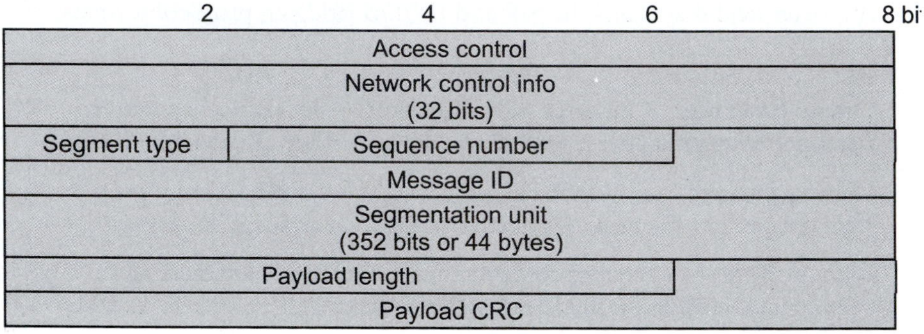

2 4 6 8 bi

Access control

Network control info
(32 bits)

Segment type | Sequence number

Message ID

Segmentation unit
(352 bits or 44 bytes)

Payload length

Payload CRC

SIP level 2 PDU

The Level 2 PDU fields are described as follows:

Access control

8-bit field that indicates whether the Level 2 PDU Access Control contains information (1) or is empty (0).

Network control info

4-octet field that determines whether Network Control Information of the Level 2 PDU contains information (FFFFF022H) or is empty (0).

Segment type

2-bit field that indicates how the receiver should process non-empty Level 2 PDUs. Possible values are:

00 Continuation of Message (COM).
01 End of Message (EOM).
10 Beginning of Message (BOM).
11 Single Segment Message (SSM).

Sequence number

4-bit number that verifies that all the Level 2 PDUs belonging to a single Level 3 PDU have been received in the correct order.

Message identifier

10-bit number that allows the various segments to be associated with a single Level 3 PDU.

Segmentation unit

44-octet field that contains a portion of the Level 3 PDU.

Payload length

6-bit field that indicates which of the 44 octets in the Segmentation Unit contain actual data. BOM and COM segments always indicate 44 octets. EOM segments indicate between 4 and 44 octets, in multiples of 4 octets. SSM segments indicate between 28 and 44 octets, in multiples of 4 octets.

Payload CRC

10-bit field that performs error detection on the Segment Type, Sequence Number, Message Identifier, Segmentation Unit, Payload Length and Payload CRC fields.

Once assembled, SIP Level 2 PDUs are passed to the PLCP and physical functions within SIP Level 1 for transmission.

SIP Level 1

The SIP Level 1 transmits the Level 2 PDUs generated at SIP Level 2. Transmission functions are divided into two sublayers, an upper PLCP sublayer and a lower transmission system sublayer. The PLCP sublayer interfaces to the SIP Level 2 functions and supports the transfer of data and control information. The transmission system sublayer defines characteristics such as the format and speed for transmitted data. The two most common implementations for this layer are based on DS-1 and DS-3 technologies and standards.

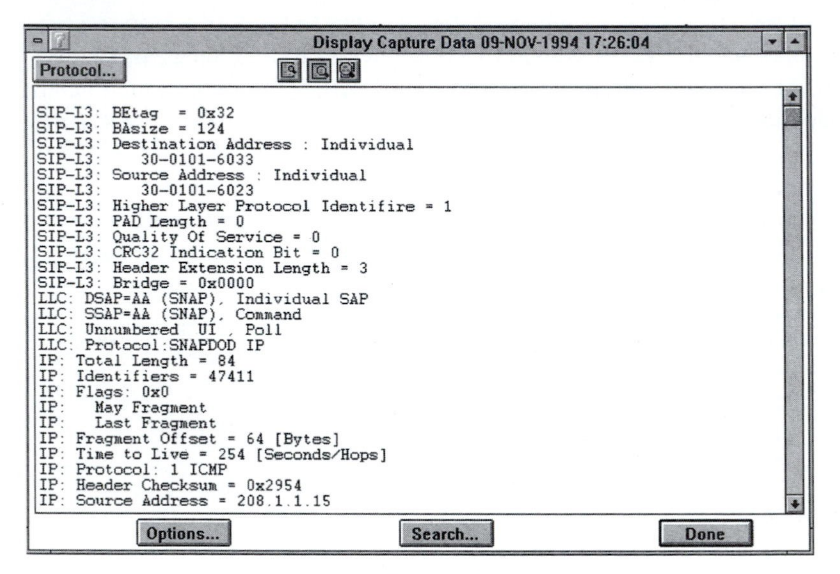

SMDS decode

27

SS7 Suite

CCITT developed the Signalling System 7 (SS7) specification. SS7 is a common channel signalling system. This means that one channel is used only for sending the signalling information, whether the system has one bearer channel or multiple bearer channels. The hardware and software functions of the SS7 protocol are divided into layers which loosely correspond to the OSI 7 layer model.

This chapter describes the following SS7 protocols:

- MTP-2: Message Transfer Part Level 2.
- MTP-3: Message Transfer Part Level 3.
- SCCP: Signalling Connection Control Part.
- DUP: Data User Part.
- ISUP: ISDN User Part.
- TUP: Telephone User Part.
- TCAP: Transaction Capabilities Application Part.
- MAP: Mobile Application Part.

The following diagram illustrates the SS7 protocol suite in relation to the OSI model:

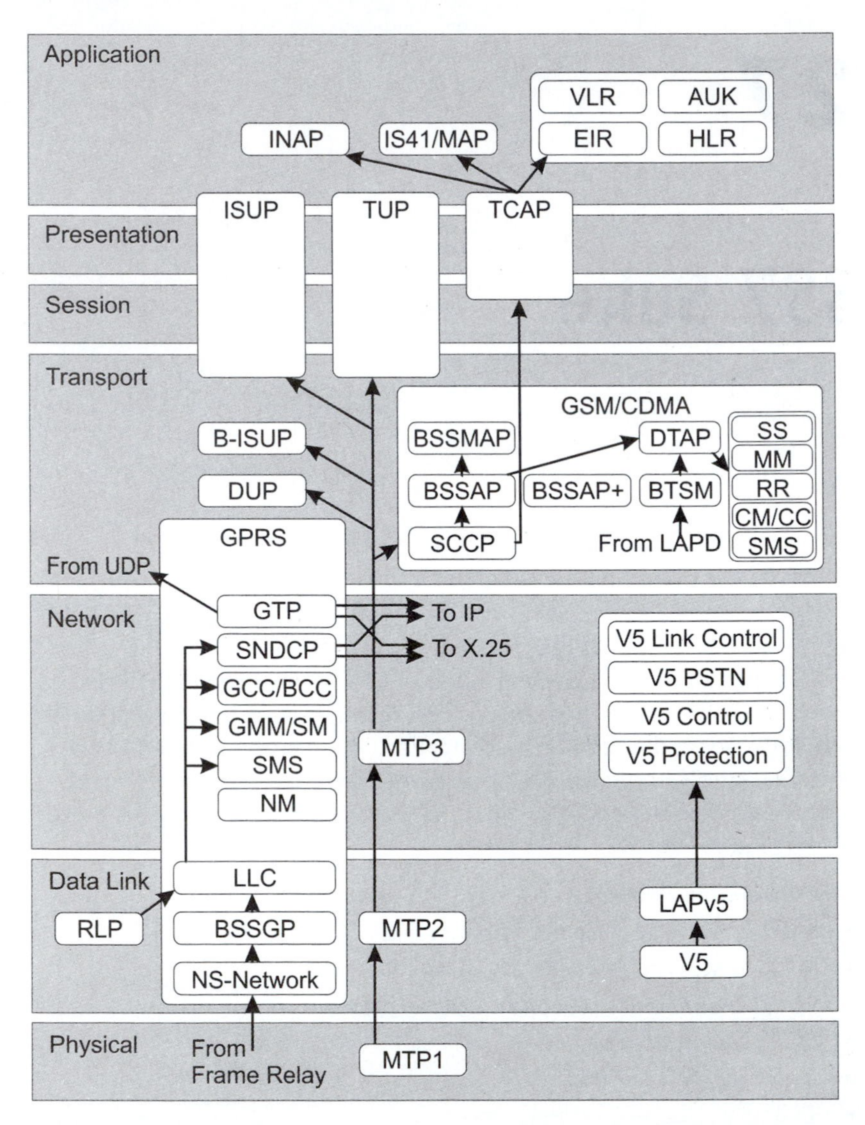

SS7 in relation to the OSI model

MTP-3

Q.704 http://www.itu.ch/itudoc/itu-t/rec/q/q500-999/q704_27792.html

Message Transfer Part - Level 3 (MTP-3) connects Q.SAAL to the users. It transfers messages between the nodes of the signalling network. MTP-3 ensures reliable transfer of the signalling messages, even in the case of the failure of the signalling links and signalling transfer points. The protocol therefore includes the appropriate functions and procedures necessary both to inform the remote parts of the signalling network of the consequences of a fault, and appropriately reconfigure the routing of messages through the signalling network.

The structure of the MTP-3 header is shown in the following illustration:

Service indicator	Subservice field
4 bits	4 bits

MTP-3 header structure

Service indicator

Used to perform message distribution and in some cases to perform message routing. The service indicator codes are used in international signalling networks for the following purposes:

- Signalling network management messages.
- Signalling network testing and maintenance messages.
- SCCP.
- Telephone user part.
- ISDN user part.
- Data user part.
- Reserved for MTP testing user part.

Sub-service field

The sub-service field contains the network indicator and two spare bits to discriminate between national and international messages.

MTP-2

Q.703 http://www.itu.int/itudoc/itu-t/rec/q/q500-999/q703_24110.html
ANSI T1.111 199

Message Transfer Part - Level 2 (MTP-2) is a signalling link which together with MTP-3 provides reliable transfer of signalling messages between two directly connected signalling points.

The format of the header is shown in the following illustration:

7	8 bits
Flag	
BSN (7 bits)	BIB
FSN (7 bits)	FIB
LI (6 + 2 bits)	
SIO	
SIF	
Checksum (16 bits)	
Flag	

MTP-2 header structure

BSN

Backward sequence number. Used to acknowledge message signal units which have been received from the remote end of the signalling link.

BIB

Backward indicator bit. The forward and backward indicator bit together with forward and backward sequence number are used in the basic error control method to perform the signal unit sequence control and acknowledgment functions.

FSN

Forward sequence number.

FIB

Forward indicator bit.

LI
Length indicator. This indicates the number of octets following the length indicator octet.

SIO
Service information octet.

SIF
Signalling information field.

Checksum
Every signal unit has 16 check bits for error detection.

SCCP

Q.713 http://www.itu.int/itudoc/itu-t/rec/q/q500-999/q713_23786.html
ANSI T1.112

The Signalling Connection Control Part (SCCP) offers enhancements to MTP level 3 to provide connectionless and connection-oriented network services, as well as to address translation capabilities. The SCCP enhancements to MTP provide a network service which is equivalent to the OSI Network layer 3.

The format of the header is shown in the following illustration:

	Octets
Routing label	3-4
Message type code	1
Mandatory fixed part	
Mandatory variable part	
Optional part	

SCCP header structure

Routing label

A standard routing label.

Message type code

A one octet code which is mandatory for all messages. The message type code uniquely defines the function and format of each SCCP message. Existing Message Type Codes are:

CR Connection Request.
CC Connection Confirm.
CREF Connection Refused.
RLSD Released.
RLC Release Complete.
DT1 Data Form 1.
DT2 Data Form 2.
AK Data Acknowledgment.
UDT Unidata.
UDTS Unidata Service.
ED Expedited Data.
EA Expedited Data Acknowledgment.

RSR Reset Request.
RSC Reset Confirm.
ERR Protocol Data Unit Error.
IT Inactivity Test.
XUDT Extended Unidata.
XUDTS Extended Unidata Service.

Mandatory fixed part

The parts that are mandatory and of fixed length for a particular message type will be contained in the mandatory fixed part.

Mandatory variable part

Mandatory parameters of variable length will be included in the mandatory variable part. The name of each parameter and the order in which the pointers are sent is implicit in the message type.

Optional part

The optional part consists of parameters that may or may not occur in any particular message type. Both fixed length and variable length parameters may be included. Optional parameters may be transmitted in any order. Each optional parameter will include the parameter name (one octet) and the length indicator (one octet) followed by the parameter contents.

DUP

TU-T recommendation X.61 (Q.741)
http://www.itv.int/itudoc/itu-t/rec/q/q500-999/q741.html

The Data User Part (DUP) defines the necessary call control, facility registration and cancellation related elements, for international common channel signalling, by using SS7 for circuit-switched data transmission services. The data signalling messages are divided into two categories:

- Call and circuit related messages: used to set up and clear a call or control and supervise the circuit state.
- Facility registration and cancellation related messages: used to exchange information between originating and destination exchanges to register and cancel information related to user facilities.

The general format of the header of call and circuit related messages is shown in the following illustration:

8 14 16 28 32 bits

DPS		OPC		BIC
BIC	TSC		HC	Parameters

DUP call and circuit related message structure

The general format of the header of Facility Registration and Cancellation Messages is shown in the following illustration:

8 14 16 28 32 bits

DPS		OPC	Spare
HC	Parameters		

DUP facility registration and cancellation message structure

Routing label

The label contained in a signalling message and used by the relevant user part to identify particulars to which the message refers. This is also used by the message transfer part to route the message towards its destination point. It contains the DPS, OPC, BIC and TSC.

DPS

Destination Point code.

OPC
Origination Point Code.

BIC
Bearer Identification Code.

TSC
Time Slot Code.

HC
Heading code, contains the message type code which is mandatory for all messages. It uniquely defines the function and format of each Data User Part message.

Parameters
Contains specific fields for each message. Variable length.

Spare bits
Not used, should be set to 0000.

ISUP

Q.763 http://www.itu.int/itudoc/itu-t/rec/q/q500-999/q763_23976.html

ISUP is the ISDN User Part of SS7. ISUP defines the protocol and procedures used to setup, manage and release trunk circuits that carry voice and data calls over the public switched telephone network. ISUP is used for both ISDN and non-ISDN calls. Calls that originate and terminate at the same switch do not use ISUP signalling. ISDN User Part messages are carried on the signalling link by means of signal units. The signalling information field of each message signal unit contains an ISDN User Part message consisting of an integral number of octets.

The format of the ISUP packet is shown in the following illustration:

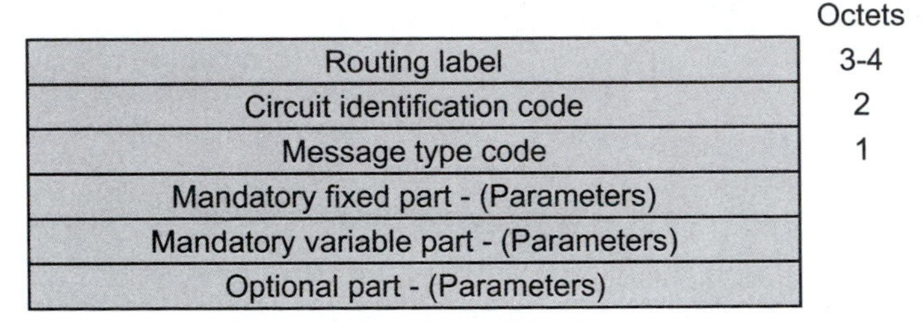

ISUP packet structure

Routing label

The label contained in a signalling message, and used by the relevant user part to identify particulars to which the message refers. It is also used by the Message Transfer Part to route the message towards its destination point.

Circuit identification code

The allocation of circuit identification codes to individual circuits is determined by bilateral agreement and/or in accordance with applicable predetermined rules.

Message type code

The message type code consists of a one octet field and is mandatory for all messages. The message type code uniquely defines the function and format of each ISDN User Part message. Each message consists of a number of parameters. Message types may be:

- Address complete.

- Answer.
- Blocking.
- Blocking acknowledgement.
- Call progress.
- Circuit group blocking.
- Circuit group blocking acknowledgement.
- Circuit group query @.
- Circuit group query response @.
- Circuit group reset.
- Circuit group reset acknowledgement.
- Circuit group unblocking.
- Circuit group unblocking acknowledgement.
- Charge information @.
- Confusion.
- Connect.
- Continuity.
- Continuity check request.
- Facility @.
- Facility accepted.
- Facility reject.
- Forward transfer.
- Identification request.
- Identification response.
- Information @.
- Information request @.
- Initial address.
- Loop back acknowledgement.
- Network resource management.
- Overload @.
- Pass-along @.
- Release.
- Release complete.
- Reset circuit.
- Resume.
- Segmentation.
- Subsequent address.

- Suspend.
- Unblocking.
- Unblocking acknowledgement.
- Unequipped CIC @.
- User Part available.
- User Part test.
- User-to-user information.

Parameters

Each parameter has a name which is coded as a single octet. The length of a parameter may be fixed or variable, and a length indicator for each parameter may be included.

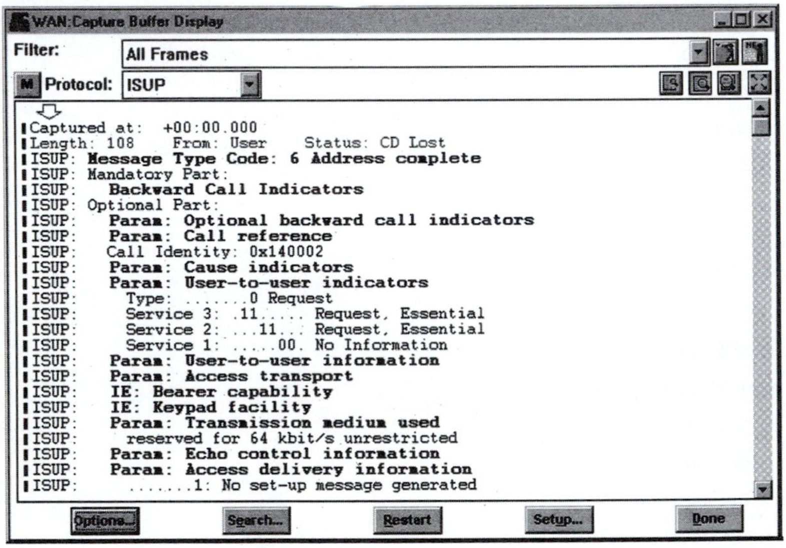

```
WAN:Capture Buffer Display
Filter: All Frames
Protocol: ISUP
Captured at:  +00:00.000
Length: 108     From: User     Status: CD Lost
ISUP: Message Type Code: 6 Address complete
ISUP: Mandatory Part:
ISUP:   Backward Call Indicators
ISUP: Optional Part:
ISUP:   Param: Optional backward call indicators
ISUP:   Param: Call reference
ISUP:   Call Identity: 0x140002
ISUP:   Param: Cause indicators
ISUP:   Param: User-to-user indicators
ISUP:     Type: .......0 Request
ISUP:     Service 3: .11..... Request, Essential
ISUP:     Service 2: ...11... Request, Essential
ISUP:     Service 1: .....00. No Information
ISUP:   Param: User-to-user information
ISUP:   Param: Access transport
ISUP:   IE: Bearer capability
ISUP:   IE: Keypad facility
ISUP:   Param: Transmission medium used
ISUP:     reserved for 64 kbit/s unrestricted
ISUP:   Param: Echo control information
ISUP:   Param: Access delivery information
ISUP:     .......1: No set-up message generated
Options...  Search...  Restart  Setup...  Done
```

ISUP decode

TUP

ITU-T recommendation q.723
http://www.itu.int/itudoc/itu-t/rec/q/q500-999/q763_23976.html

In Telephone User Part (TUP) Signalling System no.7, Telephone User messages are carried on the signalling data link by means of signal units. The signalling information of each message constitutes the signalling information field of the corresponding signal unit and consists of an integral number of octets. It consists of the label, the heading code and one or more signals and/or indications. Each of these is described here.

The format of the label is shown in the following illustration:

Field	bits
DPC	14
OPC	14
CIC	12

TUP standard label format

DPC

Destination point code indicates the signalling point for which the message is intended.

OPC

Originating point code indicates the signalling point which is the source of the message.

CIC

Circuit identification code indicates one speech circuit among those directly interconnecting the destination and the originating points.

The heading code consists of two parts, heading code H0 and heading code H1. H0 identifies a specific message group, while H1 either contains a signal code or in the case of more complex messages, identifies the format of these messages.

TCAP

ITU recommendation Q.773
http://www.itu.int/itudoc/itu-t/rec/q/q500-999/q773_24880.html

TCAP (Transaction Capabilities Application Part) enables the deployment of advanced intelligent network services by supporting non-circuit related information exchange between signalling points using the SCCP connectionless service. TCAP messages are contained within the SCCP portion of an MSU. A TCAP message is comprised of a transaction portion and a component portion.

A TCAP message is structured as a single constructor information element consisting of the following: Transaction Portion, which contains information elements used by the Transaction sub-layer; a Component Portion, which contains information elements used by the Component sub-layer related to components; and, optionally, the Dialogue Portion, which contains the Application Context and user information, which are not components. Each Component is a constructor information element.

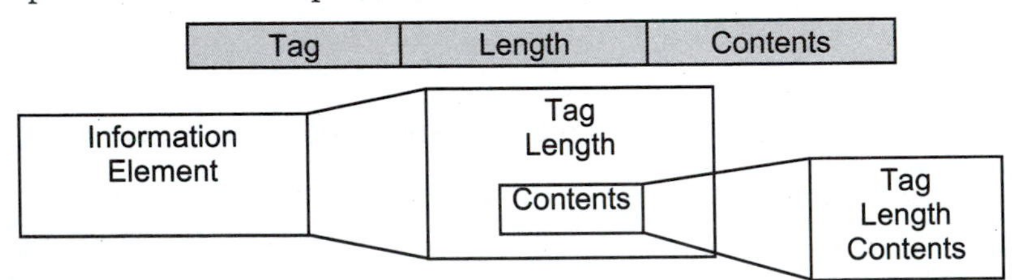

TCAP packet structure

Information element

An information element is first interpreted according to its position within the syntax of the message. Each information element within a TCAP message has the same structure. An information element consists of three fields, which always appear in the following order.

Tag

The Tag distinguishes one information element from another and governs the interpretation of the Contents. It may be one or more octets in length. The Tag is composed of Class, Form and Tag codes.

Length

Specifies the length of the contents field.

Contents

Contains the substance of the element, containing the primary information the element is intended to convey.

TCAP Packet Types

TCAP packet types are as follows:

- Unidirectional.
- Query with permission.
- Query without permission.
- Response.
- Conversation with permission.
- Conversation without permission.
- Abort.

MAP

EIA/TIA-41 http://www.tiaonline.org

Mobile Application Part (MAP) messages sent between mobile switches and databases to support user authentication, equipment identification, and roaming, are carried by TCAP in mobile networks (IS-41 and GSM). When a mobile subscriber roams into a new mobile switching center (MSC) area, the integrated visitor location register requests service profile information from the subscriber's home location register (HLR) using MAP information carried within TCAP messages.

The packet consists of a header followed by up to four information elements. The general format of the header is shown in the following illustration. The operation specifier field, when present and length field are actually part of the TCAP message which is above MAP.

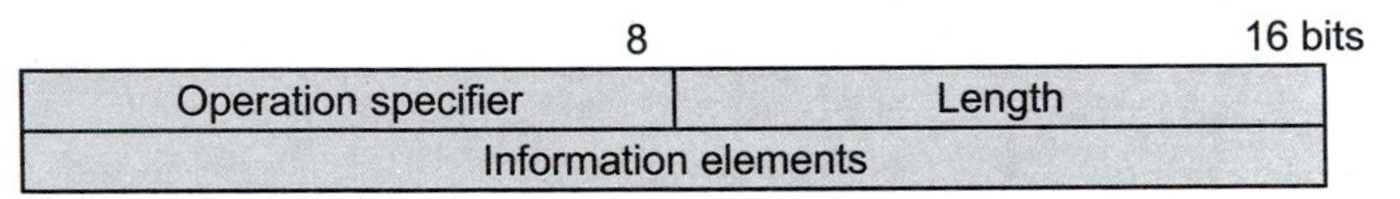

MAP header structure

Operation specifier

Optional field which specifies the type of packet. The following operations are specified:

- AuthenticationDirective.
- AuthenticationDirectiveForward.
- AuthenticationFailureReport.
- AuthenticationRequest.
- AuthenticationStatusReport.
- BaseStationChallenge.
- Blocking.
- BulkDeregistration.
- CountRequest.
- FacilitiesDirective.
- FacilitiesDirective2.
- FacilitiesRelease.
- FeatureRequest.

- FlashRequest.
- HandoffBack.
- HandoffBack2.
- HandoffMeasurementRequest.
- HandoffMeasurementRequest2.
- HandoffToThird.
- HandoffToThird2.
- InformationDirective.
- InformationForward.
- InterSystemAnswer.
- InterSystemPage.
- InterSystemPage2.
- InterSystemSetup.
- LocationRequest.
- MobileOnChannel.
- MSInactive.
- OriginationRequest.
- QualificationDirective.
- QualificationRequest.
- RandomVariableRequest.
- RedirectionDirective.
- RedirectionRequest.
- RegistrationCancellation.
- RegistrationNotification.
- RemoteUserInteractionDirective.
- ResetCircuit.
- RoutingRequest.
- SMSDeliveryBackward.
- SMSDeliveryForward.
- SMSDeliveryPointToPoint.
- SMSNotification.
- SMSRequest.
- TransferToNumberRequest.
- TrunkTest.
- TrunkTestDisconnect.
- Unblocking.

- UnreliableRoamerDataDirective.
- UnsolicitedResponse.

Length

The length of the packet.

Information elements

Various information elements which depend on the operation specified.

28

Sun Protocols

The Sun protocols are UNIX-based applications that use TCP/IP for transport. These protocols are based on the client-server model of networking, where client workstations make procedure calls to network servers where applications reside. The Sun suite includes the following protocols:

- MOUNT.
- NFS: Network File System.
- PMAP: Port Mapper.
- RPC: Remote Procedure Call.
- YP (NIS): Yellow Pages (Network Information Services).

The following diagram illustrates the Sun protocol suite in relation to the OSI model:

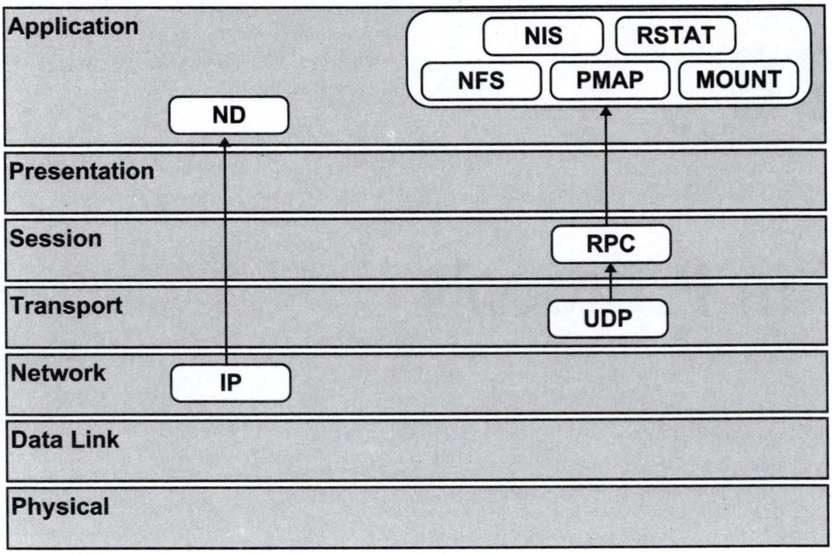

Sun protocol suite in relation to the OSI model

MOUNT

The MOUNT protocol is used to initiate client access to a server supporting the Network File System (NFS) application. It handles local operating system specifics such as path name format and user authentication. Clients desiring access to the NFS program first call the MOUNT program to obtain a file handle suitable for use with NFS.

Frames

MOUNT frames may be one of the following types:

[no operation]	Performs no operation.
[add mount entry]	Requests file handle for NFS.
[get all mounts]	Requests list of mounts for the client.
[del mount entry]	Requests deletion of mount entry.
[del all mounts]	Requests removal of mounts for client.
[get export list]	Requests group permissions for mounts.
[mount added]	Supplies file handle for use with NFS.
[give all mounts]	Supplies list of mounts for the client.
[mount deleted]	Confirms deletion of mount entry.
[mounts deleted]	Confirms deletion of all client mounts.
[give export list]	Lists group permissions for mounts.

Frame Parameters

MOUNT frames can have the following parameters:

Path name

The server path name of the mounted directory displayed in double quotes.

File handle

The 32-byte file handle used to access the mounted directory.

Failure to mount on a UNIX system can lead to the display of the error message UNIX Error xxxx, where xxxx is a standard UNIX operating system error.

NFS

The Sun Network File System (NFS) protocol is the file sharing application for the Sun protocol suite. NFS allows distributed file resources to appear as a single-file system to clients. UNIX (Sun OS) is the standard platform for NFS, but Sun designed the file access procedures in NFS to be sufficiently generic to allow porting to a wide variety of operating systems and protocols.

Frames

NFS frames may be one of the following types:

[no operation]	Performs no operation.
[get file attrib]	Request for file attributes.
[set file attrib]	Attempts to set file attributes.
[get filsys root]	Request for the root file handle.
[search for file]	Requests a search for the specified file.
[read from link]	Requests a read from a symbolic link.
[read from file]	Requests a read from a file.
[write to cache]	Requests a write to a cache.
[write to file]	Requests a write to a file.
[create file]	Requests creation of the specified file.
[delete file]	Requests deletion of a file.
[rename file]	Requests renaming of a file.
[link to file]	Requests creation of a file link.
[make symb link]	Requests creation of a symbolic link.
[make directory]	Requests creation of a directory.
[remove directry]	Requests deletion of a directory.
[read directory]	Requests a directory listing.
[get filsys attr]	Requests information on the file system.
[give file attrib]	Returns attributes for a file.
[file attrib set]	Confirms setting of file attributes.
[filesystem root]	Returns the root file handle.
[file search done]	Returns result of file search.
[link read done]	Returns result of a symbolic link read.
[file read done]	Returns result of a file read.
[cache write done]	Returns result of a cache write.
[file write done]	Returns result of a file write.
[create file rep]	Returns file creation status.

[delete file rep]	Returns file deletion status.
[rename file rep]	Returns file renaming status.
[link file reply]	Returns file linking status.
[symb link made]	Returns symbolic link creation status.
[make dir reply]	Returns directory creation status.
[remove dir reply]	Returns directory deletion status.
[dir read done]	Returns directory listing.
[filsystem attrib]	Returns file system attributes.

Frame Parameters

Sun NFS frames can contain the following parameters:

File handle
32-byte file handle used to access a file.

Number
Number of bytes read or written in file operations.

Offset
Beginning file offset for read and write operations.

File type
Includes the following file types:

NON	Non-file.
REG	Regular file.
DIR	Directory.
BLK	Block device.
CHR	Character device.
LNK	Symbolic link.

Mode
File access mode is represented in four fields as Mode/Owner/Group/Others and can include the following:

Field	Parameter	Explanation
Mode	U	Executable by specifying user ID.
	G	Executable by specifying group ID.
	S	Save as executable after use.
Owner	R	Read permission.
	W	Write/delete permission.
	X	Execute/search permission.

Field	Parameter	Explanation
Group	r	Read permission.
	w	Write/delete permission.
	x	Execute/search permission.
Others	r	Read permission.
	w	Write/delete permission.
	x	Execute/search permission.

Links
Number of hard links or filenames for the file.

Size
File size in bytes.

Block
Block size used for file storage.

File system ID
Identifying code of the file system.

File ID
Protocol ID for the file.

Cookie pointer
Pointer to first directory listing entry requested.

Maximum
Maximum number of entries to return for this request.

Read device
For block and character device, reads the number of the read device.

In addition to the above parameters, text strings such as file names are displayed in double quotes.

Response Frames

NFS response frames contain a completion status as indicated below:

{OK}	Completed with no errors.
{Ownership required}	Ownership required for operation.
{File/dir not found}	File or directory not found.
{Device error}	System or hardware error.

{Device/addr not found}	Device or I/O address error.
{Insuff access rights}	Insufficient access rights.
{File/dir already exists}	Duplicate file or directory.
{Device not found}	Device access error.
{Not a directory}	Operation only valid on directory.
{Invalid dir operation}	Operation not valid on directory.
{File too large}	File size too large.
{Out of disk space}	File device out of space.
{Write protect violation}	File system is read-only.
{Filename too long}	File name too long.
{Directory not empty}	Directory not empty.
{Disk quota exceeded}	User disk quota exceeded.
{Invalid file handle}	File handle invalid.
{Write cache was flushed}	Write cache flushed to disk.
{Unknown error}	Unknown error type.

PMAP

The Port Mapper (PMAP) protocol manages the allocation of transport layer ports to network server applications. Server applications obtain a port by requesting a port assignment. Clients wanting to access an application first call the PMAP program (on a well-known port) to obtain the transport port registered to the application. The client then calls the application directly using the registered port. Use of PMAP eliminates the need to reserve permanently a port number for each application. Only the PMAP application itself requires a reserved port.

Frames

PMAP frames may be one of the following types:

[no operation]	Performs no operation.
[set port number]	Attempt to register an application.
[unset port numb]	Attempt to unregister an application.
[get port number]	Requests registered port number.
[get all ports]	Requests all registered ports.
[call program]	Direct call to a registered application.
[port assigned]	Registers an application to a port.
[port unassigned]	Unregisters an application.
[give port number]	Informs client of registered port.
[give all ports]	Informs client of all registered ports.
[program called]	Returns information from called program.

Frame Parameters

PMAP frames can contain the following parameters:

Transport port

Transport layer port assigned to the specified application.

Program number

Application program number.

Program version

Version number of the program.

Transport protocol

Transport layer protocol the program is registered to use.

Procedure

Procedure number within the called program.

If the program number is unregistered, the message {Program is unregistered} is displayed.

RPC

RFC 1057 http://www.cis.ohio-state.edu/htbin/rfc/rfc1057.html

The Remote Procedure Call (RPC) protocol is a session layer protocol that uses the XDR language in order to establish a remote procedure call. The RPC call message includes the procedures parameters and the RPC reply message includes the results. Once the reply message is received, the results of the procedure are extracted and the callers execution is resumed.

There are 2 message types; RPC call and RPC reply. The formats of these messages are shown in the following illustrations:

32	64	Octets
Transaction ID	Message type	8
RPC version	Program	16
Version	Procedure	24
RPC call message body		

RPC call message structure

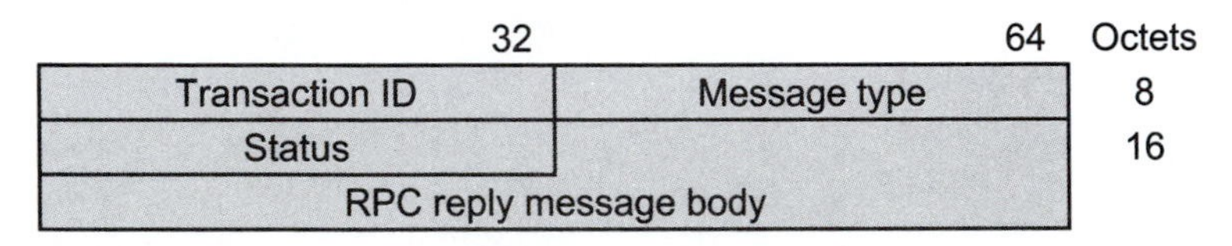

RPC reply message structure

Transaction ID
The transaction number.

Message type
The message type.

RPC version
The version of RPC being used.

Program

The name of the remote program. Programs are identified by number and, where well-known, by name. The well-known program numbers and their respective names are listed below:

Number	Program	Explanation
100000	Port_Mapper	Manages use of transport ports.
100001	RemoteStats	Remote statistics.
100002	RemoteUser	Remote users.
100003	NFS	Network file system.
100004	YellowPages	Yellow pages directory service.
100005	MountDaemon	Mount protocol.
100006	Remote_DBX	Remote DBX.
100007	YP_Binder	Yellow pages binding service.
100008	Shutdown	Shutdown messages.
100009	YP_Password	Yellow pages password server.
100010	Enet_Status	Ethernet status.
100011	Disk_Quotas	Disk quota manager.
100012	SprayPacket	Packet generator.
100013	3270_Mapper	3270 mapping service.
100014	RJE Mapper	Remote job entry mapping service.
100015	Select_Srvc	Selection service.
100016	Remote_DB	Remote database access.
100017	Remote_Exec	Remote execution.
100018	Alice_O/A	Alice office automation program.
100019	Scheduler	Scheduling service.
100020	Local_Lock	Local lock manager.
100021	NetworkLock	Network lock manager.
100022	X.25_INR	X.25 INR protocol.
100023	StatusMon_1	Status monitor 1.
100024	StatusMon_2	Status monitor 2.
100025	Select_Lib	Selection library.
100026	BootService	Boot parameters service.
100027	Mazewars	Mazewars game.
100028	YP_Updates	Yellow pages update service.
100029	Key_Server	Key server.
100030	SecureLogin	Secure login service.
100031	NFS_FwdInit	NFS network forwarding service.
100032	NFS_FwdTrns	NFS forwarding transmitter.
100033	SunLink_MAP	Sunlink MAP.
100034	Net_Monitor	Network monitor.

Number	Program	Explanation
100035	DataBase	Lightweight database.
100036	Passwd_Auth	Password authorization.
100037	TFS_Service	Translucent file service.
100038	NSE_Service	NSE server.
100039	NSE_Daemon	NSE activate process.
150001	PCPaswdAuth	PC password authorization.
200000	PyramidLock	Pyramid lock service.
200001	PyramidSys5	Pyramid Sys5 service.
200002	CADDS_Image	CV CADDS images.
300001	ADTFileLock	ADT file locking service.

Version
The version number of the remote program.

Procedure
The remote procedure being used.

Status
This field makes the identification of error messages possible.

Message body
RPC reply frames can contain the following messages:

{call successful}	Call completed with no errors.
{program unknown}	Program number was not found.
{bad program ver}	Program version was not found.
{proced unknown}	Program procedure was not found.
{bad parameters}	Invalid call parameters were found.
{bad RPC version}	RPC Version not supported.
{bad credentials}	Invalid call credentials supplied.
{restart session}	Request to begin a new session.
{bad verifier}	Invalid verifier was supplied.
{verify rejected}	Verifier has expired or was re-used.
{failed security}	Caller has insufficient privileges.

YP (NIS)

The Sun Yellow Pages (YP) protocol, now known as Network Information Service (NIS), is a directory service used for name look-up and general table enumeration. Each YP database consists of key-value pairs, maps, and domains. The key is the index, e.g., name to which YP maps values such as a phone number. Because keys and values mapped to them can be any arbitrary string of bytes, a YP server can contain a wide variety of database types.

YP defines a set of key-value pairs as a map. Each map belongs to a domain that is a category of maps. This hierarchy of key-value pairs, maps, and domains provides a generic structure for modeling a database of information.

An optional component to a YP server database implementation is the YP binder (YPbind) server. YP uses YP binder servers to provide addressing information about YP database servers to potential clients. A client desiring access to a YP database server can call the YP binder service with a domain name and receive the IP address and transport port of the YP server for that domain. If clients can more easily obtain this information elsewhere, the YP binder service is not needed.

Frames

YP and YPbind frames may be one of the following commands:

[no operation]	Performs no operation.
[domain serve query1]	Asks whether or not the specified domain is served.
[domain serve query2]	Asks only for servers that serve the specified domain.
[get key value]	Asks for value associated with the specified key.
[get first key pair]	Requests the first key-value pair in map.
[get next key pair]	Requests the next key-value pair in map.
[transfer map]	Requests a new copy of the map to be transferred.
[reset YP server]	Requests the YP server to reset its internal state.

[get all keys in map]	Requests all key-value pairs in specified map.
[get map master name]	Requests the name of master YP database server.
[get map number]	Requests the creation time of the specified map.
[get all maps]	Requests all maps in the specified domain.
[domain serve reply1]	Response to domain serve query1.
[give key value]	Returns value for specified key.
[give first key pair]	Returns the first key-value pair in map.
[give next key pair]	Returns the next key-value pair in map.
[map transferred]	Reports map transfer status.
[YP server reset]	Reports server reset status.
[give all keys in map]	Returns listing of all keys in map.
[give map master name]	Returns name of master YP database server.
[give map number]	Returns creation date of map.
[give all maps]	Returns listing of all maps in domain.
[no operation]	Performs no operation.
[get current binding]	Requests YP addressing information for the specified domain.
[set domain binding]	Installs YP addressing information for the specified domain.
[give current binding]	Returns YP addressing information for a domain.
[domain binding set]	Returns status of YP addressing installation.

Frame Parameters

YP and YPbind frames can contain the following parameters:

Bind address
IP address of the YP binder server.

Bind port
Transport port used by the YP binder server.

Created
Creation time of the map.

Domain
Domain name in use.

Key
Key index used to search for a value.

Map
Map name in use.

Master
Name of the master YP database server.

Peer
Server name of a peer YP server.

Transfer ID
Transfer ID used to reference map transfers.

Program
RPC program number used for map transfer.

Port
Transport layer port number used for map transfer.

Value
Value associated with a key.

Status
Map transfer status.

Version
YPbind protocol version.

Response Frames

YP response frames can contain the following status messages:

{OK}	Request completed successfully.
{Bad request arguments}	Request parameters invalid.
{Domain not supported}	Domain not supported by this YP server.
{General failure}	Unspecified failure.
{Invalid operation}	Request invalid.

{No more entries in map}	No more key-value pairs in map.
{No such map in domain}	Specified map not in domain.
{No such key in map}	Specified key not in map.
{Server database is bad}	Server database corrupt.
{YP server error}	Internal YP server error.
{YP version mismatch}	YP server versions do not match.

YPbind response frames can contain the following messages:

{OK}	Request completed successfully.
{Internal error}	Local YP binder error.
{No bound server for domain}	No YP database servers known for the domain.
{Can't alloc system resource}	YP binder resource error.

Transfer Status Frames

YPbind transfer status frames can contain the following messages:

{Transfer successful}	Transfer completed successfully.
{Bad request arguments}	Request parameters invalid.
{Can't clear YP server}	Cannot clear the local YP server.
{Can't find server f/map}	Cannot find YP server for map.
{Can't get master addr.}	Cannot get YP master server address.
{Domain not supported}	Domain not supported by this YP server.
{Local database failure}	Local YP server database failure.
{Local file I/O error}	Local YP server file I/O error.
{Map version mismatch}	Map versions skewed in transfer.
{Master dbase not newer}	Master database is not newer.
{Must override defaults}	Must override default settings.
{Resource alloc failure}	Resource allocation failure.
{RPC to server failed}	No RPC response from server.
{Server/map dbase error}	YP server or map database error.
{Server refused transfer}	YP server refused to transfer database.
{YP transfer error}	Error occurred during database transfer.

29

Tag Switching Protocols

The following tag switching protocols are described in this book:

- TDP - Tag Distribution Protocol.
- MPLS - Multi Protocol Label Switching *see MPLS*

TDP

IETF draft-doolan-tdp-spec-01

TDP (Tag Distribution Protocol) is a two party protocol that runs over a connection oriented transport layer with guaranteed sequential delivery. Tag switching routers use this protocol to communicate tag binding information to their peers. TDP supports multiple network layer protocols including, but not limited to IPv4, IPv6, IPX and AppleTalk. Tag Switching Routers (TSRs) create tag bindings, and then distribute the tag binding information among other TSRs. TDP provides the means for TSRs to distribute, request and release tag binding information for multiple network layer protocols. TDP also provides the means to open, monitor and close TDP sessions and to indicate errors that occur during those sessions. TCP is used as the transport for TDP.

The format of the packet is shown in the following illustration:

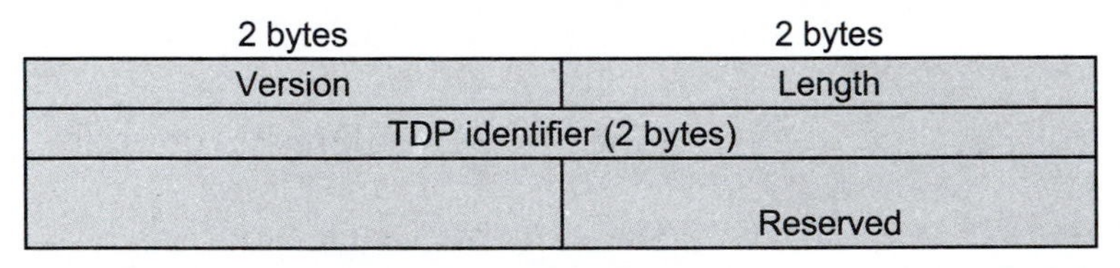

TDP packet structure

Version

The version number of the protocol.

Length

The length in octets of the data portions.

TDP identifier

A unique identifier for the TSR that generated the PDU.

Reserved

A reserved field.

30

TCP/IP Suite

The Defense Advance Research Projects Agency (DARPA) originally developed Transmission Control Protocol/Internet Protocol (TCP/IP) to interconnect various defense department computer networks. The Internet, an international Wide Area Network, uses TCP/IP to connect government and educational institutions across the world. TCP/IP is also in widespread use on commercial and private networks. The TCP/IP suite includes the following protocols:

- IP / IPv6: Internet Protocol.
- TCP: Transmission Control Protocol.
- UDP: User Datagram Protocol.

Data Link Layer

- ARP/RARP: Address Resolution Protocol/Reverse Address.
- DCAP: Data Link Switching Client Access Protocol.

Tunneling protocols

- ATMP: Ascend Tunnel Management Protocol.
- L2F: Layer 2 Forwarding Protocol.
- L2TP: Layer 2 Tunneling Protocol.
- PPTP: Point-to-Point Tunneling Protocol.

Network Layer

- DHCP / DHCPv6: Dynamic Host Configuration Protocol.
- DVMRP: Distance Vector Multicast Routing Protocol.
- ICMP / ICMPv6: Internet Control Message Protocol.
- IGMP: Internet Group Management Protocol.
- MARS: Multicast Address Resolution Server.
- PIM: Protocol Independent Multicast.
- RIP: Routing Information Protocol.
- RIPng for IPv6.
- RSVP: Resource ReSerVation setup Protocol
- VRRP: Virtual Router Redundancy Protocol.

Security

- AH: Authentication Header.
- ESP: Encapsulating Security Payload.

Routing

- BGP-4: Border Gateway Protocol.
- EGP: Exterior Gateway Protocol.
- EIGRP: Enhanced Interior Gateway Routing Protocol.
- GRE: Generic Routing Encapsulation.
- HSRP: Cisco Hot Standby Router Protocol.
- IGRP: Interior Gateway Routing.
- NARP: NBMA Address Resolution Protocol.
- NHRP: Next Hop Resolution Protocol.
- OSPF: Open Shortest Path First.

Transport Layer

- Mobile IP.
- RUDP: Reliable UDP
- TALI: Transport Adapter Layer Interface.
- Van Jacobson: compressed TCP.
- XOT: X.25 over TCP.

Session Layer

- DNS: Domain Name Service.
- NetBIOS/IP.
- LDAP: Lightweight Directory Access Protocol.

Application Layer

- COPS: Common Open Policy Service.
- FTP: File Transfer Protocol.
- TFTP: Trivial File Transfer Protocol.
- Finger: User Information Protocol.
- Gopher: Internet Gopher Protocol.
- HTTP: Hypertext Transfer Protocol.
- S-HTTP: Secure Hypertext Transfer Protocol.
- IMAP4: Internet Message Access Protocol rev 4.
- IPDC: IP Device Control.
- ISAPMP: Internet Key Exchange.
- LDP
- NTP: Network Time Protocol.
- POP3: Post Office Protocol version 3.
- Radius.
- RLOGIN: Remote Login.
- RTSP: Real-time Streaming Protocol.
- SCTP: Stream Control Transmission Protocol.
- SLP Service Location Protocol.
- SMTP: Simple Mail Transfer Protocol.
- SNMP: Simple Network Management Protocol.
- SOCKS
- TACACS+: Terminal Access Controller Access Control System.
- TELNET.
- WCCP: Web Cache Coordination Protocol
- X-Window.

The following diagram illustrates the TCP/IP suite in relation to the OSI model:

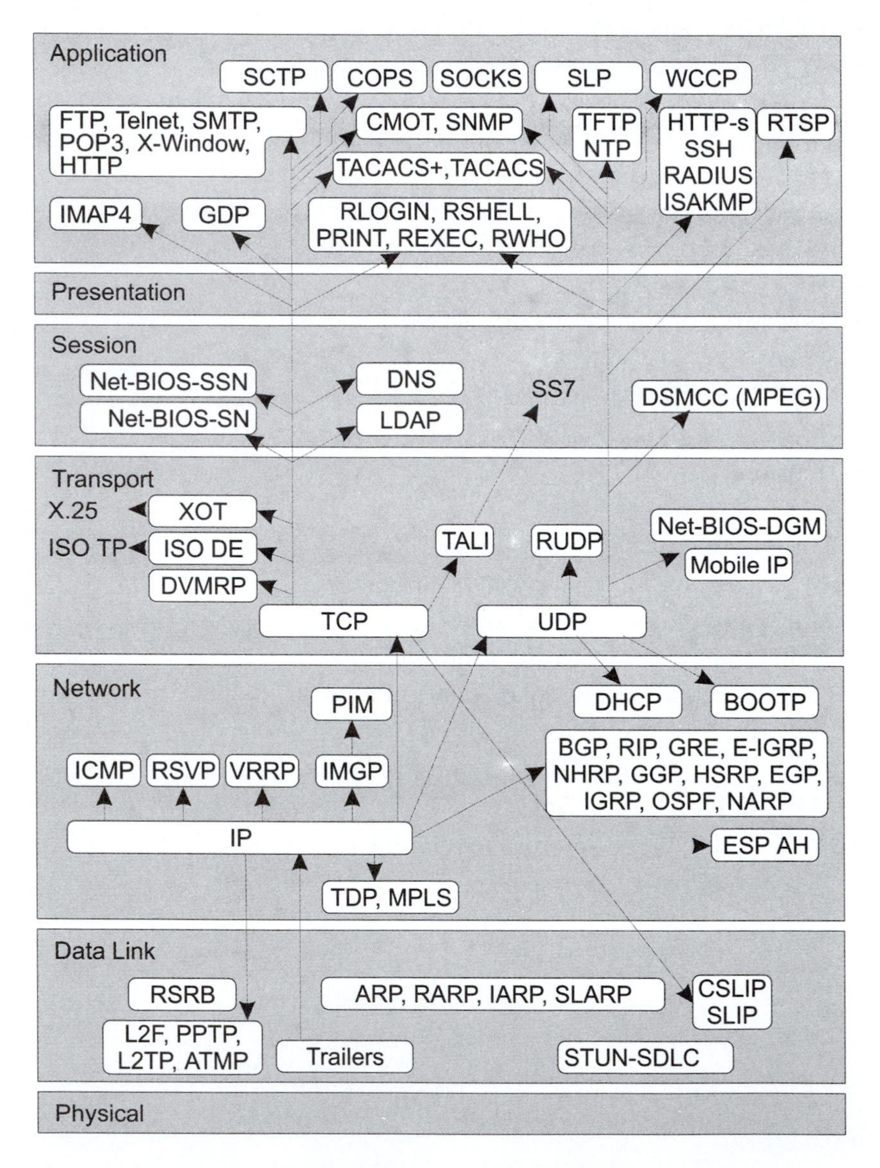

TCP/IP in relation to the OSI model

IP

IETF RFC 791 1981-09 http://www.cis.ohio-state.edu/htbin/rfc/rfc791.html
IETF RFC 1853 1995-1 http://www.cis.ohio-state.edu/htbin/rfc/rfc1853.html

The Internet Protocol (IP) is the routing layer datagram service of the TCP/IP suite. All other protocols within the TCP/IP suite, except ARP and RARP, use IP to route frames from host to host. The IP frame header contains routing information and control information associated with datagram delivery.

The IP header structure is as follows:

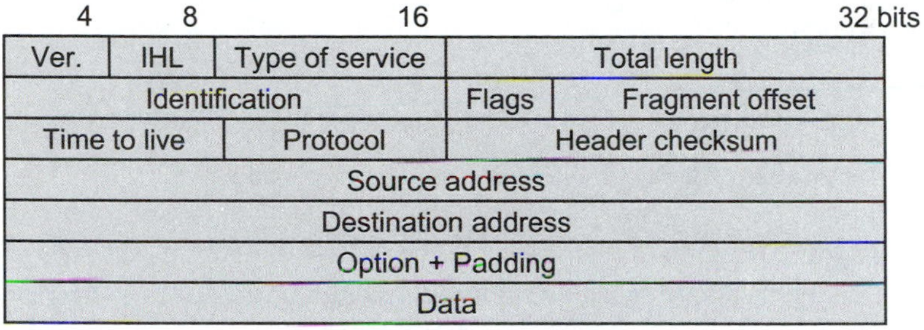

<table>
<tr><td>4</td><td>8</td><td>16</td><td></td><td>32 bits</td></tr>
<tr><td>Ver.</td><td>IHL</td><td>Type of service</td><td colspan="2">Total length</td></tr>
<tr><td colspan="3">Identification</td><td>Flags</td><td>Fragment offset</td></tr>
<tr><td colspan="2">Time to live</td><td>Protocol</td><td colspan="2">Header checksum</td></tr>
<tr><td colspan="5">Source address</td></tr>
<tr><td colspan="5">Destination address</td></tr>
<tr><td colspan="5">Option + Padding</td></tr>
<tr><td colspan="5">Data</td></tr>
</table>

IP header structure

Version

Version field indicates the format of the Internet header.

IHL

Internet header length is the length of the Internet header in 32-bit words. Points to the beginning of the data. The minimum value for a correct header is 5.

Type of service

Indicates the quality of service desired. Networks may offer service precedence, meaning that they accept traffic only above a certain precedence at times of high load. There is a three-way trade-off between low delay, high reliability and high throughput.

Bits 0-2: Precedence
111 Network control.
110 Internetwork control.

101 CRITIC/ECP.
100 Flash override.
011 Flash.
010 Immediate.
001 Priority.
000 Routine.

Bit 3: Delay
0 Normal delay.
1 Low delay.

Bit 4: Throughput
0 Normal throughput.
1 High throughput.

Bit 5: Reliability
0 Normal reliability.
1 High reliability.

Bits 6-7: Reserved for future use.

Total length

Length of the datagram measured in bytes, including the Internet header and data. This field allows the length of a datagram to be up to 65,535 bytes, although such long datagrams are impractical for most hosts and networks. All hosts must be prepared to accept datagrams of up to 576 bytes, regardless of whether they arrive whole or in fragments. It is recommended that hosts send datagrams larger than 576 bytes only if the destination is prepared to accept the larger datagrams.

Identification

Identifying value assigned by the sender to aid in assembling the fragments of a datagram.

Flags

3 bits. Control flags:

Bit 0 is reserved and must be zero.

Bit 1: Don't fragment bit:
0 May fragment.
1 Don't fragment.

Bit 2: More fragments bit:
0 Last fragment.
1 More fragments.

Fragment offset

13 bits. Indicates where this fragment belongs in the datagram. The fragment offset is measured in units of 8 bytes (64 bits). The first fragment has offset zero.

Time to live

Indicates the maximum time the datagram is allowed to remain in the Internet system. If this field contains the value zero, the datagram must be destroyed. This field is modified in Internet header processing. The time is measured in units of seconds. However, since every module that processes a datagram must decrease the TTL by at least one (even if it processes the datagram in less than 1 second), the TTL must be thought of only as an upper limit on the time a datagram may exist. The intention is to cause undeliverable datagrams to be discarded and to bound the maximum datagram lifetime.

Protocol

Indicates the next level protocol used in the data portion of the Internet datagram.

Header checksum

A checksum on the header only. Since some header fields change, e.g., Time To Live, this is recomputed and verified at each point that the Internet header is processed.

Source address / destination address

32 bits each. A distinction is made between names, addresses and routes. A *name* indicates an object to be sought. An *address* indicates the location of the object. A *route* indicates how to arrive at the object. The Internet protocol deals primarily with addresses. It is the task of higher level protocols (such as host-to-host or application) to make the mapping from names to addresses. The Internet module maps Internet addresses to local net addresses. It is the task of lower level procedures (such as local net or gateways) to make the mapping from local net addresses to routes.

Options

Options may or may not appear in datagrams. They must be implemented by all IP modules (host and gateways). What is optional is their transmission in any particular datagram, not their implementation. In some environments, the security option may be required in all datagrams.

The option field is variable in length. There may be zero or more options. There are two possible formats for an option:

- A single octet of option type.
- An option type octet, an option length octet and the actual option data octets.

The length octet includes the option type octet and the actual option data octets.

The option type octet has 3 fields:

1 bit: Copied flag. Indicates that this option is copied into all fragments during fragmentation:
0 Copied.
1 Not copied.

2 bits: Option class.
0 Control.
1 Reserved for future use.
2 Debugging and measurement.
3 Reserved for future use.

5 bits: Option number.

Data

IP data or higher layer protocol header.

IPv6

IETF RFC 1883 http://www.cis.ohio-state.edu/htbin/rfc/rfc1883.html
IETF RFC 1826 http://www.cis.ohio-state.edu/htbin/rfc/rfc1826.html
IETF RFC 1827 1995-12 http://www.cis.ohio-state.edu/htbin/rfc/rfc1827.html

IP version 6 (IPv6) is an updated version of the Internet Protocol based on IPv4. IPv4 and IPv6 are demultiplexed at the media layer. For example, IPv6 packets are carried over Ethernet with the content type 86DD (hexadecimal) instead of IPv4's 0800.

IPv6 increases the IP address size from 32 bits to 128 bits, to support more levels of addressing hierarchy, a much greater number of addressable nodes and simpler auto-configuration of addresses. Scalability of multicast addresses is introduced. A new type of address called an *anycast address* is also defined, to send a packet to any one of a group of nodes.

Improved support for extensions and options - IPv6 options are placed in separate headers that are located between the IPv6 header and the transport layer header. Changes in the way IP header options are encoded allow more efficient forwarding, less stringent limits on the length of options, and greater flexibility for introducing new options in the future. The extension headers are: Hop-by-Hop Option, Routing (Type 0), Fragment, Destination Option, Authentication, Encapsulation Payload.

Flow labeling capability - A new capability has been added to enable the labeling of packets belonging to particular traffic flows for which the sender requests special handling, such as non-default Quality of Service or real-time service.

The IPv6 header structure is as follows:

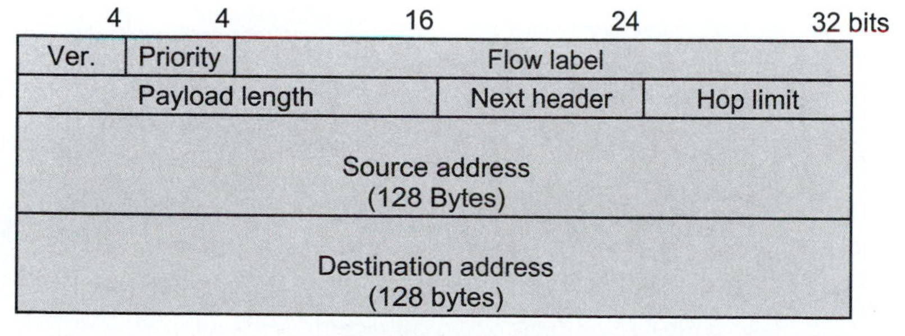

IPv6 header structure

Version

Internet Protocol version number (IPv6 is 6).

Priority

Enables a source to identify the desired delivery priority of the packets. Priority values are divided into ranges: traffic where the source provides congestion control and non-congestion control traffic.

Flow label

Used by a source to label those products for which it requests special handling by the IPv6 router. The flow is uniquely identified by the combination of a source address and a non-zero flow label.

Payload length

Length of payload (in octets).

Next header

Identifies the type of header immediately following the IPv6 header.

Hop limit

8-bit integer that is decremented by one, by each node that forwards the packet. The packet is discarded if the Hop Limit is decremented to zero.

Source address

128-bit address of the originator of the packet.

Destination address

128-bit address of the intended recipient of the packet.

TCP

IETF RFC 793 1981-09 http://www.cis.ohio-state.edu/htbin/rfc/rfc793.html
IETF RFC 1072 1988-10 http://www.cis.ohio-state.edu/htbin/rfc/rfc1072.html
IETF RFC 1693 1994-11 http://www.cis.ohio-state.edu/htbin/rfc/rfc1693.html
IETF RFC 1146 1990-03 http://www.cis.ohio-state.edu/htbin/rfc/rfc1146.html
IETF RFC 1323 1992-05 http://www.cis.ohio-state.edu/htbin/rfc/rfc1323.html

The Transmission Control Protocol (TCP) provides a reliable stream delivery and virtual connection service to applications through the use of sequenced acknowledgment with retransmission of packets when necessary.

The TCP header structure is as follows:

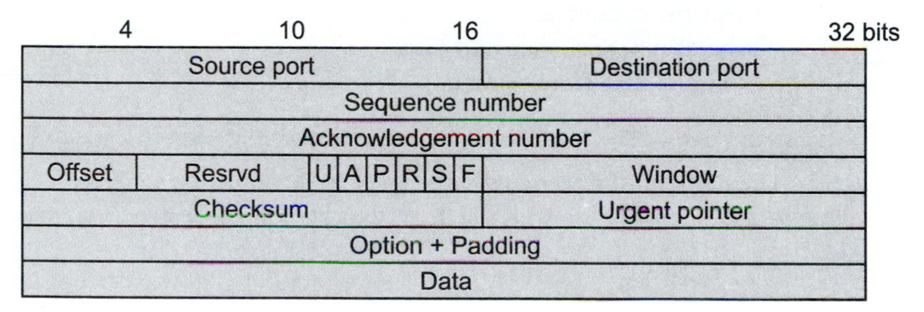

TCP header structure

Source port
Source port number.

Destination port
Destination port number.

Sequence number
The sequence number of the first data octet in this segment (except when SYN is present). If SYN is present, the sequence number is the initial sequence number (ISN) and the first data octet is ISN+1.

Acknowledgment number
If the ACK control bit is set, this field contains the value of the next sequence number which the sender of the segment is expecting to receive. Once a connection is established, this value is always sent.

Data offset

4 bits. The number of 32-bit words in the TCP header, which indicates where the data begins. The TCP header (even one including options) has a length which is an integral number of 32 bits.

Reserved

6 bits. Reserved for future use. Must be zero.

Control bits

6 bits. The control bits may be (from right to left):

U (URG) Urgent pointer field significant.
A (ACK) Acknowledgment field significant.
P (PSH) Push function.
R (RST) Reset the connection.
S (SYN) Synchronize sequence numbers.
F (FIN) No more data from sender.

Window

16 bits. The number of data octets which the sender of this segment is willing to accept, beginning with the octet indicated in the acknowledgment field.

Checksum

16 bits. The checksum field is the 16 bit one's complement of the one's complement sum of all 16-bit words in the header and text. If a segment contains an odd number of header and text octets to be checksummed, the last octet is padded on the right with zeros to form a 16-bit word for checksum purposes. The pad is not transmitted as part of the segment. While computing the checksum, the checksum field itself is replaced with zeros.

Urgent Pointer

16 bits. This field communicates the current value of the urgent pointer as a positive offset from the sequence number in this segment. The urgent pointer points to the sequence number of the octet following the urgent data. This field can only be interpreted in segments for which the URG control bit has been set.

Options

Options may be transmitted at the end of the TCP header and always have a length which is a multiple of 8 bits. All options are included in the checksum. An option may begin on any octet boundary.

There are two possible formats for an option:

- A single octet of option type.
- An octet of option type, an octet of option length, and the actual option data octets.

The option length includes the option type and option length, as well as the option data octets.

The list of options may be shorter than that designated by the data offset field because the contents of the header beyond the End-of-Option option must be header padding i.e., zero.

A TCP must implement all options.

Data

TCP data or higher layer protocol.

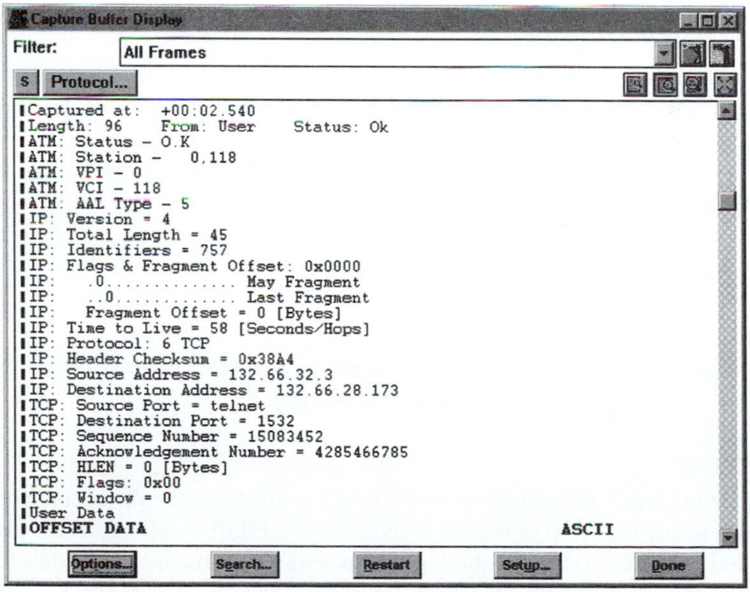

```
Capture Buffer Display
Filter: All Frames
S  Protocol...
Captured at:  +00:02.540
Length: 96     From: User     Status: Ok
ATM: Status - O.K
ATM: Station -   0,118
ATM: VPI - 0
ATM: VCI - 118
ATM: AAL Type - 5
IP: Version = 4
IP: Total Length = 45
IP: Identifiers = 757
IP: Flags & Fragment Offset: 0x0000
IP:    .0.............. May Fragment
IP:    ..0............. Last Fragment
IP:    Fragment Offset = 0 [Bytes]
IP: Time to Live = 58 [Seconds/Hops]
IP: Protocol: 6 TCP
IP: Header Checksum = 0x38A4
IP: Source Address = 132.66.32.3
IP: Destination Address = 132.66.28.173
TCP: Source Port = telnet
TCP: Destination Port = 1532
TCP: Sequence Number = 15083452
TCP: Acknowledgement Number = 4285466785
TCP: HLEN = 0 [Bytes]
TCP: Flags: 0x00
TCP: Window = 0
User Data
OFFSET DATA                                        ASCII
Options...  Search...  Restart  Setup...  Done
```

TCP/IP over ATM decode

UDP

RFC 768 1980-08 http://www.cis.ohio-state.edu/htbin/rfc/rfc768.html

The User Datagram Protocol (UDP) provides a simple, but unreliable message service for transaction-oriented services. Each UDP header carries both a source port identifier and destination port identifier, allowing high-level protocols to target specific applications and services among hosts.

The UDP header structure is shown as follows:

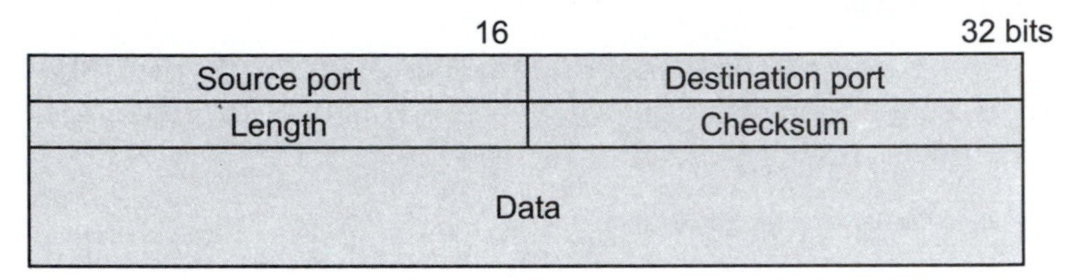

UDP header structure

Source port

Source port is an optional field. When used, it indicates the port of the sending process and may be assumed to be the port to which a reply should be addressed in the absence of any other information. If not used, a value of zero is inserted.

Destination port

Destination port has a meaning within the context of a particular Internet destination address.

Length

The length in octets of this user datagram, including this header and the data. The minimum value of the length is eight.

Checksum

The 16-bit one's complement of the one's complement sum of a pseudo header of information from the IP header, the UDP header and the data, padded with zero octets at the end (if necessary) to make a multiple of two octets.

Data
UDP data field.

ARP/RARP

IETF RFC 826 1982-11 http://www.cis.ohio-state.edu/htbin/rfc/rfc826.html
IETF RFC 1390 1993-01 http://www.cis.ohio-state.edu/htbin/rfc/rfc1390.html
IETF RFC 1293 1992-01 http://www.cis.ohio-state.edu/htbin/rfc/rfc1293.html

TCP/IP uses the Address Resolution Protocol (ARP) and the Reverse Address Resolution Protocol (RARP) to initialize the use of Internet addressing on an Ethernet or other network that uses its own media access control (MAC). ARP allows a host to communicate with other hosts when only the Internet address of its neighbors is known. Before using IP, the host sends a broadcast ARP request containing the Internet address of the desired destination system.

The ARP/RARP header structure is shown in the illustration below.

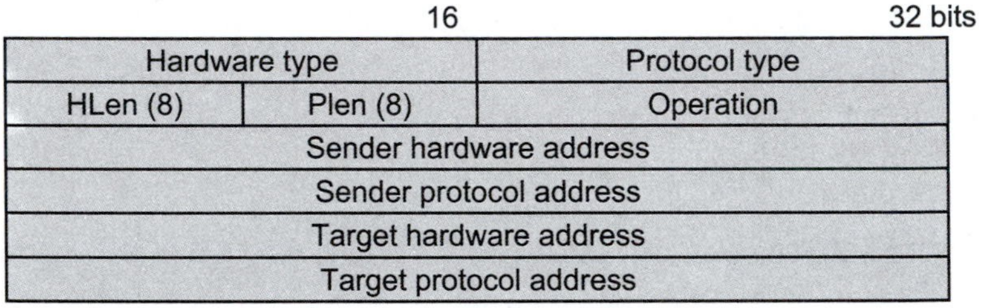

16		32 bits
Hardware type		Protocol type
HLen (8)	Plen (8)	Operation
Sender hardware address		
Sender protocol address		
Target hardware address		
Target protocol address		

ARP/RARP header structure

Hardware type

Specifies a hardware interface type for which the sender requires a response.

Protocol type

Specifies the type of high-level protocol address the sender has supplied.

HLen

Hardware address length.

PLen

Protocol address length.

Operation

The values are as follows:

1 ARP request.
2 ARP response.
3 RARP request.
4 RARP response.
5 Dynamic RARP request.
6 Dynamic RARP reply.
7 Dynamic RARP error.
8 InARP request.
9 InARP reply.

Sender hardware address

HLen bytes in length.

Sender protocol address

PLen bytes in length.

Target hardware address

HLen bytes in length.

Target protocol address

PLen bytes in length.

DCAP

http://info.internet.isi.edu/in-notes/rfc/files/rfc2114.txt

The (DLSw) Data Link Switching Client Access Protocol (DCAP) is used between workstations and routers to transport SNA/NetBIOS traffic over TCP sessions.

Since the Data Link Switching Protocol, RFC 1795, was published, some software vendors have begun implementing DLSw on workstations. The implementation of DLSw on a large number of workstations raises the important issues of scalability and efficiency. Since DLSw is a switch-to-switch protocol, it is not efficient when implemented on workstations. DCAP addresses these issues. It introduces a hierarchical structure to resolve the scalability problems. All workstations are clients to the router (server) rather than peers to the router. This creates a client/server model. It also provides a more efficient protocol between the workstation (client) and the router (server).

DCAP Packet Header

The DCAP packet header is used to identify the message type and length of the frame. This is a general purpose header used for each frame that is passed between the DCAP server and the client.

Protocol ID/version number (1 byte)
Message type (1 byte)
Packet length (2 bytes)

DCAP Header Format

Protocol ID

The Protocol ID uses the first 4 bits of this field and is set to 1000.

Version number

The Version number uses the next 4 bits in this field and is set to 0001.

Message type

The message type is the DCAP message type.

The following message types exist:

Code	DCAP Frame Name	Function
0x01	CAN U REACH	Find if the station given is reachable
0x02	I CAN REACH	Positive response to CAN U REACH
0x03	I CANNOT REACH	Negative response to CAN U REACH
0x04	START DL	Setup session for given addresses
0x05	DL STARTED	Session started
0x06	START DL FAILED	Session Start failed
0x07	XID FRAME	XID frame
0x08	CONTACT STN	Contact destination to establish SABME
0x09	STN CONTACTED	Station contacted - SABME mode set
0x0A	DATA FRAME	Connectionless Data Frame for a link
0x0B	INFO FRAME	Connection oriented I-Frame
0x0C	HALT DL	Halt Data Link session
0x0D	HALT DL NOACK	Halt Data Link session without ack
0x0E	DL HALTED	Session halted
0x0F	FCM FRAME	Data Link Session Flow Control Message
0x11	DGRM FRAME	Connectionless Datagram Frame for circuit
0x12	CAP XCHANGE	Capabilities Exchange Message
0x13	CLOSE PEER REQUEST	Disconnect Peer Connection Request
0x14	CLOSE PEER RESPONSE	Disconnect Peer Connection Response
0x1D	PEER TEST REQ	Peer keepalive test request
0x1E	PEER TEST RSP	Peer keepalive test response

Packet length

The total packet length is the length of the packet including the DCAP header, DCAP data and user data. The minimum size of the packet is 4, which is the length of the header.

ATMP

RFC 2107 http://www.cis.ohio-state.edu/htbin/rfc/rfc2107.html

The Ascend Tunnel Management Protocol (ATMP) is a protocol currently being used in Ascend Communication products to allow dial-in client software to obtain virtual presence on a user's home network from remote locations. A user calls into a remote NAS but instead of using an address belonging to a network directly supported by the NAS, the client software uses an address belonging to the user's "Home Network". This address can be either provided by the client software or assigned from a pool of addresses from the Home Network address space. In either case, this address belongs to the Home Network and therefore special routing considerations are required in order to route packets to and from these clients. A tunnel between the NAS and a special "Home Agent" (HA) located on the Home Network is used to carry data to and from the client.

The format of the ATMP header is shown in the following illustration:

Version	Message type	Identifier

ATMP packet structure

Version

The ATMP protocol version must be 1.

Message type

ATMP defines a set of request and reply messages sent with UDP. There are 7 different ATMP message types represented by the following values.

MessageType	*Type Code*
Registration Request	1
Challenge Request	2
Challenge Reply	3
Registration Reply	4
Deregister Request	5
Deregister Reply	6
Error Notification	7

Identifier

A 16 bit number used to match replies with requests. A new value should be provided in each new request. Retransmissions of the same request should use the same identifier.

L2F

RFC 2341 http://www.cis.ohio-state.edu/htbin/rfc/rfc2341.html

The Layer 2 Forwarding protocol (L2F) permits the tunneling of the link layer of higher layer protocols. Using such tunnels it is possible to divorce the location of the initial dial-up server from the location at which the dial-up protocol connection is terminated and access to the network provided.

The format of the packet is shown in the following illustration:

13	16	24	32
F K P S 0 0 0 0 0 0 0 0 C	Ver	Protocol	Sequence (opt)
Multiplex ID		Client ID	
Length		Payload offset	
Packet key (optional)			
Payload			
		Checksum	

L2F packet structure

Version

The major version of the L2F software creating the packet.

Protocol

The protocol field specifies the protocol carried within the L2F packet.

Sequence

The sequence number is present if the S bit in the L2F header is set to 1.

Multiplex ID

The packet multiplex ID identifies a particular connection within a tunnel.

Client ID

The client ID (CLID) assists endpoints in demultiplexing tunnels.

Length

The length is the size in octets of the entire packet, including the header, all the fields and the payload.

Payload offset

This field specifies the number of bytes past the L2F header at which the payload data is expected to start. This field is present if the F bit in the L2F header is set to 1.

Packet key

The key field is present if the K bit is set in the L2F header. This is part of the authentication process.

Checksum

The checksum of the packet. The checksum field is present if the C bit in the L2F header is set to 1.

Option Messages

When the link is initiated, the endpoints communicate to verify the presence of L2F on the remote end, and to permit any needed authentication. The protocol for such negotiation is always 1, indicating L2F management. The message itself is structured as a sequence of single octets indicating an option. When the protocol field of an L2F specifies L2F management, the body of the packet is encoded as zero or more options. An option is a single octet message type, followed by zero or more sub-options. Each sub-option is a single byte sub-option value, and followed by additional bytes as appropriate for the sub-option.

Possible option messages are:

Invalid	Invalid message.
L2F CONF	Request configuration.
L2F CONF NAME	Name of peer sending L2F CONF.
L2F CONF CHAL	Random number peer challenges.
L2F CONF CLID	Assigned CLID for peer to use.
L2F OPEN	Accept configuration.
L2F OPEN NAME	Name received from client.
L2F OPEN CHAL	Challenge client received.
L2F OPEN RESP	Challenge response from client.
L2F ACK LCP1 LCP	CONFACK accepted from client.
L2F ACK LCP2 LCP	CONFACK sent to client.
L2F OPEN TYPE	Type of authentication used.
L2F OPEN ID	ID associated with authentication.
L2F REQ LCP0	First LCP CONFREQ from client.
L2F CLOSE	Request disconnect.

L2F CLOSE WHY	Reason code for close.
L2F CLOSE STR	ASCII string description.
L2F ECHO	Verify presence of peer.
L2F ECHO RESP	Respond to L2F_ECHO.

L2TP

IETF draft
http://info.internet.isi.edu:80/in-drafts/files/draft-ietf-pppext-l2tp-11.txt
RFC 2661

The L2TP Protocol is used for integrating multi-protocol dial-up services into existing Internet Service Providers Point of Presence (hereafter referred to as ISP and POP, respectively). This protocol may also be used to solve the "multilink hunt-group splitting" problem. Multilink PPP, often used to aggregate ISDN B channels, requires that all channels composing a multilink bundle be grouped at a single Network Access Server (NAS). Because L2TP makes a PPP session appear at a location other than the physical point at which the session was physically received, it can be used to make all channels appear at a single NAS, allowing for a multilink operation even when the physical calls are spread across distinct physical NASs.

The format of the L2TP packet is shown in the following illustration:

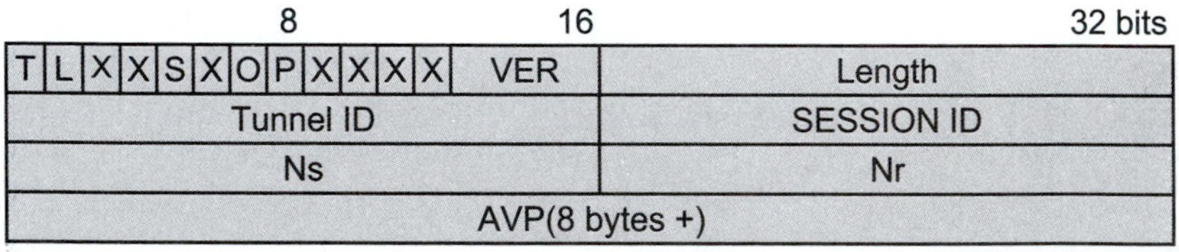

8	16	32 bits
T L X X S X O P X X X X	VER	Length
Tunnel ID		SESSION ID
Ns		Nr
AVP(8 bytes +)		

L2TP packet structure

T
The T bit indicates the type of message. It is set to 0 for data messages and 1 for control messages.

L
When set, this indicates that the Length field is present, indicating the total length of the received packet. Must be set for control messages.

X
The X bits are reserved for future extensions. All reserved bits are set to 0 on outgoing messages and are ignored on incoming messages.

S

If the S bit is set, both the Nr and Ns fields are present. S must be set for control messages.

O

When set, this field indicates that the Offset Size field is present in payload messages. This bit is set to 0 for control messages.

P

If the Priority (P) bit is 1, this data message receives preferential treatment in its local queuing and transmission. LCP echo requests used as a keepalive for the link, for instance, are generally sent with this bit set to 1. Without it, a temporary interval of local congestion could result in interference with keepalive messages and unnecessary loss of the link. This feature is only for use with data messages. The P bit has a value of 0 for all control messages.

Ver

The value of the ver bit is always 002. This indicates a version 1 L2TP message.

Length

Overall length of the message, including header, message type AVP, plus any additional AVP's associated with a given control message type.

Tunnel ID

Identifies the tunnel to which a control message applies. If an Assigned Tunnel ID has not yet been received from the peer, Tunnel ID must be set to 0. Once an Assigned Tunnel ID is received, all further packets must be sent with Tunnel ID set to the indicated value.

Call ID

Identifies the user session within a tunnel to which a control message applies. If a control message does not apply to a single user session within the tunnel (for instance, a Stop-Control-Connection-Notification message), Call ID must be set to 0.

Nr

The sequence number expected in the next control message to be receivec.

Ns

The sequence number for this data or control message.

Data messages have two additional fields before the AVP as follows:

Offset size (16 bits)	Offset pad (16 bits)

Additional fields in L2TP payload message

Offset size

This field specifies the number of bytes past the L2TP header at which the payload data is expected to start. It is recommended that data thus skipped be initialized to 0s. If the offset size is 0, or the O bit is not set, the first byte following the last byte of the L2TP header is the first byte of payload data.

AVP

The AVP (Attribute-Value Pair) is a uniform method used for encoding message types and bodies throughout L2TP. The format of the AVP is given below:

						16	32 bits
M	H	0	0	0	0	Overall length	Vendor ID
Attribute							
Value							

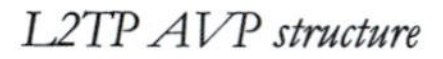

L2TP AVP structure

M

The first six bits are a bit mask, describing the general attributes of the AVP. The M bit, known as the mandatory bit, controls the behavior required of an implementation which receives an AVP which it does not recognize.

H

The hidden bit controls the hiding of the data in the value field of an AVP. This capability can be used to avoid the passing of sensitive data, such as user passwords, as cleartext in an AVP.

Overall length

Encodes the number of octets (including the overall length field itself) contained in this AVP. It is 10 bits, permitting a maximum of 1024 bytes of data in a single AVP.

Vendor ID

The IANA assigned SMI Network Management Private Enterprise Codes value, encoded in network byte order.

Attribute

The actual attribute, a 16-bit value with a unique interpretation across all AVP's defined under a given Vendor ID.

Value

The value field follows immediately after the Attribute field, and runs for the remaining octets indicated in the overall length (i.e., overall length minus six octets of header).

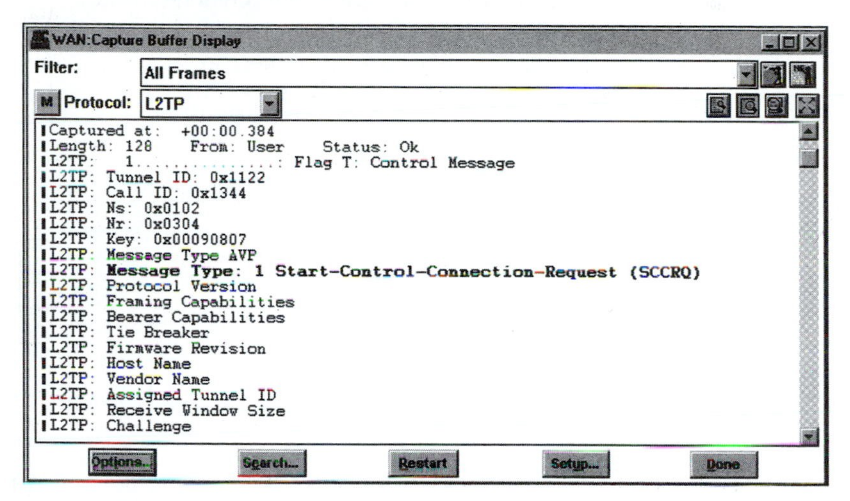

L2TP decode

PPTP

IETF draft
http://info.internet.isi.edu:80/in-drafts/files/draft-ietf-pppext-pptp-04.txt

PPTP (Point to Point Tunneling Protocol) allows PPP to be channeled through an IP network. It uses a client-server architecture to decouple functions which exist in current Network Access Servers and support Virtual Private Networks. It specifies a call-control and management protocol which allows the server to control access for dial-in circuit switched calls originating from a PSTN or ISDN, or to initiate outbound circuit switched connections. PPTP uses a GRE-like (Generic Routing Encapsulation) mechanism to provide a flow- and congestion-controlled encapsulated datagram service for carrying PPP packets.

The format of the header is shown in the following illustration:

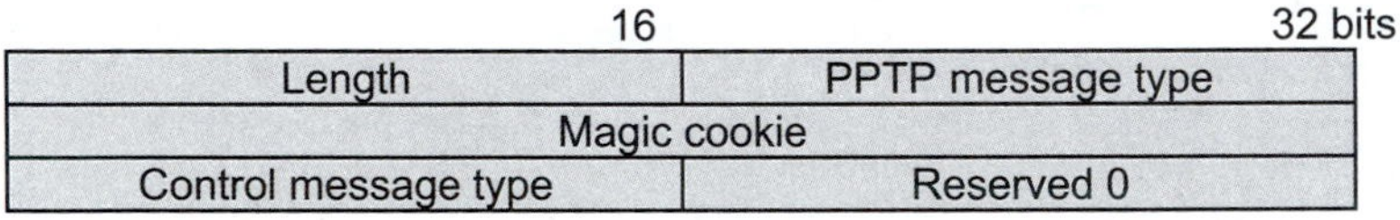

16	32 bits
Length	PPTP message type
Magic cookie	
Control message type	Reserved 0

PPTP header structure

Length

Total length in octets of this PPTP message including the entire PPTP header.

PPTP message type

The message type. Possible values are:

1 Control message.
2 Management message.

Magic cookie

The magic cookie is always sent as the constant 0x1A2B3C4D. Its basic purpose is to allow the receiver to ensure that it is properly synchronized with the TCP data stream.

Control Message Type

Values may be:

1 Start-Control-Connection-Request.
2 Start-Control-Connection-Reply.
3 Stop-Control-Connection-Request.
4 Stop-Control-Connection-Reply.
5 Echo-Request.
6 Echo-Reply.

Call Management

7 Outgoing-Call-Request.
8 Outgoing-Call-Reply.
9 Incoming-Call-Request.
10 Incoming-Call-Reply.
11 Incoming-Call-Connected.
12 Call-Clear-Request.
13 Call-Disconnect-Notify.

Error Reporting

14 WAN-Error-Notify.

PPP Session Control

15 Set-Link-Info.

Reserved

A reserved field, must be set to 0.

DHCP

RFC 1531 http://www.cis.ohio-state.edu/htbin/rfc/rfc1531.html

The Dynamic Host Configuration Protocol (DHCP) provides Internet hosts with configuration parameters. DHCP is an extension of BOOTP. DHCP consists of two components: a protocol for delivering host-specific configuration parameters from a DHCP server to a host and a mechanism for allocation of network addresses to hosts.

The format of the header is shown in the following illustration:

8	16	24	32 bits
Op	Htype	Hlen	Hops
XID			
Secs		Flags	
Ciaddr			
Yiaddr			
Siaddr			
Giaddr			
Chaddr (16 bytes)			

DHCP header structure

Op
The message operation code. Messages can be either BOOTREQUEST or BOOTREPLY.

Htype
The hardware address type.

Hlen
The hardware address length.

XID
The transaction ID.

Secs
The seconds elapsed since the client began the address acquisition or renewal process.

Flags
The flags.

Ciaddr
The client IP address.

Yiaddr
The "Your" (client) IP address.

Siaddr
The IP address of the next server to use in bootstrap.

Giaddr
The relay agent IP address used in booting via a relay agent.

Chaddr
The client hardware address.

DHCPv6

http://www.ietf.org/internet-drafts/draft-ietf-dhc-dhcpv6-14.txt

The Dynamic Host Configuration Protocol for IPv6 (DHCPv6) enables DHCP servers to pass configuration information, via extensions, to IPv6 nodes. It offers the capability of automatic allocation of reusable network addresses and additional configuration flexibility. This protocol is a stateful counterpart to the IPv6 Stateless Address Autoconfiguration protocol, and can be used separately or together with the latter to obtain configuration information.

DHCPv6 has 6 different message types: Solicit, Advertise, Request, Reply, Release and Reconfigure.

DHCP Solicit message

A client transmits a DHCP Solicit message over the interface to be configured, to obtain one or more server addresses. Unless otherwise noted, the value of all fields are set by the client.

8		16 24	25 32 bits
Message type	C	reserved	Prefix-size
Client link local address (16 octets)			
Relay address (16 octets)			
Saved agent address (16 octets)			

DHCP Solicit message structure

Message type

Value of 1 specifies a Solicit message.

C

Indicates that the client requests that all servers receiving the message deallocate the resources associated with the client. When set, the client should provide a saved agent address to locate the clients binding by a server.

Prefix size

When non-zero, indicates the number of left-most bits of the agent's IPv6 address which conprise the routing prefix.

Reserved
Set to zero.

Client link local address
IP link local address of the client interface from which the client issued the DHCP Request message.

Relay address
Set by the client to zero. If received by a DHCP relay this is set by the relay to the IP address of the interface on which the relay received the client's DHCP Solicit message.

Saved agent address
When present, indicates the IP address of an agent's interface retained by the client from a previous DHCP transaction.

DHCP Advertise message

A DHCP agent sends a DHCP Advertise message to inform a prospective client about the IP address of a server to which a DHCP Request message may be sent. When the client and server are on different links, the server sends the advertisement back through the relay whence the solicitation came. The value of all fields in the DHCP Advertise message are filled in by the DHCP server and not changed by any DHCP relay.

8		16 / 24 25	32 bits
Message type	S	reserved	Preference
Client link local address (16 octets)			
Agent address (16 octets)			
Server address (16 octets)			
Extensions			

DHCP Advertise message structure

Message type
Value of 2 specifies an Advertise message.

S
If set, specifies that the server address is present.

Preference
Indicates a server's willingness to provide service to the client.

Client link local address
IP link local address of the client interface from which the client issued the DHCP Request message.

Agent address
IP address of a DHCP agent interface on the same link as the client.

Server address
When present, the IP address of the DHCP server.

Extensions
Described in the standard.

DHCP Request message

In order to request configuration parameters from a server, a client sends a DHCP Request message, and may append extensions. If the client does not know any server address, it must first obtain one by multicasting a DHCP Solicit message. Typically, when a client reboots, it does not have a valid IP address of sufficient scope for the server to communicate with the client. In such cases, the client cannot send the message directly to the server because the server could not return any response to the client. In this case, the client must send the message to the local relay and insert the relay address as the agent address in the message header.

8				16	24 25	32 bits
Message type	C	S	R	rsvd	Transaction ID	
Client link local address (16 octets)						
Agent address (16 octets)						
Server address (16 octets)						
Extensions						

DHCP Request message structure

Message type
Value of 3 specifies a Request message.

R
If set, specifies that the client has rebooted and requests that all of its previous transaction IDs be expunged and made available for reuse.

Transaction ID
Unsigned integer identifier used to identify this request.

The remaining fields are described in the Solicit and Advertise messages.

DHCP Reply message

The server sends one DHCP Reply message in response to every DHCP Request or DHCP Release received. If the request comes with the S bit set, the client could not directly send the Request to the server and had to use a neighboring relay agent. In that case, the server sends back the DHCP Reply with the L bit set, and the DHCP Reply is addressed to the agent-address found in the DHCP Request message. All the fields in the DHCP Reply message are set by the DHCP server.

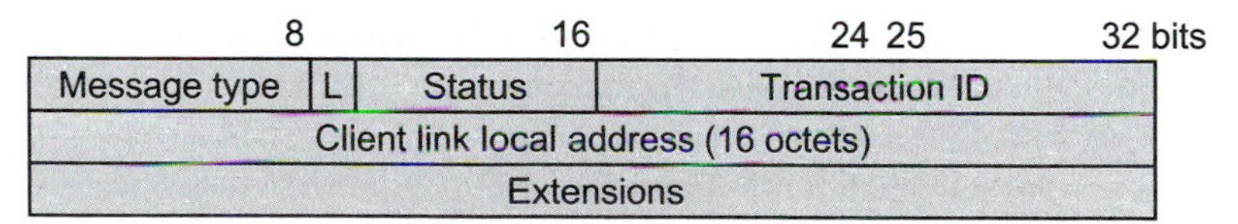

DHCP Reply message structure

Message type
Value of 4 specifies a Reply message.

L
If set, the client link local address is present.

Status
May have the following values:

0	Success
16	Failure, reason unspecified
17	Authentication failed or nonexistent
18	Poorly formed Request or Release
19	Resources unavailable
20	Client record unavailable
21	Invalid client IP address in Release

23 Relay cannot find server address
64 Server unreachable (ICMP error)

Transaction ID
Unsigned integer identifier used to identify this Reply, copied from the client Request.

Client link local address
If present, the IP address of the client interface which issued the corresponding DHCP Request message. If the L bit is set, the client's link-local address is present in the Reply message. Then the Reply is sent by the server to the relay's address which was specified as the agent-address in the DHCP Request message, and the relay uses the link-local address to deliver the Reply message to the client. The transaction-ID in the DHCP Reply is copied by the server from the client Request message.

DHCP Release message

The DHCP Release message is sent without the assistance of any DHCP relay. When a client sends a Release message, it is assumed to have a valid IP address with sufficient scope to allow access to the target server. If parameters are specified in the extensions, only those parameters are released. The values of all fields of the DHCP Release message are entered by the Client. The DHCP server acknowledges the Release message by sending a DHCP Reply.

8		16	24 25	32 bits
Message type	D	Reserved	Transaction ID	
Client link local address (16 octets)				
Agent address (16 octets)				
Client address (16 octets)				
Extensions				

DHCP Release message structure

Message type
Value of 5 specifies a Release message.

D

When set, the client instructs the server to send the DHCP Reply directly back to the client instead of using the given agent address and link local address to relay the Reply message.

Transaction ID

Unsigned integer identifier used to identify this Release, and copied into the Reply.

The remaining fields are described in the other DHCP messages.

DHCP Reconfigure message

DHCP Reconfigure messages can only be sent to clients which have established an IP address which routes to the link at which they are reachable, hence, the DHCP Reconfigure message is sent without the assistance of any DHCP relay. When a server sends a Reconfigure message, the receivers are assumed to have a valid IP address with sufficient scope to be accessible by the server. Only the parameters which are specified in the extensions to the Reconfigure message need be requested again by the client. A Reconfigure message can either be unicast or multicast by the server. The client extracts the extensions provided by the server and sends a DHCP Request message to the server using those extensions.

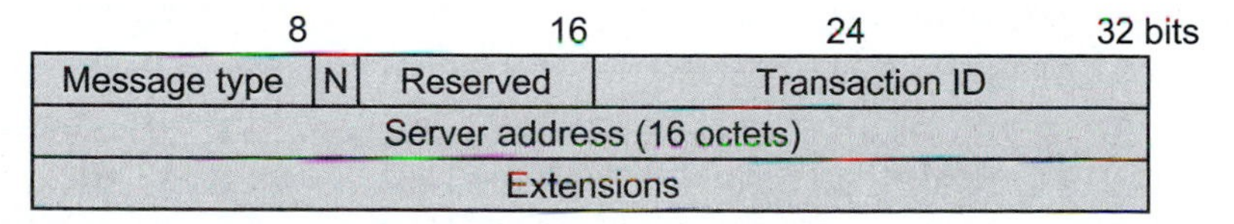

DHCP Reconfigure message structure

Message type

Value of 6 specifies a Reconfigure message.

N

Indicates that the client should not expect a DHCP Reply in response to the DHCP Request it sends as a result of the DHCP Reconfigure message.

The remaining fields are described in the other DHCP messages.

DVMRP

RFC 1075: http://www.cis.ohio-state.edu/htbin/rfc/rfc1075.html
IETF draft:
http://www.ietf.org/internet-drafts/draft-ietf-idmr-dvmrp-v3-08.txt

Distance Vector Multicast Routing Protocol (DVMRP) is an Internet routing protocol that provides an efficient mechanism for connectionless datagram delivery to a group of hosts across an internetwork. It is a distributed protocol that dynamically generates IP multicast delivery trees using a technique called Reverse Path Multicasting

DVMRP combines many of the features of RIP with the Truncated Reverse Path Broadcasting (TRPB) algorithm. DVMRP is developed based upon RIP because an implementation was available and distance vector algorithms are simple, as compared to link-state algorithms. In addition, to allow experiments to traverse networks that do not support multicasting, a mechanism called *tunneling* was developed.

DVMRP differs from RIP in one very important way. RIP routes and forwards datagrams to a particular destination. The purpose of DVMRP is to keep track of the return paths to the source of multicast datagrams. To make the explanation of DVMRP more consistent with RIP, the term *destination* is used instead of the more proper term *source*, however, datagrams are not forwarded to these destinations, but rather, originate from them.

DVMRP packets are encapsulated in IP datagrams, with an IP protocol number of 2 (IGMP). All fields are transmitted in Network Byte Order. DVMRP packets use a common protocol header that specifies the IGMP Packet Type as DVMRP. DVMRP protocol packets should be sent with the Precedence field in the IP header set to Internetwork Control (hexadecimal 0xc0 for the Type of Service Octet). The common protocol header is as shown in the following illustration:

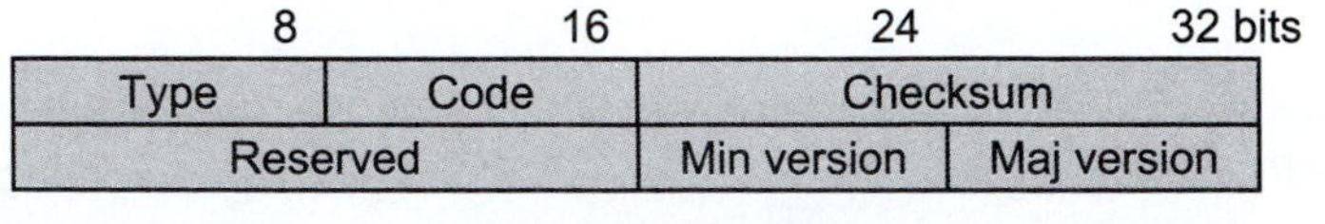

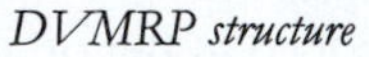
DVMRP structure

Type

Packet type. 0x13 indicates a DVMRP packet.

Code

Determines the type of DVMRP packet. Currently, there are codes for DVMRP protocol message types as well as protocol analysis and troubleshooting packets. The protocol message codes may be as follows:
Probe Neighbor discovery.
Report Route exchange.
Prune Pruning multicast delivery trees.
Graft Grafting multicast delivery trees.
Graft ack Acknowledging graft messages.

Checksum

16-bit one's complement of the one's complement sum of the DVMRP message. The checksum must be calculated upon transmission and must be validated on reception of a packet. The checksum of the DVMRP message should be calculated with the checksum field set to zero.

Reserved

Reserved for later use.

Min version

Minor version. Value must be 0xFF for this version of DVMRP.

Maj version

Major version. Value must be 3 for this version of DVMRP.

ICMP

RFC 792 http://www.cis.ohio-state.edu/htbin/rfc/rfc792.html

Internet Control Message Protocol (ICMP) messages generally contain information about routing difficulties with IP datagrams or simple exchanges such as time-stamp or echo transactions.

The ICMP header structure is shown as follows:

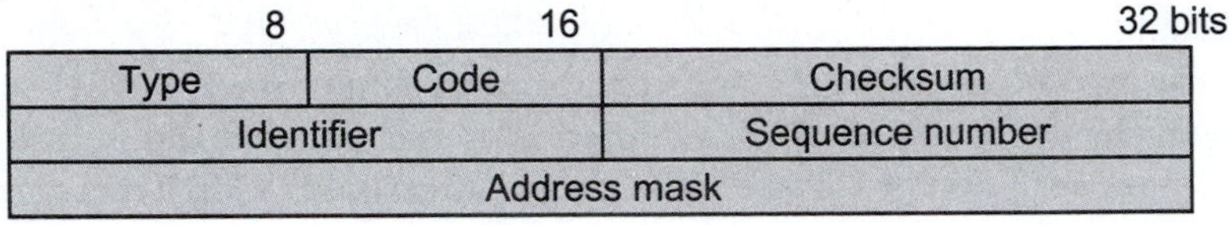

8	16	32 bits
Type	Code	Checksum
Identifier		Sequence number
Address mask		

ICMP header structure

Type	Code	Description
0		Echo reply.
3		Destination unreachable.
3	0	Net unreachable.
3	1	Host unreachable.
3	2	Protocol unreachable.
3	3	Port unreachable.
3	4	Fragmentation needed and DF set.
3	5	Source route failed.
4		Source quench.
5		Redirect.
5	0	Redirect datagrams for the network.
5	1	Redirect datagrams for the host.
5	2	Redirect datagrams for the type of service and network.
5	3	Redirect datagrams for the type of service and host.
8		Echo.
11		Time exceeded.
11	0	Time to live exceeded in transit.
11	1	Fragment reassemble time exceeded.
12		Parameter problem.
13		Timestamp.
14		Timestamp reply.

Type	Code	Description
15		Information request.
16		Information reply.

Checksum

The 16-bit one's complement of the one's complement sum of the ICMP message starting with the ICMP Type. For computing the checksum, the checksum field should be zero.

Identifier

An identifier to aid in matching requests/replies; may be zero.

Sequence number

Sequence number to aid in matching requests/replies; may be zero.

Address mask

A 32-bit mask.

ICMPv6

IETF RFC 1885 1970, 1995-12
http://www.cis.ohio-state.edu/htbin/rfc/rfc1885.html

The Internet Control Message Protocol (ICMP) was revised during the definition of IPv6. In addition, the multicast control functions of the IPv4 Group Membership Protocol (IGMP) are now incorporated with the ICMPv6.

The structure of the ICMPv6 header is shown in the following illustration.

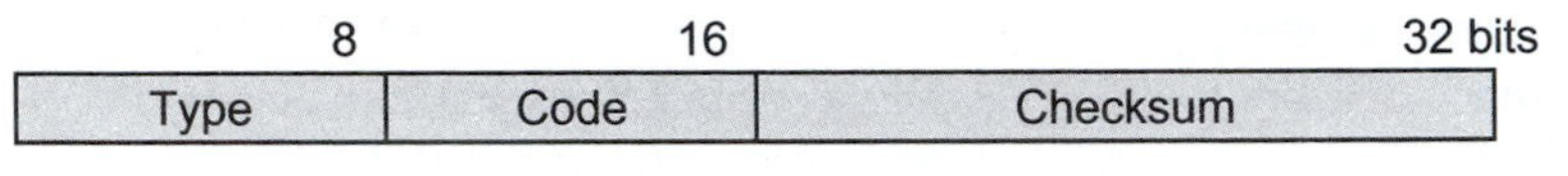

ICMPv6 header structure

Type

The type of the message. Messages can be error or informational messages. Error messages can be Destination unreachable, Packet too big, Time exceed, Parameter problem. The possible informational messages are, Echo Request, Echo Reply, Group Membership Query, Group Membership Report, Group Membership Reduction.

Code

For each type of message several different codes are defined.
An example of this is the Destination Unreachable message, where possible messages are: no route to destination, communication with destination administratively prohibited, not a neighbor, address unreachable, port unreachable. For further details, refer to the standard.

Checksum

Used to check data corruption in the ICMPv6 message and parts of the IPv6 header.

IGMP

IETF RFC 1112 1989-08 http://www.cis.ohio-state.edu/htbin/rfc/rfc1885.html

The Internet Group Management Protocol (IGMP) is used by IP hosts to report their host group memberships to any immediately neighboring multicast routers. IGMP is a integral part of IP. It must be implemented by all hosts conforming to level 2 of the IP multicasting specification. IGMP messages are encapsulated in IP datagrams, with an IP protocol number of 2.

The format of the IGMP packet is shown in the following illustration:

4	8	16	32 bits
Ver	Type	Unused	Checksum
Group address			

IGMP packet structure

Version
The protocol version.

Type
The message type:

1 Host Membership Query.

2 Host Membership Report.

Unused
An unused field.

Checksum
The checksum.

Group address
In a Host Membership Report Message this field holds the IP host group address of the group being reported.

MARS

IETF RFC 2022 1996-11 http://www.cis.ohio-state.edu/htbin/rfc/rfc2022.html

Multicasting is the process whereby a source host or protocol entity sends a packet to multiple destinations simultaneously using a single, local 'transmit' operation. ATM is being utilized as a new link layer technology to support a variety of protocols, including IP. The MARS protocol has two broad goals: to define a group address registration and membership distribution mechanism that allows UNI 3.0/3.1 based networks to support the multicast service of protocols such as IP and to define specific endpoint behaviors for managing point to multipoint VCs to achieve multicasting of layer 3 packets. The Multicast Address Resolution Server (MARS) is an extended analog of the ATM ARP Server. It acts as a registry, associating layer 3 multicast group identifiers with the ATM interfaces representing the group's members. MARS messages support the distribution of multicast group membership information between MARS and endpoints (hosts or routers). Endpoint address resolution entities query the MARS when a layer 3 address needs to be resolved to the set of ATM endpoints making up the group at any one time. Endpoints keep the MARS informed when they need to join or leave particular layer 3 groups. To provide for asynchronous notification of group membership changes, the MARS manages a point to multipoint VC out to all endpoints desiring multicast support. Each MARS manages a cluster of ATM-attached endpoints.

The format of the header is shown in the following illustration:

Field	Octets
Address family	1-2
Protocol identification	3-9
Reserved	10-12
Checksum	13-14
Extensions offset	15-16
Operation code	17-18
Type and length of source ATM number	19
Type and length of source ATM subaddress	20

MARS header structure

Address family
Defines the type of link layer addresses being carried.

Protocol ID
Contains 2 subfields:
16-bit protocol type.
40-bit optional SNAP extension to protocol type.

Reserved
This reserved field may be subdivided and assigned specific meanings for other control protocols indicated by the version number.

Checksum
This field carries a standard IP checksum calculated across the entire message.

Extension offset
This field identifies the existence and location of an optional supplementary parameters list.

Operation code
This field is divided into 2 sub fields: version and type. Version indicates the operation being performed, within the context of the control protocol version indicated by mar$op.version.

Type and length of ATM source number
Information regarding the source hardware address.

Type and length of ATM source subaddress
Information regarding the source hardware subaddress.

PIM

RFC 2362 http://www.cis.ohio-state.edu/htbin/rfc/rfc2362.html

Protocol Independent Multicast-Sparse Mode (PIM-SM) is a protocol for efficiently routing to multicast groups that may span wide-area (and inter-domain) internets. The protocol is not dependent on any particular unicast routing protocol, and is designed to support sparse groups.

The format of the PIM packet is shown in the following illustration:

PIM version	Type	Address length	Checksum

PIM header structure

PIM version
Current PIM version is 2.

Type
Types for specific PIM messages.

Address length
Address length in bytes. The length of the address field throughout, in the specific message.

Checksum
The 16-bit one's complement, of the one's complement sum of the entire PIM message (excluding the data portion in the register message). For computing the checksum, the checksum field is zeroed.

RIP

IETF RFC 1058 1988-06 http://www.cis.ohio-state.edu/htbin/rfc/rfc1058.html
IETF RFC 1723 1996-05 http://www.cis.ohio-state.edu/htbin/rfc/rfc1723.html
IETF RFC 1528 1994-02 http://www.cis.ohio-state.edu/htbin/rfc/rfc1528.html

RIP (Routing Information Protocol) is used by Berkeley 4BSD UNIX systems to exchange routing information. Implemented by a UNIX program, RIP derives from an earlier protocol of the same name developed by Xerox.

RIP is an extension of the Routing Information Protocol (RIP) intended to expand the amount of useful information carried in the RIP messages and to add a measure of security.

RIP is a UDP-based protocol. Each host that uses RIP has a routing process that sends and receives datagrams on UDP port number 520. The packet format of RIP is shown in the illustration below.

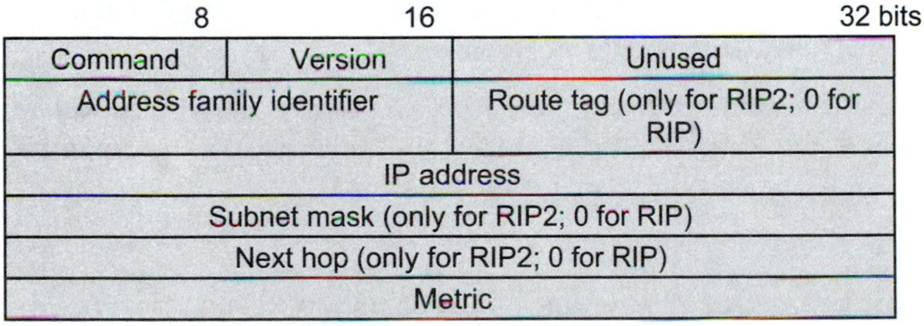

8	16	32 bits
Command	Version	Unused
Address family identifier		Route tag (only for RIP2; 0 for RIP)
IP address		
Subnet mask (only for RIP2; 0 for RIP)		
Next hop (only for RIP2; 0 for RIP)		
Metric		

RIP packet structure

The portion of the datagram from Address Family Identifier through Metric may appear up to 25 times.

Command

The command field is used to specify the purpose of this datagram:

1 Request: A request for the responding system to send all or part of its routing table.

2 Response: A message containing all or part of the sender's routing table. This message may be sent in response to a request or poll, or it may be an update message generated by the sender.

3 Traceon: Obsolete. Messages containing this command are to be ignored.

4 Traceoff: Obsolete. Messages containing this command are to be ignored.

5 Reserved: Used by Sun Microsystems for its own purposes.

Version

The RIP version number. Datagrams are processed according to version number, as follows:

0 Datagrams whose version number is zero are to be ignored.

1 Datagrams whose version number is one are processed. All fields that are to be 0, are to be checked. If any such field contains a non-zero value, the entire message is ignored.

2 Specifies RIP messages which use authentication or carry information in any of the newly defined fields.

>2 Datagrams whose version numbers are greater than 1 are processed. All fields that are 0 are ignored.

Address family identifier

Indicates what type of address is specified in this particular entry. This is used because RIP may carry routing information for several different protocols. The address family identifier for IP is 2.

When authentication is in use, the Address Family Identifier field will be set to 0xFFFF, the Route Tag field contains the authentication type and the remainder of the message contains the password.

Route tag

Attribute assigned to a route which must be preserved and readvertised with a route. The route tag provides a method of separating internal RIP routes (routes for networks within the RIP routing domain) from external RIP routes, which may have been imported from an EGP or another IGP.

IP address

The IP address of the destination.

Subnet mask

Value applied to the IP address to yield the non-host portion of the address. If zero, then no subnet mask has been included for this entry.

Next hop

Immediate next hop IP address to which packets to the destination specified by this route entry should be forwarded.

Metric

Represents the total cost of getting a datagram from the host to that destination. This metric is the sum of the costs associated with the networks that would be traversed in getting to the destination.

RIPng for IPv6

IETF RFC 2080 1997-01 http://www.cis.ohio-state.edu/htbin/rfc/rfc2080.html

RIPng for IPv6 is a routing protocol for the IPv6 Internet. It is based on protocols and algorithms used extensively in the IPv4 Internet.

The format of the header is shown in the following illustration:

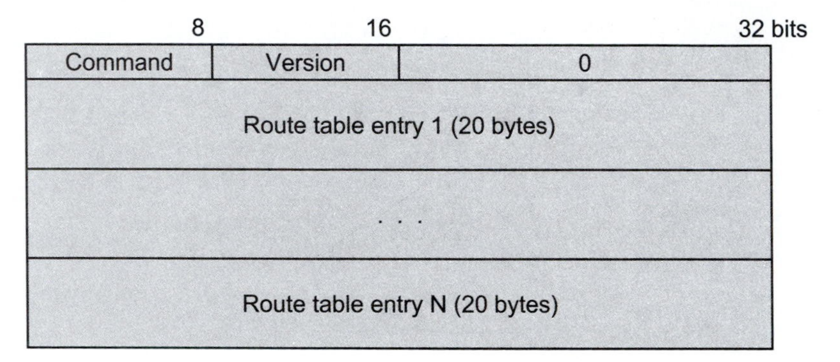

RIPng for IPv6 header structure

Command

The purpose of the message. Possible commands are:

Request A request for the responding system to send all or part of its routing table

Response A message containing all or part of the sender's routing table.

Version

The version of the protocol. The current version is version 1.

Route table entry

Each route table entry contains a destination prefix, the number of significant bits in the prefix and the cost of reaching that destination.

RSVP

IETF draft-ietf-rsvp-spec-13.txt 08-1996, IETF draft-ietf-rsvp-md5-02.txt 06-199 draft-ietf-mpls-rsvp-lsp-tunnel-05.txt

RSVP is a Resource ReSerVation setup Protocol designed for an integrated services Internet. It is used by a host on behalf of an application data stream to request a specific quality of service from the network for particular data streams or flows. It is also used by routers to deliver QoS control requests to all nodes.

The format of the header is shown in the following illustration:

4	8	16	32 bits
Ver	Flags	Message type	RSVP checksum
Send TTL		(Reserved)	RSVP length

RSVP header structure

Version
The protocol version number, this is version 1.

Flags
No flag bits are defined yet.

Message type
Possible values are:

1 Path.
2 Resv.
3 PathErr.
4 ResvErr.
5 PathTear.
6 ResvTear.
7 ResvConf.

RSVP checksum
The checksum.

Send TTL
The IP TTL value with which the message was sent.

RSVP length

The total length of the RSVP message in bytes, including the common header and the variable length objects that follow.

Protocol Object Format

The protocol object follows the header.

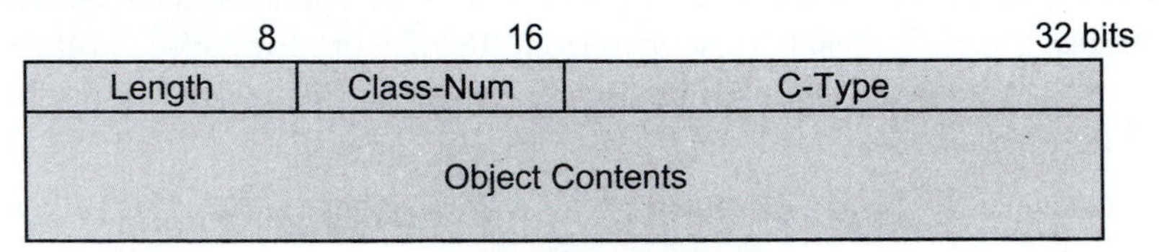

Protocol Object Format

The following optional object types exist:

0 Null
1 Session
3 RSVP Hop
4 Internity
5 Time Values
6 Error Spec
7 Scope
8 Style
9 FlowSpec
10 FilterSpec
11 Sender Template
12 SenderTSpec
13 ADSpec
14 Policy Data
15 Resv Confirm

VRRP

Virtual Router Redundancy Protocol (VRRP) specifies an election protocol that dynamically assigns responsibility for a virtual router to one of the VRRP routers on a LAN. The VRRP router controlling the IP address(es) associated with a virtual router is called the Master, and forwards packets sent to these IP addresses. The election process provides dynamic fail-over in the forwarding responsibility should the Master become unavailable. This allows any of the virtual router IP addresses on the LAN to be used as the default first hop router by end-hosts. The advantage gained from using VRRP is a higher availability default path without requiring configuration of dynamic routing or router discovery protocols on every end-host. This protocol is intended for use with IPv4 routers only. VRRP packets are sent encapsulated in IP packets.

The structure of the VRRP packet is:

4	8	16		32 bits
Ver	Type	Virtual Rtr. ID	Prority	Count IP Address
Auth Type		Adver Int	Checksum	
IP address 1				
⇓ . . .				
IP adresss N				
Authentication Data (1)				
Authentication Data (2)				

VRRP header structure

Version

The version field specifies the VRRP protocol version of this packet. This version is version 2.

Type

The type field specifies the type of this VRRP packet. The only packet type defined in this version of the protocol is: 1 ADVERTISEMENT.
A packet with an unknown type must be discarded.

Virtual Rtr ID

The Virtual Router Identifier (VRID) field identifies the virtual router this packet is reporting status for.

Priority

Specifies the sending VRRP router's priority for the virtual router. VRRP routers backing up a virtual router must use priority values between 1-254 (decimal).

Count IP addresses

The number of IP addresses contained in this VRRP advertisement.

Auth Type

Identifies the authentication method being utilized.

Authentication Methods

0 No Authentication
1 Simple Text Password
2 IP Authentication Header

Advertisement interval

Indicates the time interval (in seconds) between advertisements.

Checksum

Used to detect data corruption in the VRRP message. The checksum is the 16-bit one's complement of the one's complement sum of the entire VRRP message starting with the version field. For computing the checksum, the checksum field is set to zero.

IP address(es)

One or more IP addresses that are associated with the virtual router. The number of addresses included is specified in the "Count IP Addrs" field. These fields are used for troubleshooting misconfigured routers.

Authentication data

The authentication string is currently only utilized for simple text authentication, similar to the simple text authentication found in the Open Shortest Path First routing protocol (OSPF). It is up to 8 characters of plain text.

AH

RFC 1826 http://www.cis.ohio-state.edu/htbin/rfc/rfc1826.html
RFC 1827 http://www.cis.ohio-state.edu/htbin/rfc/rfc1827.html

The IP Authentication Header seeks to provide security by adding authentication information to an IP datagram. This authentication information is calculated using all of the fields in the IP datagram (including not only the IP Header but also other headers and the user data) which do not change in transit. Fields or options which need to change in transit (e.g., hop count, time to live, ident, fragment offset or routing pointer) are considered to be zero for the calculation of the authentication data. This provides significantly more security than is currently present in IPv4 and might be sufficient for the needs of many users. When used with IPv6, the Authentication Header normally appears after the IPv6 Hop-by-Hop Header and before the IPv6 Destination Options. When used with IPv4, the Authentication Header normally follows the main IPv4 header.

The format of the header is shown in the following illustration:

Next header (8)	Length (16)	Reserved (32 bits)
SPI		
Authentication data		

AH header structure

Next header
The next payload after the authentication payload.

Length
The length of the authentication data field.

Reserved
Reserved for future use, must be set to zero.

Security parameters index (SPI)
Identifies the security association for this datagram.

Authentication data

Variable number of 32-bit words.

ESP

RFC 1826 http://www.cis.ohio-state.edu/htbin/rfc/rfc1826.html
RFC 1827 http://www.cis.ohio-state.edu/htbin/rfc/rfc2406.html

The IP Encapsulating Security Payload (ESP) seeks to provide confidentiality and integrity by encrypting data to be protected and placing the encrypted data in the data portion of the IP ESP. Depending on the user's security requirements, this mechanism may be used to encrypt either a transport-layer segment (e.g., TCP, UDP, ICMP, IGMP) or an entire IP datagram. Encapsulating the protected data is necessary to provide confidentiality for the entire original datagram.

ESP may appear anywhere after the IP header and before the final transport-layer protocol. The Internet Assigned Numbers Authority has assigned Protocol Number 50 to ESP. The header immediately preceding an ESP header will always contain the value 50 in its Next Header (IPv6) or Protocol (IPv4) field. ESP consists of an unencrypted header followed by encrypted data. The encrypted data includes both the protected ESP header fields and the protected user data, which is either an entire IP datagram or an upper-layer protocol frame (e.g., TCP or UDP).

The format of the header is shown in the following illustration:

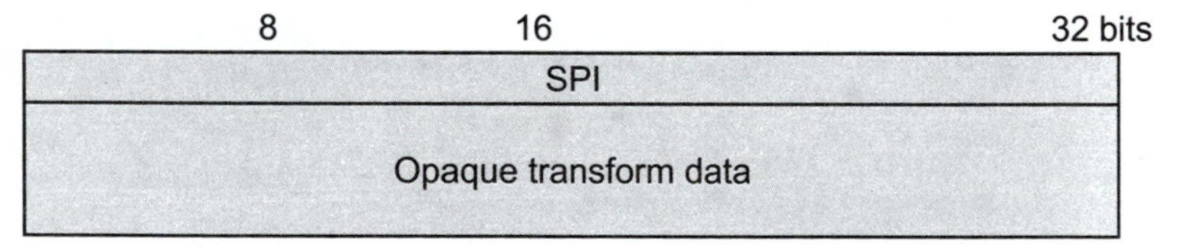

ESP header structure

Security parameters index (SPI)

A 32-bit pseudo-random value identifying the security association for this datagram. If no security association has been established, the value of the SPI field is 0x00000000. An SPI is similar to the SAID used in other security protocols. The name has been changed because the semantics used here are not exactly the same as those used in other security protocols.

Opaque transform data

Variable length data field.

BGP-4

IETF RFC 1654 http://www.cis.ohio-state.edu/htbin/rfc/rfc1654.html
Draft-ietf-idr-BGP-4-muiltiprotocol-v2-05

The Border Gateway Protocol (BGP) is an inter-Autonomous System routing protocol. The primary function of a BGP speaking system is to exchange network reachability information with other BGP systems. BGP-4 provides a new set of mechanisms for supporting classes interdomain routing.

The format of the header is shown in the following illustration:

Marker	Length	Type	
16	2	1	bytes

BGP-4 header structure

Marker
A 16-byte message containing a value predictable by the receiver of the message.

Length
The length of the message including the header.

Type
The message type. Possible messages are:
Open, Update, Notification, KeepAlive.

EGP

RFC 904 http://www.cis.ohio-state.edu/htbin/rfc/rfc904.html

The Exterior Gateway Protocol EGP exists in order to convey net-reachability information between neighboring gateways, possibly in different autonomous systems. The protocol includes mechanisms to acquire neighbors, monitor neighbor reachability and exchange net-reachability information in the form of Update messages. The protocol is based on periodic polling using Hello/I-Heard-You (I-H-U) message exchanges to monitor neighbor reachability and Poll commands to solicit Update responses.

The format of the header is shown in the following illustration:

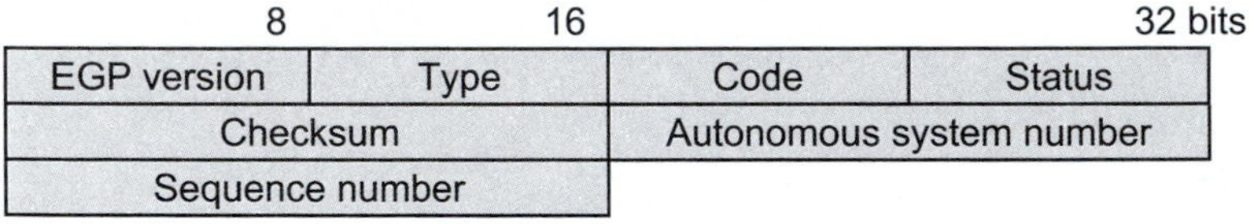

8	16		32 bits
EGP version	Type	Code	Status
Checksum		Autonomous system number	
Sequence number			

EGP header structure

EGP Version
The version number.

Type
Identifies the message type. Possible types are as follows:

1 Update response/indication.
2 Poll command.
3 Neighbor acquisition message.
5 Neighbor reachability message.
8 Error response/indication.

Code
Identifies the message code.

Status
Contains message-dependent status information.

Checksum
The 16-bit one's complement of the one's complement sum of the EGP message starting with the EGP version number field. When computing the checksum the checksum field itself should be zero.

Autonomous system number
Assigned number identifying the particular autonomous system.

Sequence number
Send state variable (commands) or receive state variable (responses and indications).

EIGRP

Enhanced Interior Gateway Routing Protocol (EIGRP) is an enhanced version of IGRP. IGRP is Cisco's Interior Gateway Routing Protocol used in TCP/IP and OSI internets. It is regarded as an interior gateway protocol (IGP) but has also been used extensively as an exterior gateway protocol for inter-domain routing. IGRP uses distance vector routing technology. The same distance vector technology found in IGRP is also used in EIGRP, and the underlying distance information remains unchanged. The convergence properties and the operating efficiency of this protocol have improved significantly.

The format of the EIGRP header is shown in the following illustration.

8	16	32 bits
Version	Opcode	Checksum
Flags		
Sequence number		
Acknowledge number		
Autonomous system number		
Type		Length

EIGRP header structure

Version

The version of the protocol.

Opcode

1 Update.
2 Reserved.
3 Query.
4 Hello.
5 IPX-SAP.

Type

1 EIGRP Parameters.
2 Reserved.
3 Sequence.
4 Software version.
5 Next Multicast sequence.

Length
Length of the frame.

GRE

RFC 1701 http://www.cis.ohio-state.edu/htbin/rfc/rfc1701.html
RFC 1702 http://www.cis.ohio-state.edu/htbin/rfc/rfc1702.html

The Generic Routing Encapsulation (GRE) protocol provides a mechanism for encapsulating arbitrary packets within an arbitrary transport protocol. In the most general case, a system has a packet that needs to be encapsulated and routed (the payload packet). The payload is first encapsulated in a GRE packet, which possibly also includes a route. The resulting GRE packet is then encapsulated in some other protocol and forwarded (the delivery protocol).

GRE is also used with IP, using IP as the delivery protocol or the payload protocol. The GRE header used in PPTP is enhanced slightly from that specified in the current GRE protocol specification.

The format of the header is shown in the following illustration:

16	32 bits
Flags	Protocol type
Checksum (optional)	Offset (optional)
Key (optional)	
Sequence number (optional)	
Routing (optional)	

GRE header structure

Flags

The GRE flags are encoded in the first two octets. Bit 0 is the most significant bit, bit 12 is the least significant bit. Flags are as follows:

Checksum present (bit 0). When set to 1, the Checksum field is present and contains valid information.

Routing present (bit 1). When set to 1, the Offset and Routing fields are present and contain valid information.

Key present (bit 2). When set to 1, the Key field is present in the GRE header.

Sequence number present (bit 3). When set to 1, the Sequence number field is present.

Strict Source Route (bit 4). It is recommended that this bit only be set to 1 if all of the Routing Information consists of Strict Source Routes.

Recursion Control (bits 5-7). Contains a three bit unsigned integer which contains the number of additional encapsulations which are permissible.

Version Number (bits 13-15). Contains the value 0.

Protocol type

Contains the protocol type of the payload packet. In general, the value will be the Ethernet protocol type field for the packet.

Offset

Indicates the octet offset from the start of the Routing field to the first octet of the active Source Route Entry to be examined.

Checksum

Contains the IP (one's complement) checksum of the GRE header and the payload packet.

Key

Contains a four octet number which was inserted by the encapsulator. It may be used by the receiver to authenticate the source of the packet.

Sequence number

Contains an unsigned 32 bit integer which is inserted by the encapsulator. It may be used by the receiver to establish the order in which packets have been transmitted from the encapsulator to the receiver.

Routing

Contains data which may be used in routing this packet.

The format of the enhanced GRE header is as follows:

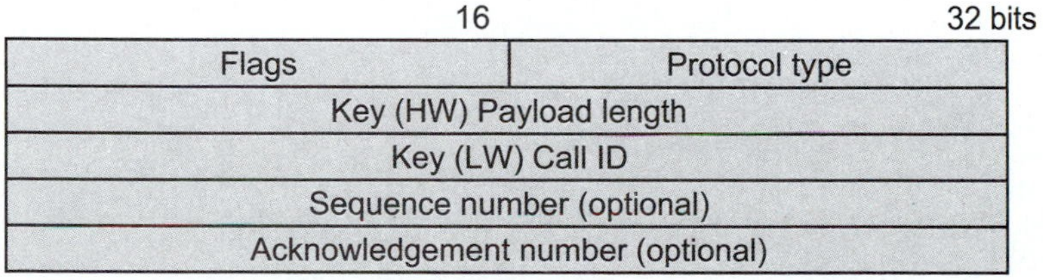

16	32 bits
Flags	Protocol type
Key (HW) Payload length	
Key (LW) Call ID	
Sequence number (optional)	
Acknowledgement number (optional)	

GRE header structure

Flags

Flags are defined as follows:
C (bit 0). Checksum present.
R (bit 1). Routing present.
K (bit 2). Key present.
S (bit 3). Sequence Number present.
s (bit 4). Strict source route present.
Recur (bits 5-7). Recursion control.
A (bit 8). Acknowledgment sequence number present.
Flags (bits 9-12). Must be set to zero.
Ver (bits 13-15). Must contain 1 (enhanced GRE).

Protocol type

Set to hex 880B.

Key

Use of the Key field is up to the implementation.

Sequence number

Contains the sequence number of the payload.

Acknowledgment number

Contains the sequence number of the highest numbered GRE packet received by the sending peer for this user session.

HSRP

RFC 2281 http://www.cis.ohio-state.edu/htbin/rfc/rfc2281.html

The Cisco Hot Standby Router Protocol (HSRP) provides a mechanism which is designed to support non-disruptive failover of IP traffic in certain circumstances. In particular, the protocol protects against the failure of the first hop router when the source host cannot learn the IP address of the first hop router dynamically. The protocol is designed for use over multi-access, multicast, or broadcast capable LANs (e.g., Ethernet). A large class of legacy host implementations that do not support dynamic discovery are capable of configuring a default router. HSRP provides failover services to those hosts.

HSRP runs on top of UDP, and uses port number 1985. Packets are sent to multicast address 224.0.0.2 with TTL 1. Routers use their actual IP address as the source address for protocol packets, not the virtual IP address. This is necessary so that the HSRP routers can identify each other.

The format of the data portion of the UDP datagram is shown in the following illustration:

8	16		32 bits
Version	OpCode	State	Hellotime
Holdtime	Priority	Group	Reserved
Authentication data			
Virtual IP address			

HSRP header structure

Version

HSRP version number, 0 for this version.

OpCode

Type of message contained in the packet. Possible values are:

0 Hello, sent to indicate that a router is running and is capable of becoming the active or standby router.

1 Coup, sent when a router wishes to become the active router.

2 Resign, sent when a router no longer wishes to be the active router.

State

Internally, each router in the standby group implements a state machine. The State field describes the current state of the router sending the message. Possible values are:

0 Initial
1 Learn
2 Listen
4 Speak
8 Standby
16 Active

Hellotime

Approximate period between the Hello messages that the router sends (for Hello messages only). The time is given in seconds. If the Hellotime is not configured on a router, then it may be learned from the Hello message from the active router. The Hellotime should only be learned if no Hellotime is configured and the Hello message is authenticated. A router that sends a Hello message must insert the Hellotime that it is using in the Hellotime field in the Hello message. If the Hellotime is not learned from a Hello message from the active router and it is not manually configured, a default value of 3 seconds is recommended.

Holdtime

The amount of time, in seconds, that the current Hello message should be considered valid. (For Hello messages only.)

Priority

Used to elect the active and standby routers. When comparing priorities of two different routers, the router with the numerically higher priority wins. In the case of routers with equal priority the router with the higher IP address wins.

Group

Identifies the standby group. For Token Ring, values between 0 and 2 inclusive are valid. For other media, values between 0 and 255 inclusive are valid.

Authentication date

Clear-text 8 character reused password. If no authentication data is configured, the recommended default value is 0x63 0x69 0x73 0x63 0x6F 0x00 0x00 0x00.

Virtual IP Address

Virtual IP address used by this group. If the virtual IP address is not configured on a router, then it may be learned from the Hello message from the active router. An address should only be learned if no address was configured and the Hello message is authenticated.

IGRP

The Interior Gateway Routing Protocol (IGRP) was developed by the Cisco company. It is used to transfer routing information between routers.

IGRP is sent using IP datagrams with IP 9 (IGP). The packet begins with a header which starts immediately after the IP header.

Field	Octets
Version	1
Opcode	1
Edition	1
ASystem	1
Ninterior	1
Nsystem	1
Nexterior	1
Checksum	1

IGRP header structure

Version
Protocol version number (currently 1).

Opcode
Operation code indicating the message type:

1 Update.
2 Request.

Edition
Serial number which is incremented whenever there is a change in the routing table. The edition number allows gateways to avoid processing updates containing information that they have already seen.

ASystem
Autonomous system number. A gateway can participate in more than one autonomous system where each system runs its own IGRP. For each autonomous system, there are completely separate routing tables. This field allows the gateway to select which set of routing tables to use.

Ninterior, Nsystem, Nexterior

Indicate the number of entries in each of these three sections of update messages. The first entries (Ninterior) are taken to be interior, the next entries (Nsystem) as being system, and the final entries (Nexterior) as exterior.

Checksum

IP checksum which is computed using the same checksum algorithm as a UDP checksum. The checksum is computed on the IGRP header and any routing information that follows it. The checksum field is set to zero when computing the checksum. The checksum does not include the IP header and there is no virtual header as in UDP and TCP.

An IGRP request asks the recipient to send its routing table. The request message has only a header. Only the Version, Opcode and ASystem fields are used; all other fields are zero.

An IGRP update message contains a header, immediately followed by routing entries. As many routing entries as possible are included to fit into a 1500-byte datagram (including the IP header). With current structure declarations, this allows up to 104 entries. If more entries are needed, several update messages are sent.

NARP

RFC 1735 http://www.cis.ohio-state.edu/htbin/rfc/rfc1735.html

The NBMA Address Resolution Protocol (NARP) allows a source terminal (a host or router), wishing to communicate over a Non-Broadcast, Multi-Access link layer (NBMA) network, to find out the NBMA addresses of a destination terminal if the destination terminal is connected to the same NBMA network as the source.

The general format of the header is shown in the following illustration. The configuration varies according to the value of the type field to request type and reply type:

8	16	32 bits
Version	Hop count	Checksum
Type	Code	Unused
Destination IP address		
Source IP address		
NBMA length	NBMA address (variable length)	

NARP header structure

Version

The NARP version number. Currently this value is 1.

Hop count

Indicates the maximum number of NASs that a request or reply is allowed to traverse before being discarded.

Checksum

The standard IP checksum over the entire NARP packet (starting with the fixed header).

Type

The NARP packet type. The NARP Request has a Type code 1, NARP Reply has a Type code 2.

Code

A response to an NARP request may contain cached information. If an authoritative answer is desired, then code 2 (NARP Request for Authoritative Information) should be used. Otherwise, a code value of 1 (NARP Request) should be used. NARP replies may be positive or negative. A Positive, Non- authoritative Reply carries a code of 1, while a Positive, Authoritative Reply carries a code of 2. A Negative, Non- authoritative Reply carries a code of 3 and a Negative, Authoritative reply carries a code of 4.

Source and destination IP addresses

Respectively, these are the IP addresses of the NARP requestor and the target terminal for which the NBMA address is desired.

NBMA length and NBMA address

The NBMA length field is the length of the NBMA address of the source terminal in bits. The NBMA address itself is zero-filled to the nearest 32-bit boundary.

NHRP

RFC 2332 http://www.cis.ohio-state.edu/htbin/rfc/rfc2332.html
draft http://info.internet.isi.edu:80/in-drafts/files/draft-ietf-rolc-nhrp-15.txt

The NBMA Next Hop Resolution Protocol (NHRP) allows a source station (a host or router), wishing to communicate over a Non-Broadcast, Multi-Access (NBMA) subnetwork, to determine the internetworking layer addresses and NBMA addresses of suitable *NBMA next hops* toward a destination station.

The format of the header is shown in the following illustration:

8	16	24	32 bits
ar$afn		ar$pro.type	
ar$pro.snap			
ar$pro.snap	ar$hopcnt	ar$pkstz	
ar$chksum		ar$extoff	
ar$op.version	ar$op.type	ar$shtl	ar$sstl

NHRP header structure

ar$afn
Defines the type of link layer address being carried.

ar$pro.type
This field is a 16 bit unsigned integer.

ar$pro.snap
When ar$pro.type has a value of 0x0080, a snap encoded extension is being used to encode the protocol type. This snap extension is placed in the ar$pro.snap field; otherwise this field should be set to 0.

ar$hopcnt
The hop count. This indicates the maximum number of NHSs that an NHRP packet is allowed to traverse before being discarded.

ar$pktsz
The total length of the NHRP packet in octets.

ar$chksum

The standard IP checksum over the entire NHRP packet.

ar$extoff

This field identifies the existence and location of NHRP extensions.

ar$op.version

This field indicates what version of generic address mapping and management protocol is represented by this message.

ar$op.type

If the ar$op.version is 1 then this field represents the NHRP packet type. Possible values for packet types are:

1 NHRP Resolution Request.
2 NHRP Resolution Reply.
3 NHRP Registration Request.
4 NHRP Registration Reply.
5 NHRP Purge Request.
6 NHRP Purge Reply.
7 NHRP Error Indication.

ar$shtl

The type and length of the source NBMA address interpreted in the context of the *address family number*.

ar$sstl

The type and length of the source NBMA subaddress interpreted in the context of the "address family number".

OSPF

RFC1583 http://www.cis.ohio-state.edu/htbin/rfc/rfc1583.html
draft-yeung-ospf-traffic-00

OSPF (Open Shortest Path First) protocol is a link-state routing protocol used for routing IP. It is an interior gateway protocol which is used for routing within a group of routers. It uses link-state technology in which routers send each other information about the direct connections and links which they have to other routers.

The OSPF header structure is shown in the illustration below.

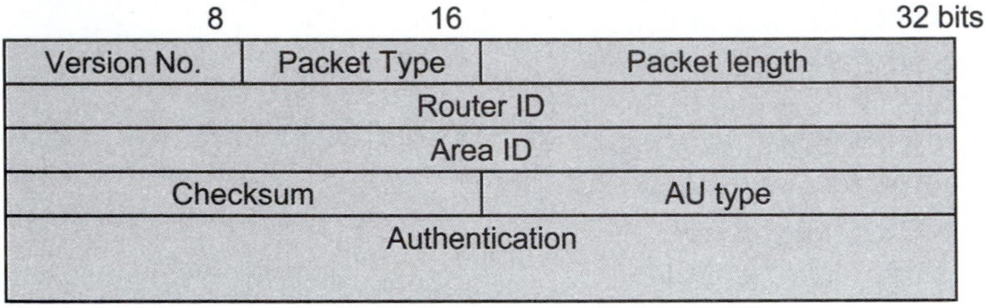

8	16	32 bits
Version No.	Packet Type	Packet length
Router ID		
Area ID		
Checksum		AU type
Authentication		

OSPF header structure

Version number

Protocol version number (currently 1).

Packet type

Valid types are as follows:

1 Hello
2 Database Description
3 Link State Request
4 Link State Update
5 Link State Acknowledgment.

Packet length

The length of the protocol packet in bytes. This length includes the standard OSPF header.

Router ID

The router ID of the packet's source. In OSPF, the source and destination of a routing protocol packet are the two ends of an (potential) adjacency.

Area ID

A 32-bit number identifying the area that this packet belongs to. All OSPF packets are associated with a single area. Most travel a single hop only. Packets traveling over a virtual link are labeled with the back bone area ID of 0.0.0.0.

Checksum

The standard IP checksum of the entire contents of the packet, starting with the OSPF packet header, but excluding the 64-bit authentication field. This checksum is calculated as the 16-bit one's complement of the one's complement sum of all the 16-bit words in the packet, except for the authentication field. If the packet length is not an integral number of 16-bit words, the packet is padded with a byte of zero before checksumming.

AU type

Identifies the authentication scheme to be used for the packet.

Authentication

A 64-bit field for use by the authentication scheme.

Mobile IP

RFC 2002: http://www.cis.ohio-state.edu/htbin/rfc/rfc2002.html
RFC 2290: http://www.cis.ohio-state.edu/htbin/rfc/rfc2290.html
RFC 2344: http://www.isi.edu/in-notes/rfc2344.txt

The Mobile IP protocol enables nodes to move from one IP subnet to another. Each mobile node is always identified by its home address, regardless of its current point of attachment to the Internet. While situated away from its home, a mobile node is also associated with a care-of address, which provides information about its current point of attachment to the Internet. The protocol allows registration of the care-of address with a home agent. The home agent sends datagrams destined for the mobile node through a tunnel to the care-of address. After arriving at the end of the tunnel, each datagram is then delivered to the mobile node. It can be used for mobility across both homogeneous and heterogeneous media. Mobile IP defines a set of new control messages, sent with UDP, Registration Request and Registration Reply.

The IP packet consists of the IP source and destination addresses, followed by the UDP source and destination ports, followed by the Mobile IP fields. Mobile IP packets can be either registration request or registration reply.

The format of the Mobile IP registration request message is shown in the following illustration:

	8	9	10	11	12	13	14	15	16	Octet
Type		S	B	D	M	G	V	T	Rsv	2
Lifetime										4
Home address										8
Home agent										12
Care of address										16
Identification										20
Extensions										…

Mobile IP registration request message structure

Type
1 signifies a registration request.

S
Simultaneous bindings. When set, the mobile node is requesting that the home agent retain its prior mobility bindings.

B
Broadcast datagrams. When set, the mobile node requests that the home agent tunnel to it any broadcast datagrams that it receives on the home network.

D
Decapsulation by mobile node. When set, the mobile node will itself decapsulate datagrams which are sent to the care-of address. In other words, the mobile node is using a co-located care-of address.

M
Minimal encapsulation. When set, the mobile node requests that its home agent use minimal encapsulation for datagrams tunneled to the mobile node.

G
GRE encapsulation. When set, the mobile node requests that its home agent use GRE encapsulation for datagrams tunneled to the mobile node.

V
The mobile node requests that its mobility agent use Van Jacobson header compression over its link with the mobile node.

T
When set, the mobile node asks its home agent to accept a reverse tunnel from the care-of address. Mobile nodes using a foreign agent care-of address ask the foreign agent to reverse-tunnel its packets.

Rsv
Reserved bit, set to zero.

Lifetime
The number of seconds remaining before the registration expires.

Home address
IP address of the mobile node.

Home agent
IP address of the mobile node's home agent.

Care-of address
IP address for the end of the tunnel.

Identification
A 64-bit number, constructed by the mobile node, used for matching registration requests with registration replies, and for protecting against replay attacks of registration messages.

Extensions
The fixed portion of the registration request is followed by one or more of the extensions listed in Section 3.5 of RFC2002. The Mobile-Home Authentication Extension must be included in all registration requests.

The format of the Mobile IP registration reply message is shown in the following illustration:

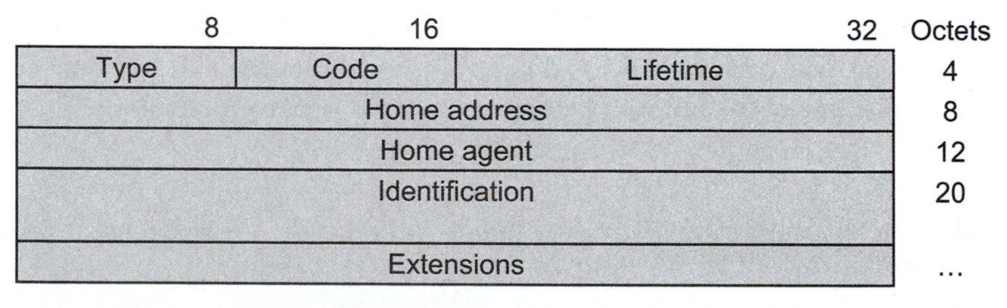

Mobile IP registration reply message structure

Type
3 indicates a registration reply.

Code
A value indicating the result of the Registration Request. Values may be as follows:

Registration successful:

0	Registration accepted.
1	Registration accepted, but simultaneous mobility bindings unsupported.

Registration denied by the foreign agent:

64	Reason unspecified.
65	Administratively prohibited.

66 Insufficient resources.
67 Mobile node failed authentication.
68 Home agent failed authentication.
69 Requested Lifetime too long.
70 Poorly formed Request.
71 Poorly formed Reply.
72 Requested encapsulation unavailable.
73 Requested Van Jacobson compression unavailable.

Service denied by the foreign agent:

74 Requested reverse tunnel unavailable.
75 Reverse tunnel is mandatory and T bit not set.
76 Mobile node too distant

Registration denied by the home agent:

80 Home network unreachable (ICMP error received).
81 Home agent host unreachable (ICMP error received).
82 Home agent port unreachable (ICMP error received).
88 Home agent unreachable (other ICMP error received).

Service denied by the home agent:

137 Requested reverse tunnel unavailable.
138 Reverse tunnel is mandatory and T bit not set.
139 Requested encapsulation unavailable.

Lifetime

If the Code field indicates that the registration was accepted, the Lifetime field is set to the number of seconds remaining before the registration expires. A value of zero indicates that the mobile node has been deregistered. A value of 0xffff indicates infinity. If the Code field indicates that the registration was denied, the contents of the Lifetime field are unspecified and are ignored on reception.

RUDP

Draft-ietg-sigtran-reliable-udp-00.txt (1999)

The Reliable UDP protocol is a simple packet based transport protocol, based on RFCs 1151 and 908. A reliable transport protocol is needed to transport telephony signalling across IP networks. This should provide architecture for a variety of signalling protocols needing transport over IP. RUDP is designed to allow characteristics of each connection to be individually configured so that a number of protocols with different transport requirement can be implemented simultaneously not on the same platform. It is layered on the UDP/IP protocols and provides reliable in-order delivery (up to a maximum number of retransmissions) for virtual connections. RUDP has a very flexible design that would make it suitable for a variety of transport uses. One such use would be to transport telecommunication-signalling protocols.

RUDP Header

Each UDP packet sent by RUDP must start with at least a 6-octet header. The first octet contains a series of single bit flags. The next 3 fields are one octet in size. They are followed by a 2-octet checksum.

SYN	ACK	EAK	RST	NUL	CHK	TCS	0	Header Length
Sequence number								Ack number
Checksum								

RUDP header

Control bits

The 8 control bits are altogether one byte in length and indicate what is present in the packet.

SYN

The SYN bit indicates a synchronization segment is present.

ACK

The ACK bit indicates the acknowledgment number in the header is valid.

EACK

The EACK bit indicates an extended acknowledge segment is present.

RST

The RST bit indicates the packet is a reset segment.

NUL

The NUL bit indicates the packet is a null segment.

CHK

The CHK bit indicates whether the Checksum field contains the checksum of just the header or the header and the body (data).

TCS

The TCS bit indicates the packet is a transfer connection state segment.

0

The value of this field must be zero.

Header length

It is one byte in length and indicates where user data begins in the packet.

Sequence number

When a connection is first opened, each peer randomly picks an initial sequence number. This sequence number is used in the SYN segments to open the connection. Each transmitter increments the sequence number before sending a data, null, or reset segment. It is one byte in length.

Acknowledgement number

The acknowledgment number field indicates to a transmitter the last in-sequence packet the receiver has received. It is one byte in length.

Checksum

The checksum is always calculated on the RUDP header to ensure integrity. The checksum here is the same algorithm used in UDP and TCP headers.

Segments

The following segments can appear in the packet: SYN, Ack EACK, RST, NUL and TCS segments:

The Syn segment

The SYN is used to establish a connection and synchronize sequence numbers between two hosts. The SYN segment also contains the negotiable parameters of the connection. All configurable parameters that the peer must know about are contained in this segment. This includes the maximum number of segments the local RUDP is willing to accept and option flags that indicate the features of the connection being established.

Ack segment

The Ack segment is used to acknowledge in-sequence segments. It contains both the next send sequence number and the acknowledgement sequence number in the RUDP header.

The Eack segment:

The EACK segment is used to acknowledge segments received out of sequence. It contains the sequence numbers of one or more segments received out of sequence. The EACK is always combined with an ACK in the segment, giving the sequence number of the last segment received in sequence. The header length is variable for the EACK segment. Its minimum value is seven and its maximum value depends on the maximum receive queue length.

RST segment

This is used to close or reset a connection. Upon receipt of an RST segment, the sender must stop sending new packets, but continue to attempt delivery of packets already accepted from the API.

NUL segment

This segment is used to determine whether the other side of a connection is still active. When a NUL segment is received, an RUDP implementation must immediately acknowledge the segment if a valid connection exists and the segment acknowledge number is the next one in sequence.

TCS segment

The TCS is used to transfer the state of connection.

TALI

http://search.ietf.org/internet-drafts/draft-benedyk-sigtran-tali-00.txt

This protocol proposes the interfaces of a Signalling Gateway, which provides interworking between the Switched Circuit Network (SCN) and an IP network. Since the Gateway is the central point of signalling information, not only does it provide transportation of signalling from one network to another, but can also provide additional functions such as protocol translation, security screening, routing information, and seamless access to Intelligent Network (IN) services on both networks.

The Transport Adapter Layer Interface (TALI) protocol provides TCAP, ISUP, and MTP messaging over TCP/IP and is used to support reliable communication between the SS7 Signalling Network and applications residing within the IP network.

This version of TALI provides 3 SS7 signalling transport methods and provides functionality for MTP over TCP/IP (lmtp3), SCCP/TCAP over TCP/IP (sccp), and ISUP over TCP/IP (isot). These three methods comprise the service messages.

The following is a description of the TALI payload:

2 bytes	2 bytes
SYNC	
OpCode	
Length	Sevice message data

TALI Payload structure

SYNC

Four bytes must be (54 41 4C 49) TALI in ASCII.

OpCode

The kind of TALI frame.

The following types of frame exist

Type of frame	Ascii OpCode
Test Service on this Socket	test
Allow Service messages on this socket	allo
Prohibit Service messages on this socket	proh

Type of frame	Ascii OpCode
Prohibit Service messages Ack	proa
Monitor Socket message on this socket	moni
Monitor Socket message Ack	mona
SCCP Service message	sccp
ISUP Service message	isot
MTP3 Service message	mtp3
MTP Primitives	mtpp
SCCP Primitives	scpp
Routing Key Registration	rkrg
Routing Key De-Registration	rkdr
Special Service Message	spcl

Length

The length of the frame. Non zero if message contains a Service or Monitor Socket message.

Service message data:

The service message data.

Van Jacobson

IETF RFC 1144 http://www.cis.ohio-state.edu/htbin/rfc/rfc1144.html

Van Jacobson is a compressed TCP protocol which thereby improves the TCP/IP performance over low speed (300 to 19,200 bps) serial links.

The format of the compressed TCP is as follows:

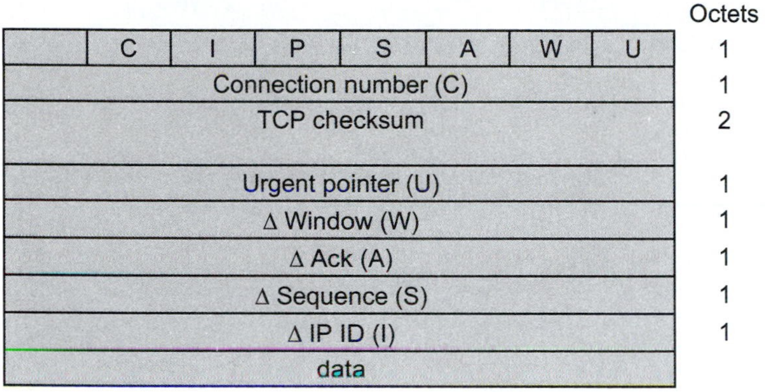

								Octets
	C	I	P	S	A	W	U	1
Connection number (C)								1
TCP checksum								2
Urgent pointer (U)								1
Δ Window (W)								1
Δ Ack (A)								1
Δ Sequence (S)								1
Δ IP ID (I)								1
data								

Compressed TCP structure

C, I, P, S, A, W, U
Change mask. Identifies which of the fields expected to change per-packet actually changed.

Connection number
Used to locate the saved copy of the last packet for this TCP connection.

TCP checksum
Included so that the end-to-end data integrity check will still be valid.

Urgent pointer
This is sent if URG is set.

Δ values for each field
Represent the amount the associated field changed from the original TCP (for each field specified in the change mask).

XOT

IETF RFC 1613 http://www.cis.ohio-state.edu/htbin/rfc/rfc1613.html

XOT is Cisco Systems X.25 over TCP. The format of the header is shown in the following illustration:

Version	Length
2 bytes	2 bytes

XOT header structure

Version
The version number.

Length
The length of the packet.

DNS

RFC 1035 1987-11 http://www.cis.ohio-state.edu/htbin/rfc/rfc1035.html
RFC 1706 1994-01 http://www.cis.ohio-state.edu/htbin/rfc/rfc1706.html

The Domain Name Service (DNS) protocol searches for resources using a database distributed among different name servers.

The DNS message header structure is shown in the following illustration:

16		21					28	32 bits
ID	Q	Query	A	T	R	V	B	Rcode
Question count	Answer count							
Authority count	Additional count							

DNS message header structure

ID
16-bit field used to correlate queries and responses.

Q
1-bit field that identifies the message as a query or response.

Query
4-bit field that describes the type of message:
0 Standard query (name to address).
1 Inverse query (address to name).
2 Server status request.

A
Authoritative Answer. 1-bit field. When set to 1, identifies the response as one made by an authoritative name server.

T
Truncation. 1-bit field. When set to 1, indicates the message has been truncated.

R
1-bit field. Set to 1 by the resolve to request recursive service by the name server.

V

1-bit field. Signals the availability of recursive service by the name server.

B

3-bit field. Reserved for future use. Must be set to 0.

RCode

Response Code. 4-bit field that is set by the name server to identify the status of the query:

0	No error condition.
1	Unable to interpret query due to format error.
2	Unable to process due to server failure.
3	Name in query does not exist.
4	Type of query not supported.
5	Query refused.

Question count

16-bit field that defines the number of entries in the question section.

Answer count

16-bit field that defines the number of resource records in the answer section.

Authority count

16-bit field that defines the number of name server resource records in the authority section.

Additional count

16-bit field that defines the number of resource records in the additional records section.

NetBIOS/IP

IETF RFC 1002 http://www.cis.ohio-state.edu/htbin/rfc/rfc1002.html

NetBIOS/IP is a standard protocol to support NetBIOS services in a TCP/IP environment. Both local network and Internet operations are supported. Various node types are defined to accommodate local and Internet topologies and to allow operation with or without the use of IP broadcast.

NetBIOS types may be Name Service, Session or Datagram.

The format of the header is shown in the following illustration:

16	21	28	32 bits
Name_trn_id	Opcode	Nm_flags	Rcode
Qdcount (16 bits)	Ancount (16 bits)		
Nscount (16 bits)	Arcount (16 bits)		

NetBIOS/IP header structure

Name_trn_id
Transaction ID for the Name Service Transaction.

Opcode
Packet type code: Possible values are:

0 Query.
5 Registration.
6 Release.
7 WACK.
8 Refresh.

Nm_flags
Flags for operation.

Rcode
Result codes of request.

Qdcount
Unsigned 16 bit integer specifying the number of entries in the question section of a name.

Ancount

Unsigned 16 bit integer specifying the number of resource records in the answer section of a name service packet.

Nscount

Unsigned 16 bit integer specifying the number of resource records in the authority section of a name service packet.

Arcount

Unsigned 16 bit integer specifying the number of resource records in the additional records section of a name service packet.

LDAP

RFC 1777.

The LDAP (Lightweight Directory Access Protocol.) provides access to X.500 directories without using the DAP (Directory Access Protocol). It is used for simple management applications and browser applications that provide simple read/write interactive access to the X.500 directory and should complement the DAP. X.500 technology has proved to be highly popular, and therefore led to efforts to reduce the high "cost of entry" associated with it. Until now methods suggested were based on specific applications and, as such, were limited. The LDAP is also a directory protocol alternative, but it is not dependant on a particular application. As such it is intended to be simpler and less expensive than existing ones.

Main features:

- Protocol elements are carried directly over TCP or any other transport layer protocol.
- Protocol data elements are encoded in ordinary strings.
- Lightweight BER encoding is used to encode all protocol elements.

LDAP works by a client transmitting a request to a server. In the request the client specifies the operation to be performed. The server must then perform the required operation on the directory. After this, the server returns a response containing the results, or any errors.

LADP messages are PDUs mapped directly onto the TCP bytestream and use port 389.

The LDAP messages do not have their own header and are text based messages based on ASN.1

COPS

RFC 2748

The COPS (Common Open Policy Service) protocol describes a simple query and response protocol that can be used to exchange policy information between a policy server (Policy Decision Point or PDP) and its clients (Policy Enforcement Points or PEPs). It is designed to be extensible so that other kinds of policy clients may be supported in the future. The model does not make any assumptions about the methods of the policy server, but is based on the server returning decisions to policy requests. Each message consists of the COPS header followed by a number of typed objects.

The structure of the COPS header is:

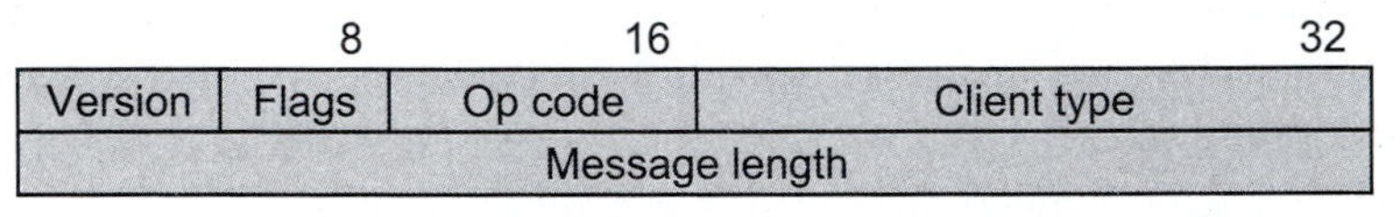

COPS Header structure

Version

The version field specifies the COPS version number. The current version is 1.

Flags

The defined flag value is 1 a Solicited Message Flag Bit. This flag is set when the message is solicited by another COPS message (all other flags are set to 0).

Op code

Code identifying the COPS operations:

1 Request (REQ)
2 Decision (DEC)
3 Report State (RPT)
4 Delete Request State (DRQ)
5 Synchronize State Req (SSQ)
6 Client-Open (OPN)
7 Client-Accept (CAT)

8	Client-Close (CC)
9	Keep-Alive (KA)
10	Synchronize Complete (SSC)

Client-type

The Client-type identifies the policy client. Interpretation of all encapsulated objects is relative to the client-type.

Message length

Size of message in octets, which includes the standard COPS header and all encapsulated objects. Messages are aligned on 4 octet intervals.

COPS specific object formats

After the COPS header come all encapsulated objects that follow the same object format.

Each object consists of one or more 32-bit words with a four-octet header, using the following format:

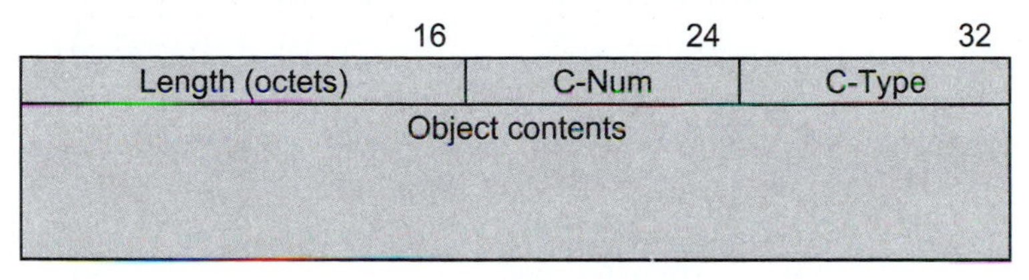

COPS specific object formats

Length

The length is a two-octet value that describes the number of octets (including the header) that compose the object. If the length in octets does not fall on a 32-bit word boundary, padding is added to the end of the object so that it is aligned to the next 32- bit boundary before the object can be sent on the wire. On the receiving side, a subsequent object boundary can be found by simply rounding up the previous stated object length to the next 32-bit boundary.

C-Num

Identifies the class of information contained in the object.

The possible values for the C-number are:

C-Num	Object Contents
1	Handle

2	Context
3	In Interface
4	Out Interface
5	Reason code
6	Decision
7	LDP Decision
8	Error
9	Client Specific Info
10	Keep-Alive Timer
11	PEP Identification
12	Report Type
13	PDP Redirect Address
14	Last PDP Address
15	Accounting Timer
16	Message Integrity

C-type

Identifies the subtype or version of the information contained in the object.

Object contents

The value appearing in the C-Num fields, defines the type of object contents. See the list above for possible object contents.

FTP

IETF RFC 959 1985-10 http://www.cis.ohio-state.edu/htbin/rfc/rfc959.html

The File Transfer Protocol (FTP) provides the basic elements of file sharing between hosts. FTP uses TCP to create a virtual connection for control information and then creates a separate TCP connection for data transfers. The control connection uses an image of the TELNET protocol to exchange commands and messages between hosts.

Commands

FTP control frames are TELNET exchanges and can contain TELNET commands and option negotiation. However, most FTP control frames are simple ASCII text and can be classified as FTP commands or FTP messages. The standard FTP commands are as follows:

Command	*Description*
ABOR	Abort data connection process.
ACCT <account>	Account for system privileges.
ALLO <bytes>	Allocate bytes for file storage on server.
APPE <filename>	Append file to file of same name on server.
CDUP <dir path>	Change to parent directory on server.
CWD <dir path>	Change working directory on server.
DELE <filename>	Delete specified file on server.
HELP <command>	Return information on specified command.
LIST <name>	List information if name is a file or list files if name is a directory.
MODE <mode>	Transfer mode (S=stream, B=block, C=compressed).
MKD <directory>	Create specified directory on server.
NLST <directory>	List contents of specified directory.
NOOP	Cause no action other than acknowledgement from server.
PASS <password>	Password for system log-in.
PASV	Request server wait for data connection.
PORT <address>	IP address and two-byte system port ID.
PWD	Display current working directory.
QUIT	Log off from the FTP server.
REIN	Reinitialize connection to log-in status.

Command	*Description*
REST <offset>	Restart file transfer from given offset.
RETR <filename>	Retrieve (copy) file from server.
RMD <directory>	Remove specified directory on server.
RNFR <old path>	Rename from old path.
RNTO <new path>	Rename to new path.
SITE <params>	Site specific parameters provided by server.
SMNT <pathname>	Mount the specified file structure.
STAT <directory>	Return information on current process or directory.
STOR <filename>	Store (copy) file to server.
STOU <filename>	Store file to server name.
STRU <type>	Data structure (F=file, R=record, P=page).
SYST	Return operating system used by server.
TYPE <data type>	Data type (A=ASCII, E=EBCDIC, I=binary).
USER <username>	User name for system log-in.

Messages

FTP messages are responses to FTP commands and consist of a response code followed by explanatory text. Standard FTP messages are as follows:

Response Code	*Explanatory Text*
110	Restart marker at MARK yyyy=mmmm (new file pointers).
120	Service ready in nnn minutes.
125	Data connection open, transfer starting.
150	Open connection.
200	OK.
202	Command not implemented.
211	(System status reply).
212	(Directory status reply).
213	(File status reply).
214	(Help message reply).
215	(System type reply).
220	Service ready.
221	Log off network.
225	Data connection open.
226	Close data connection.
227	Enter passive mode (IP address, port ID).
230	Log on network.

Response Code	*Explanatory Text*
250	File action completed.
257	Path name created.
331	Password required.
332	Account name required.
350	File action pending.
421	Service shutting down.
425	Cannot open data connection.
426	Connection closed.
450	File unavailable.
451	Local error encountered.
452	Insufficient disk space.
500	Invalid command.
501	Bad parameter.
502	Command not implemented.
503	Bad command sequence.
504	Parameter invalid for command.
530	Not logged onto network.
532	Need account for storing files.
550	File unavailable.
551	Page type unknown.
552	Storage allocation exceeded.
553	File name not allowed.

TFTP

IETF RFC 1350 1992-07 http://www.cis.ohio-state.edu/htbin/rfc/rfc1350.html
IETF RFC 783 http://www.cis.ohio-state.edu/htbin/rfc/rfc783.html

The Trivial File Transfer Protocol (TFTP) uses UDP. TFTP supports file writing and reading; it does not support directory service of user authorization.

Commands

The following are TFTP commands:

Command	*Description*
Read Request	Request to read a file.
Write Request	Request to write to a file.
File Data	Transfer of file data.
Data Acknowledge	Acknowledgement of file data.
Error	Error indication.

Parameters

TFTP Read and Write Request commands use the following parameters:

Parameter	*Description*
Filename	The name of the file, expressed in quotes, where the protocol is to perform the read or write operation.
Mode	Datamode. The format of the file data that the protocol is to transfer. The following formats are possible: NetASCII Standard ASCII character format. Octet Eight-bit binary data. Mail Standard ASCII character format with username in place of filename.

TFTP data and data acknowledge commands use the following parameters:

Command	*Description*
Block	Block number or sequence number of the current frame of file data.
Data	First part of the file data displayed for TFTP data frames.

Command	*Description*
TFTP Errors	TFTP error frames contain an error code in parentheses followed by the error message, as follows: (0000) Unknown Error. (0001) File not found. (0002) Access violation. (0003) Out of disk space. (0004) Illegal TFTP operation. (0005) Unknown Transfer ID. (0006) Filename already exists. (0007) Unknown user.

Finger

RFC 1288 http://www.cis.ohio-state.edu/htbin/rfc/rfc1288.html

The Finger user information protocol is a simple protocol which provides an interface to a remote user information program. It is a protocol for the exchange of user information, based on the Transmission Control Protocol, using TCP port 79 decimal (117 octal). The local host opens a TCP connection to a remote host on the Finger port. An RUIP becomes available on the remote end of the connection to process the request. The local host sends the RUIP a one line query based upon the Finger query specification, and waits for the RUIP to respond. The RUIP receives and processes the query, returns an answer, then initiates the close of the connection. The local host receives the answer and the close signal, then proceeds closing its end of the connection.

The Finger protocol displays data. Any data transferred must be in ASCII format, with no parity, and with lines ending in CRLF (ASCII 13 followed by ASCII 10). This excludes other character formats such as EBCDIC, etc. This also means that any characters between ASCII 128 and ASCII 255 should truly be international data, not 7-bit ASCII with the parity bit set. Note: if ASCII 13 followed by ASCII 10 transferred, the character won't display (because the only meaning is to end the line).

Gopher

RFC 1436 http://www.cis.ohio-state.edu/htbin/rfc/rfc1436.html

The Internet Gopher protocol and software follow a client-server model. This protocol assumes a reliable data stream; TCP is assumed. Gopher servers listen on port 70 (port 70 is assigned to Internet Gopher by IANA). Documents reside on many autonomous servers on the Internet. Users run client software on their desktop systems, connecting to a server and sending the server a selector (a line of text, which may be empty) via a TCP connection at a well-known port. The server responds with a block of text terminated by a period on a line by itself and closes the connection. No state is retained by the server.

The first character on each line tells whether the line describes a document, directory, or search service (characters '0', '1', '7'; there are a handful more of these characters described later). The succeeding characters up to the tab form a user display string to be shown to the user for use in selecting this document (or directory) for retrieval. The first character of the line is really defining the type of item described on this line. In nearly every case, the Gopher client software will give the users some sort of idea about what type of item this is (by displaying an icon, a short text tag, or the like).

The characters following the tab, up to the next tab form a selector string that the client software must send to the server to retrieve the document (or directory listing). The selector string should mean nothing to the client software; it should never be modified by the client. In practice, the selector string is often a pathname or other file selector used by the server to locate the item desired. The next two tab delimited fields denote the domain-name of the host that has this document (or directory), and the port at which to connect. If there are yet other tab delimited fields, the basic Gopher client should ignore them. A CR LF denotes the end of the item.

Item type characters

The client software decides what items are available by looking at the first character of each line in a directory listing. Augmenting this list can extend the protocol. A list of defined item-type characters follows:

0 Item is a file.
1 Item is a directory.
2 Item is a CSO phone-book server.
3 Error.

4	Item is a BinHexed Macintosh file.
5	Item is DOS binary archive of some sort. The client must read until the TCP connection closes.
6	Item is a UNIX uuencoded file.
7	Item is an Index-Search server.
8	Item points to a text-based telnet session.
9	Item is a binary file. The client must read until the TCP connection closes.
+	Item is a redundant server
T	Item points to a text-based tn3270 session.
g	Item is a GIF format graphics file.
I	Item is some kind of image file. The client decides how to display.

Characters '0' through 'Z' are reserved. Local experiments should use other characters. Machine-specific extensions are not encouraged. Note that for type 5 or type 9 the client must be prepared to read until the connection closes. There will be no period at the end of the file; the contents of these files are binary and the client must decide what to do with them based perhaps on the .xxx extension.

HTTP

RFC 1945 http://www.cis.ohio-state.edu/htbin/rfc/rfc1945.html

The Hypertext Transfer Protocol (HTTP) is an application-level protocol with the lightness and speed necessary for distributed, collaborative, hypermedia information systems. Messages are passed in a format similar to that used by Internet Mail and the Multipurpose Internet Mail Extensions (MIME).

Request Packet

The format of the Request packet header is shown in the following illustration:

Method	Request URI	HTTP version

HTTP request packet structure

Method
The method to be performed on the resource.

Request-URI
The Uniform Resource Identifier, the resource upon which to apply the request, i.e. the network resource.

HTTP version
The HTTP version being used.

Response Packet

The format of the Response packet header is shown in the following illustration:

HTTP version	Status code	Reason phrase

HTTP response packet structure

HTTP version
The HTTP version being used.

Status-code
A 3 digit integer result code of the attempt to understand and satisfy the request.

Reason-phrase
A textual description of the status code.

S-HTTP

draft-ietf-wts-shttp-06

Secure HTTP (S-HTTP) provides secure communication mechanisms between an HTTP client-server pair in order to enable spontaneous commercial transactions for a wide range of applications. S-HTTP provides a flexible protocol that supports multiple orthogonal operation modes, key management mechanisms, trust models, cryptographic algorithms and encapsulation formats through option negotiation between parties for each transaction. Syntactically, S-HTTP messages are the same as HTTP, consisting of a request or status line followed by headers and a body. However, the range of headers is different and the bodies are typically cryptographically enhanced.

IMAP4

RFC 2060 http://www.cis.ohio-state.edu/htbin/rfc/rfc2060.html

The Internet Message Access Protocol, Version 4 revision 1 (IMAP4) allows a client to access and manipulate electronic mail messages on a server. IMAP4 permits manipulation of remote message folders, called mailboxes, in a way that is functionally equivalent to local mailboxes. IMAP4 also provides the capability for an offline client to resynchronize with the server.

IMAP4 includes operations for creating, deleting, and renaming mailboxes; checking for new messages; permanently removing messages; setting and clearing flags; parsing; searching; and selective fetching of message attributes, texts, and portions thereof. Messages in IMAP4 are accessed by the use of numbers. These numbers are either message sequence numbers or unique identifiers.

IMAP4 consists of a sequence of textual messages which contain commands, status messages, etc. Each message ends with <crlf>(carriage return and line feed). For example:

```
Server Message: "a002 OK [READ-WRITE] SELECT
completed<crlf>"
```

```
Client Message: "a001 login mrc secret<crlf>"
```

There are no other predefined fields.

IPDC

Internet Drafts: – draft-taylor-ipdc-00.txt and draft-calhoun-diameter-07.txt.
http://www.ietf.org/internet-drafts/draft-taylor-ipdc-00.txt
http://www.ietf.org/internet-drafts/draft-calhoun-diameter-07.txt

The IP Device Control (IPDC) is a family of protocols which is proposed as a protocol suite, components of which can be used individually or together to perform connection control, media control, and signalling transports. It fulfils a need for one or more protocols to control gateway devices which sit at the boundary between the circuit- switched telephone network and the internet and terminate circuit- switched trunks. Examples of such devices include network access servers and voice-over-IP gateways. The need for a control protocol separate from call signalling, arises when the service control logic needed to process calls lies partly or wholly outside the gateway devices.

IPDC was built on the base structure provided by the DIAMETER protocol which was specifically written for authentication, authorization and accounting applications.

There are two different types of IPDC/DIAMETER messages: header-only messages and messages containing Attribute-Value Pairs (AVPs) in addition to headers. Header-only messages are used for explicitly acknowledging packets to the peer. An AVP is a data object encapsulated in a header. The general format of the header is shown in the following illustration:

8	13	16	32 bits
Radius PCC	Pkt flags	Ver	Packet length
Identifier			
Next sent		Next received	
Attributes			

IPDC header structure

Radius PCC
Radius packet compatibility code, used for Radius backward compatibility. In order to easily distinguish DIAMETER/IPDC messages from Radius, a special value has been reserved and allows an implementation to support

both protocols concurrently using the first octet in the header. The Radius PCC field must be set to 254 for DIAMETER/IPDC messages.

Pkt flags

Packet flags. Used to identify any options. This field must be initialized to zero. The Window-Present flag may be set (0x1), thus indicating that the Next Send and Next Received fields are present. This flag must be set unless the underlying layer provides reliability (i.e., TCP).

Version

Indicates the version number associated with the packet received. This field is set to 1 to indicate IPDC version 1.

Packet length

Indicates the length of the message including the header fields. Thus the message AVP content cannot exceed 65,528 octets. For messages received via UDP, octets outside the range of the length field should be treated as padding and are ignored upon receipt.

Identifier

Aids in matching requests and replies.

Next sent (Ns)

Present when the Window-Present bit is set in the header flags. The Next Send (Ns) is copied from the send sequence number state variable, Ss, at the time the message is transmitted.

Next received

This field is present when the Window-Present bit is set in the header flags. Nr is copied from the receive sequence number state variable, Sr, and indicates the sequence number, Ns, +1 of the highest (modulo 2^16) in-sequence message received.

Attributes

IPDC Attributes carry the specific commands and parameters which must be exchanged between IPDC protocol endpoints to perform the tasks associated with Media Gateway control.

ISAKMP

RFC2408 http://www.cis.ohio-state.edu/htbin/rfc/rfc2408.html

The Internet Security Association and Key Management Protocol, version 4rev1 (ISAKMP), defines procedures and packet formats to establish, negotiate, modify and delete Security Associations (SA). SAs contain all the information required for execution of various network security services, such as the IP layer services (such as header authentication and payload encapsulation), transport or application layer services, or self-protection of negotiation traffic. ISAKMP defines payloads for exchanging key generation and authentication data. These formats provide a consistent framework for transferring key and authentication data which is independent of the key generation technique, encryption algorithm and authentication mechanism.

The format of the header is shown in the following illustration:

8	12	16	24	32 bits
Initiator cookie (8 bytes)				
Responder cookie (8 bytes)				
Next payload	MjVer	MnVer	Exchange type	Flags
Message ID				
Length				

ISAKMP header structure

Initiator cookie

Cookie of entity that initiated SA establishment, SA notification, or SA deletion.

Responder cookie

Cookie of entity that is responding to an SA establishment, SA notification, or SA deletion.

Next payload

Indicates the type of the first payload in the message. Possible types are:

0 None.

1 Security Association (SA).

2 Proposal (P).

3 Transform (T).
4 Key Exchange (KE).
5 Identification (ID).
6 Certificate (CERT).
7 Certificate Request (CR).
8 Hash (HASH).
9 Signature (SIG).
10 Nonce (NONCE).
11 Notification (N).
12 Delete (D).
13 Vendor ID (VID).
14 - 127 Reserved.
128 - 255 Private use.

MjVer

Major Version, indicates the major version of the ISAKMP protocol in use. Implementations based on RFC2408 must set the Major Version to 1. Implementations based on previous versions of ISAKMP Internet- Drafts must set the Major Version to 0. Implementations should never accept packets with a major version number larger than its own.

MnVer

Minor Version - indicates the minor version of the ISAKMP protocol in use. Implementations based on RFC2408 must set the minor version to 0. Implementations based on previous versions of ISAKMP Internet- Drafts must set the minor version to 1. Implementations should never accept packets with a minor version number larger than its own.

Exchange Type

The type of exchange being used. This dictates the message and payload orderings in the ISAKMP exchanges. Possible values are:
0 None
1 Base
2 Identity Protection
3 Authentication Only
4 Aggressive
5 Informational
6 - 31 ISAKMP Future Use
32 - 239 DOI Specific Use
240 - 255 Private Use

Flags

Specific options that are set for the ISAKMP exchange.
E(ncryption bit) (bit 0) - Specifies that all payloads following the header are encrypted using the encryption algorithm identified in the ISAKMP SA.
C(ommit bit) (bit 1) - Signals key exchange synchronization. It is used to ensure that encrypted material is not received prior to completion of the SA establishment.
A(uthentication Only Bit) (bit 2) - Intended for use with the Informational Exchange with a Notify payload and will allow the transmission of information with integrity checking, but no encryption.
All remaining bits are set to 0 before transmission.

Message ID

Unique Message Identifier used to identify protocol state during Phase 2 negotiations. This value is randomly generated by the initiator of the Phase 2 negotiation. In the event of simultaneous SA establishments (i.e., collisions), the value of this field will likely be different because they are independently generated and, thus, two security associations will progress toward establishment. However, it is unlikely there will be absolute simultaneous establishments. During Phase 1 negotiations, the value must be set to 0.

Length

Length of total message (header + payloads) in octets. Encryption can expand the size of an ISAKMP message.

NTP

RFC 1305 http://www.cis.ohio-state.edu/htbin/rfc/rfc1305.html

The Network Time Protocol (NTP) is a time synchronization system for computer clocks through the Internet network. It provides the mechanisms to synchronize time and coordinate time distribution in a large, diverse internet operating at rates from mundane to light wave. It uses a returnable time design in which a distributed sub network of time servers, operating in a self-organizing, hierarchical master-slave configuration, synchronize logical clocks within the sub network and to national time standards via wire or radio.

The format of the header is shown in the following illustration:

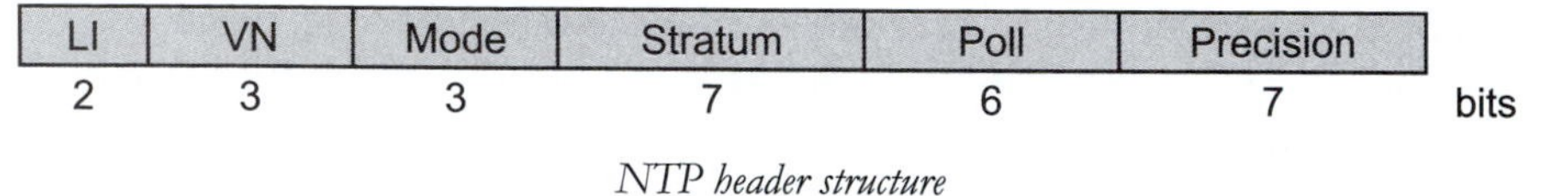

NTP header structure

LI Leap Indicator

A 2-bit code warning of impending leap-second to be inserted at the end of the last day of the current month. Bits are coded as follows:

00 No warning.
01 +1 second (following minute has 61 seconds).
10 -1 second (following minute has 59 seconds).
11 Alarm condition (clock not synchronized).

VN

Version number 3 bit code indicating the version number.

Mode

The mode: This field can contain the following values:

0 Reserved.
1 Symmetric active.
2 Symmetric passive.
3 Client.
4 Server.
5 Broadcast.
6 NTP control message.

Stratum

An integer identifying the stratum level of the local clock. Values are defined as follows:

0	Unspecified.
1	Primary reference (e.g. radio clock).
2...n	Secondary reference (via NTP).

Poll

Signed integer indicating the maximum interval between successive messages, in seconds to the nearest power of 2.

Precision

Signed integer indicating the precision of the local clock, in seconds to the nearest power of 2.

POP3

RFC 1939 http://www.cis.ohio-state.edu/htbin/rfc/rfc1939.html

The Post Office Protocol version 3 (POP3) is intended to permit a workstation to dynamically access a maildrop on a server host. It is usually used to allow a workstation to retrieve mail that the server is holding for it.

POP3 transmissions appear as data messages between stations. The messages are either command or reply messages.

Radius

RFC 2138 http://www.cis.ohio-state.edu/htbin/rfc/rfc2138.html
RFC 2139 http://www.cis.ohio-state.edu/htbin/rfc/rfc2139.html

Radius is a protocol which manages dispersed serial line and modem pools for large numbers of users. Since modem pools are by definition a link to the outside world, they require careful attention to security, authorization and accounting. This is achieved by managing a single database of users, which allows for authentication (verifying user name and password) as well as configuration information detailing the type of service to deliver to the user (for example, SLIP, PPP, telnet, rlogin).

Key features of RADIUS include:

- Client/server model.
- Network security.
- Flexible authentication mechanisms.
- Extensible protocol.

The format of the header is shown in the following illustration:

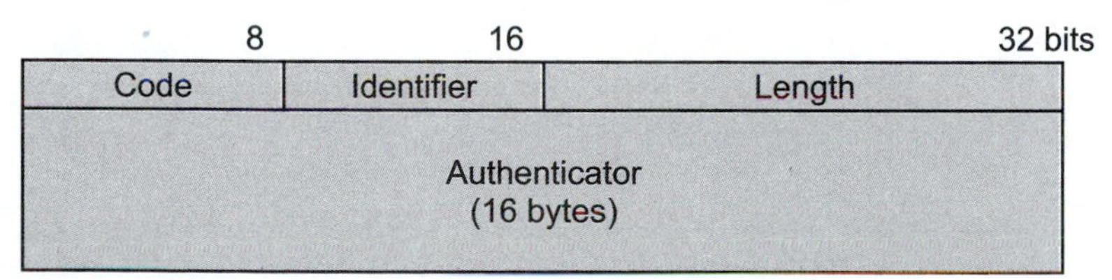

Radius header structure

Code
The message type.

Identifier
The identifier matches requests and replies.

Length
The message length including the header.

Authenticator
A field used to authenticate the reply from the radius server and in the password hiding algorithm.

RLOGIN

Remote LOGIN (RLOGIN) allows UNIX users of one machine to connect to other UNIX systems across an Internet and interact as if their terminals are directly connected to the machines. This protocol offers essentially the same services as TELNET.

RTSP

RFC 2326 http://www.cis.ohio-state.edu/htbin/rfc/rfc2326.html

RTSP (Real Time Streaming Protocol) is an application level protocol for control over the delivery of data with real-time properties. RTSP provides an extensible framework to enable controlled, on-demand delivery of real-time data, such as audio and video. Sources of data can include both live data feeds and stored clips. This protocol is intended to control multiple data delivery sessions, provide a means for choosing delivery channels such as UDP, multicast UDP and TCP, and provide a means for choosing delivery mechanisms based upon RTP.

The streams controlled by RTSP may use RTP, but the operation of RTSP does not depend on the transport mechanism used to carry continuous media. The protocol is intentionally similar in syntax and operation to HTTP/1.1 so that extension mechanisms to HTTP can in most cases also be added to RTSP. However, RTSP differs in a number of important aspects from HTTP:

- RTSP introduces a number of new methods and has a different protocol identifier.
- An RTSP server needs to maintain state by default in almost all cases, as opposed to the stateless nature of HTTP.
- Both an RTSP server and client can issue requests.
- Data is carried out-of-band by a different protocol.
- RTSP is defined to use ISO 10646 (UTF-8) rather than ISO 8859-1, consistent with current HTML internationalization efforts.
- The Request-URI always contains the absolute URI. Because of backward compatibility with an historical blunder, HTTP/1.1 carries only the absolute path in the request and puts the host name in a separate header field.

This makes virtual hosting easier, where a single host with one IP address hosts several document trees.

SCTP

draft-ietf-sigtran-sctp-03.txt 11/99

The Stream Control Transmission Protocol (SCTP) is designed to transport PSTN signalling messages over IP networks, but is capable of broader applications. SCTP is an application-level datagram transfer protocol operating on top of an unreliable datagram service such as UDP. It offers the following services:

- Acknowledged error-free non-duplicated transfer of user data.
- Application-level segmentation to conform to discovered MTU size.
- Sequenced delivery of user datagrams within multiple streams, with an option for order-of-arrival delivery of individual datagrams.
- Optional multiplexing of user datagrams into SCTP datagrams, subject to MTU size restrictions.
- Enhanced reliability through support of multi-homing at either or both ends of the association.

The design of SCTP includes appropriate congestion avoidance behaviour and resistance to flooding and masquerade attacks. The SCTP datagram is comprised of a common header and chunks. The chunks contain either control information or user data.

The following is the format of the SCTP header.

2 bytes	2 bytes
Source Port Number	Destination Port Number
Verification Tag	
Adler 32 Checksum	

Source Port Number

This is the SCTP sender's port number. It can be used by the receiver, in combination with the source IP Address, to identify the association to which this datagram belongs.

Destination Port Number

This is the SCTP port number to which this datagram is destined. The receiving host will use this port number to de-multiplex the SCTP datagram to the correct receiving endpoint/application.

Verification Tag

The receiver of this 32 bit datagram uses the Verification tag to identify the association. On transmit, the value of this Verification tag must be set to the value of the Initiate tag received from the peer endpoint during the association initialization.

For datagrams carrying the INIT chunk, the transmitter sets the Verification tag to all 0's. If the receiver receives a datagram with an all-zeros Verification tag field, it checks the Chunk ID immediately following the common header. If the chunk type is not INIT or SHUTDOWN ACK, the receiver drops the datagram. For datagrams carrying the SHUTDOWN-ACK chunk, the transmitter sets the Verification tag to the Initiate tag received from the peer endpoint during the association initialization, if known. Otherwise the Verification tag is set to all 0's.

Adler 32 Checksum

This field contains an Adler-32 checksum on this SCTP datagram.

Chunk Field Descriptions

The following is the field format for the chunks transmitted in the SCTP datagram. Each chunk has a chunk ID field, a chunk specific Flag field, a Length field and a Value field.

1 byte	1 byte	2 bytes
Chunk ID	Chunk Flags	Chunk Length
Chunk Value (variable)		

Chunk ID

The type of information contained in the chunk value field. The values of the chunk ID are defined as follows:

ID Value	**Chunk Type**
00000000	Payload Data (DATA)
00000001	Initiation (INIT)
00000010	Initiation Acknowledgment (INIT ACK)
00000011	Selective Acknowledgment (SACK)
00000100	Heartbeat Request (HEARTBEAT)
00000101	Heartbeat Acknowledgment (HEARTBEAT ACK)
00000110	Abort (ABORT)

00000111	Shutdown (SHUTDOWN)
00001000	Shutdown Acknowledgment (SHUTDOWN ACK)
00001001	Operation Error (ERROR)
00001010	State Cookie (COOKIE)
00001011	Cookie Acknowledgment (COOKIE ACK)
00001100	Reserved for Explicit Congestion Notification Echo (ECNE)
00001101	Reserved for Congestion Window Reduced (CWR)
00001110 to 11111101 - reserved by IETF	
11111110	Vendor-specific Chunk Extensions
11111111 -	IETF-defined Chunk Extensions

Chunk Flags

The type of chunk flag as defined in the chunk ID defines whether these bits will be used. Their value is generally 0 unless otherwise specified.

Chunk Length

The size of the chunk in octets including the Chunk ID, Flags, Length and Value fields.

Chunk Value

This field contains the actual information to be transferred in the chunk. This is dependent on the chunk ID.

Chunk Types

Initiation (INIT)

This chunk is used to initiate a SCTP association between two endpoints. The INIT chunk contains the following parameters. Unless otherwise noted, each parameter is only be included once in the INIT chunk.

Fixed Parameters	Status
Initiate Tag	Mandatory
Receiver Window Credit	Mandatory
Number of Outbound Streams	Mandatory
Number of Inbound Streams	Mandatory
Initial TSN	Mandatory

Variable Parameters	Status
IPv4 Address/Port	Optional
IPv6 Address/Port	Optional

Cookie Preservative	Optional
Reserved For ECN Capable	Optional
Host Name Address	Optional
Supported Address Types	Optional

Initiate Acknowledgement (INIT ACK)

The INIT ACK chunk is used to acknowledge the initiation of a SCTP association. The parameter part of INIT ACK is formatted similarly to the INIT chunk. It uses two extra variable parameters: The Responder Cookie and the Unrecognized Parameter.

Selective Acknowledgement (SACK)

This chunk is sent to the remote endpoint to acknowledge received *Data* chunks and to inform the remote endpoint of gaps in the received subsequences of *Data* chunks as represented by their TSNs.

The selective acknowledgement chunk contains the highest consecutive TSN ACK and Rcv Window Credit (rwnd) parameters. By definition, the value of the highest consecutive TSN ACK parameter is the last TSN received at the time the Selective ACK is sent, before a break in the sequence of received TSNs occurs; the next TSN value following this one has not yet been received at the reporting end. This parameter therefore acknowledges receipt of all TSNs up to and including the value given.

The Selective ACK also contains zero or more fragment reports. Each fragment report acknowledges a sub-sequence of TSNs received following a break in the sequence of received TSNs. By definition, all TSNs acknowledged by fragment reports are higher than the value of the Highest Consecutive TSN ACK.

Heartbeat Request (HEARTBEAT)

An endpoint should send this chunk to its peer endpoint of the current association to probe the reachability of a particular destination transport address defined in the present association. The parameter fields contain the time values.

Heartbeat Acknowledgement (HEARTBEAT ACK)

An endpoint should send this chunk to its peer endpoint as a response to a Heartbeat Request. The parameter field contains the time values.

Abort Association (ABORT)

The Abort Association chunk is sent to the peer of an association to terminate the association. The Abort chunk may contain cause parameters to inform the receiver the reason for the abort. Data chunks are not bundled with the abort, control chunks may be bundled with an abort, but must be placed before the abort in the SCTP datagram or they will be ignored.

SHUTDOWN

An endpoint in an association uses this chunk to initiate a graceful termination of the association with its peer.

Shutdown Acknowledgement (SHUTDOWN ACK)

This chunk is used to acknowledge the receipt of the SHUTDOWN chunk at the completion of the shutdown process. The SHUTDOWN ACK chunk has no parameters.

Operation Error (ERROR)

This chunk is sent to the other endpoint in the association to notify certain error conditions. It contains one or more error causes.

State Cookie (COOKIE)

This chunk is used only during the initialization of an association. It is sent by the initiator of an association to its peer to complete the initialization process. This chunk precedes any Data chunk sent within the association, but may be bundled with one or more Data chunks in the same datagram.

Cookie Acknowledgement (COOKIE ACK)

This chunk is used only during the initialization of an association. It is used to acknowledge the receipt of a COOKIE chunk. This chunk precedes any Data chunk sent within the association, but may be bundled with one or more Data chunks in the same SCTP datagram.

Payload Data (DATA)

This contains the user data.

Vendor Specific Chunk Extensions

This chunk type is available to allow vendors to support their own extended data formats not defined by the IETF. It must not affect the operation of SCTP. Endpoints not equipped to interpret the vendor-specific chunk sent by a remote endpoint must ignore it. Endpoints that do not receive desired vendor specific information should make an attempt to operate without it, although they may do so (and report they are doing so) in a degraded mode.

SLP

RFC 2165

The Service Location Protocol (SLP) provides a scalable framework for the discovery and selection of network services. Using this protocol, computers using the Internet no longer need so much static configuration for network services for network-based applications. This is especially important as computers become more portable and users less tolerant or able to fulfill the demands of network system administration.

Traditionally, users find services by using the name of a network host (a human readable text string), which is an alias for a network address. The Service Location Protocol eliminates the need for a user to know the name of a network host supporting a service. Rather, the user names the service and supplies a set of attributes, which describe the service. The Service Location Protocol allows the user to bind this description to the network address of the service.

Service Location provides a dynamic configuration mechanism for applications in local area networks. It is not a global resolution system for the entire Internet; rather it is intended to serve enterprise networks with shared services. Applications are modeled as clients that need to find servers attached to the enterprise network at a possibly distant location. For cases where there are many different clients and/or services available, the protocol is adapted to make use of nearby Directory Agents that offer a centralized repository for advertised services. The basic operation in Service Location is that a client attempts to discover the location for a service. In small installations, each service is configured to respond individually to each client. In larger installations, service will register their services with one or more directory agents and clients contact the directory agent to fulfill request for service location information. This is intended to be similar to URL specifications and make user of URL technology.

The header is used in all the SLP messages

2 bytes		2 bytes
Version	Function	Length
O M U A F rsvd	Dialect	Language Code
Char encoding		XID

SLP header structure

Version

The current version is version 1

Function

The function field describes the operation of the Service location datagram. The following message types exist:

Function	Message Type
1	Service Request
2	Service Reply
3	Service Registration
4	Service Deregister
5	Service Acknowledge
6	Attribute Request
7	Attribute Reply
8	DA Advertisement
9	Service Type Request
10	Service Type Reply

Length

Number of bytes in the message including the Service location header.

O

The overflow bit.

M

The monolingual bit.

U

URL Authentication bit present.

A

Attribute authentication bit present.

F

The F bit is set. If the F bit is set in a Service Acknowledgement, the directory agent has registered the service as a new entry.

Rsvd

These bits are reserved and must have a value of 0.

Dialect
To be use by future versions of the SLP. Must be set to zero.

Language Code
The language encoded in this field indicates the language in which the remainder of the message should be interpreted.

Character Encoding
The characters making up strings within the remainder of this message may be encoded in any standardized encoding

Transaction Identifier (XID)
Allows matching replies to individual requests.

SMTP

RFC 821 http://www.cis.ohio-state.edu/htbin/rfc/rfc821.html

The Simple Mail Transfer Protocol (SMTP) is a mail service modeled on the FTP file transfer service. SMTP transfers mail messages between systems and provides notification regarding incoming mail.

Commands

SMTP commands are ASCII messages sent between SMTP hosts. Possible commands are as follows:

Command	*Description*
DATA	Begins message composition.
EXPN <string>	Returns names on the specified mail list.
HELO <domain>	Returns identity of mail server.
HELP <command>	Returns information on the specified command.
MAIL FROM <host>	Initiates a mail session from host.
NOOP	Causes no action, except acknowledgement from server.
QUIT	Terminates the mail session.
RCPT TO <user>	Designates who receives mail.
RSET	Resets mail connection.
SAML FROM <host>	Sends mail to user terminal and mailbox.
SEND FROM <host>	Sends mail to user terminal.
SOML FROM <host>	Sends mail to user terminal or mailbox.
TURN	Switches role of receiver and sender.
VRFY <user>	Verifies the identity of a user.

Messages

SMTP response messages consist of a response code followed by explanatory text, as follows:

Response Code	*Explanatory Text*
211	(Response to system status or help request).
214	(Response to help request).
220	Mail service ready.
221	Mail service closing connection.
250	Mail transfer completed.
251	User not local, forward to <path>.
354	Start mail message, end with <CRLF><CRLF>.
421	Mail service unavailable.
450	Mailbox unavailable.
451	Local error in processing command.
452	Insufficient system storage.
500	Unknown command.
501	Bad parameter.
502	Command not implemented.
503	Bad command sequence.
504	Parameter not implemented.
550	Mailbox not found.
551	User not local, try <path>.
552	Storage allocation exceeded.
553	Mailbox name not allowed.
554	Mail transaction failed.

SNMP

RFC 1157: http://www.cis.ohio-state.edu/htbin/rfc/rfc1157.html

The Internet community developed the Simple Network Management Protocol (SNMP) to allow diverse network objects to participate in a global network management architecture. Network managing systems can poll network entities implementing SNMP for information relevant to a particular network management implementation. Network management systems learn of problems by receiving traps or change notices from network devices implementing SNMP.

SNMP Message Format

SNMP is a session protocol which is encapsulated in UDP. The SNMP message format is shown below:

Version	Community	PDU

SNMP message format

Version

SNMP version number. Both the manager and agent must use the same version of SNMP. Messages containing different version numbers are discarded without further processing.

Community

Community name used for authenticating the manager before allowing access to the agent.

PDU

There are five different PDU types: GetRequest, GetNextRequest, GetResponse, SetRequest, and Trap. A general description of each of these is given in the next section.

PDU Format

The format for GetRequest, GetNext Request, GetResponse and SetRequest PDUs is shown here.

PDU type	Request ID	Error status	Error index	Object 1, value 1	Object 2, value 2	...

SNMP PDU format

PDU type

Specifies the type of PDU:

0 GetRequest.
1 GetNextRequest.
2 GetResponse.
3 SetRequest.

Request ID

Integer field which correlates the manager's request to the agent's response.

Error status

Enumerated integer type that indicates normal operation or one of five error conditions. The possible values are:

0 noError: Proper manager/agent operation.
1 tooBig: Size of the required GetResponse PDU exceeds a local limitation.
2 noSuchName: The requested object name does not match the names available in the relevant MIB View.
3 badValue: A SetRequest contains an inconsistent type, length and value for the variable.
4 readOnly: Not defined in RFC1157.
5 genErr: Other errors, which are not explicitly defined, have occurred.

Error index

Identifies the entry within the variable bindings list that caused the error.

Object/value

Variable binding pair of a variable name with its value.

Trap PDU Format

The format of the Trap PDU is shown below:

PDU type	Enterp	Agent addr	Gen trap	Spec trap	Time stamp	Obj 1, Val 1	Obj 1, Val 1	...

SNMP trap PDU

PDU type
Specifies the type of PDU (4=Trap).

Enterprise
Identifies the management enterprise under whose registration authority the trap was defined.

Agent address
IP address of the agent, used for further identification.

Generic trap type
Field describing the event being reported. The following seven values are defined:

0 coldStart: Sending protocol entity has reinitialized, indicating that the agent's configuration or entity implementation may be altered.
1 warmStart: Sending protocol has reinitialized, but neither the agent's configuration nor the protocol entity implementation has been altered.
2 linkDown: A communication link has failed.
3 linkUp: A communication link has come up.
4 authenticationFailure: The agent has received an improperly authenticated SNMP message from the manager, i.e., community name was incorrect.
5 egpNeighborLoss: An EGP peer neighbor is down.
6 enterpriseSpecific: A non-generic trap has occurred which is further identified by the Specific Trap Type and Enterprise fields.

Specific trap type
Used to identify a non-generic trap when the Generic Trap Type is enterpriseSpecific.

Timestamp
Value of the sysUpTime object, representing the amount of time elapsed between the last (re-)initialization and the generation of that Trap.

Object/value

Variable binding pair of a variable name with its value.

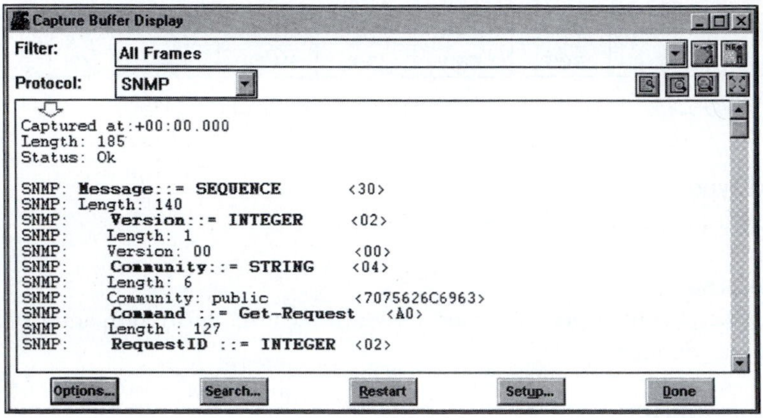

```
Capture Buffer Display
Filter: All Frames
Protocol: SNMP

Captured at:+00:00.000
Length: 185
Status: Ok

SNMP: Message::= SEQUENCE        <30>
SNMP: Length: 140
SNMP:     Version::= INTEGER      <02>
SNMP:     Length: 1
SNMP:     Version: 00             <00>
SNMP:     Community::= STRING     <04>
SNMP:     Length: 6
SNMP:     Community: public       <7075626C6963>
SNMP:     Command ::= Get-Request    <A0>
SNMP:     Length : 127
SNMP:     RequestID ::= INTEGER   <02>

Options...   Search...   Restart   Setup...   Done
```

SNMP decode

SOCKS V5

ftp://ftp.isi.edu/in-notes/rfc1928.txt

This protocol provides a framework for client-server applications in both the TCP and UDP domains to conveniently and securely use the services of a network firewall. The protocol is conceptually a "shim-layer" between the application layer and the transport layer, and as such does not provide network layer gateway services, such as forwarding of ICMP messages. SOCKS Version 4 provides unsecured firewall traversal for TCP-based client-server applications, including TELNET, FTP, and protocols such as HTTP, WAIS and GOPHER. This version of SOCKS extends the SOCKS Version 4 model to include UDP, and extends the framework to include provisions for generalized strong authentication schemes. It also adapts the addressing scheme to encompass domain-name and V6 IP addresses.

The implementation of the SOCKS protocol typically involves the recompilation or relinking of TCP-based client applications to use the appropriate encapsulation routines in the SOCKS library.

Protocol Structure for TCP-based Clients

Version identifier/method selection message:

1 byte	1 byte	1-225 bytes
Version	NMethods	Methods

Version
The version is 05.

Nmethod
The NMETHODS field contains the number of method identifier octets that appear in the METHODS field.

The method selection message:

1 byte	1 byte
Version	Method

Methods
Possible values for methods are:

00 No authentication required

01	GSSAPI
02	Username/Password
3	IANA assigned
4 to FE	Reserved for private methods
FF	No acceptable methods

Socks Request Message

1 byte	1 byte	Value of 0	1 byte	Variable	2 bytes
Version	CMD	Rsv	ATYP	DST addr	DST Port

Version

The Protocol version is 5.

CMD

Possible values for the cmnd field are:

01	CONNECT1
02	BIND
03	UDP ASSOCIATE

Reserved

The value of this field is 0.

ATYP

Address type of the following address:

01	IP V4 address
03	DOMAINNAME
04	IP V6 address: X'04'

Destination address

The destination address desired.

Destination port

The desired destination port in network octet order.

Socks Reply Message

1 byte	1 byte	Value of 0	1 byte	Variable	2 bytes
Version	REP	RSV	ATYP	BND addr	BND Port

Version

The protocol version is 5.

REP

The reply field.

Possible values for the reply field are:

00	Succeeded
01	General SOCKS server failure
02	Connection not allowed by ruleset
03	Network unreachable
04	Host unreachable
05	Connection refused
06	TTL expired
07	Command not supported
08	Address type not supported
09 to FF	Unassigned

RSV

Reserved, the value of this field is 0.

ATYP

Address type of the following address:

01	IP V4 address
03	DOMAINNAME
04	IP V6 address: X'04'

BND address

Server bound address.

BND Port

Server bound port in network octet order.

Protocol Structure for UDP-based Clients

Each UDP datagram carries a UDP request header with it:

UDP Request Header

2byte	1 byte	1 byte	Variable	2	Variable
RSV	FRAG	ATYP	DST Addr	DST Port	Data

RSV
This field is reserved. Its value is 0000.

FRAG
This field contains the current fragment number, and indicates whether the datagram is one of a number of fragments.

ATYP
Address type of the following address:

01 IP V4 address
03 DOMAINNAME
04 IP V6 address: X'04'

DST addr
Desired destination address.

DST Port
Desired destination port.

Data
User data.

TACACS+

draft-grant-tacacs-02.txt
http://www.ietf.org/internet-drafts/draft-grant-tacacs-02.txt
RFC 1492 http://www.cis.ohio-state.edu/htbin/rfc/rfc1492.html

TACACS+ (Terminal Access Controller Access Control System) is a protocol providing access control for routers, network access servers and other networked computing devices via one or more centralized servers. TACACS+ provides separate authentication, authorization and accounting services.

The format of the header is shown in the following illustration:

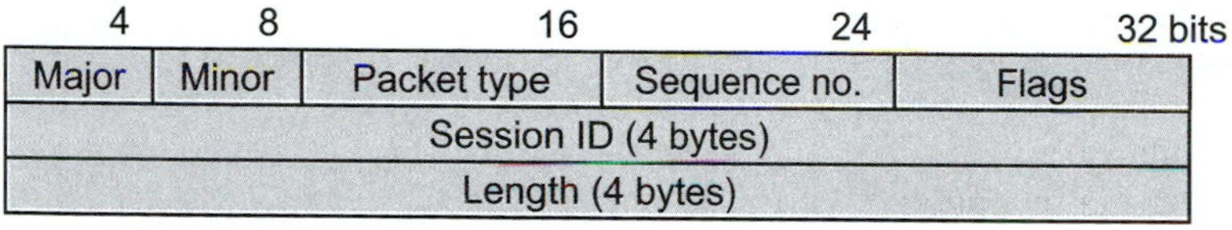

4	8	16	24	32 bits
Major	Minor	Packet type	Sequence no.	Flags
Session ID (4 bytes)				
Length (4 bytes)				

TACACS+ header structure

Major version
The major TACACS+ version number.

Minor version
The minor TACACS+ version number. This is intended to allow revisions to the TACACS+ protocol while maintaining backwards compatibility.

Packet type
Possible values are:
TAC_PLUS_AUTHEN:= 0x01 (Authentication).
TAC_PLUS_AUTHOR:= 0x02 (Authorization).
TAC_PLUS_ACCT:= 0x03 (Accounting).

Sequence number
The sequence number of the current packet for the current session. The first TACACS+ packet in a session must have the sequence number 1 and each subsequent packet will increment the sequence number by one. Thus clients only send packets containing odd sequence numbers, and TACACS+ daemons only send packets containing even sequence numbers.

Flags
This field contains various flags in the form of bitmaps. The flag values signify whether the packet is encrypted.

Session ID
The ID for this TACACS+ session.

Length
The total length of the TACACS+ packet body (not including the header).

TELNET

IETF RFC 854 1983-05 http://www.cis.ohio-state.edu/htbin/rfc/rfc854.html
IETF RFC 855 1983-05 http://www.cis.ohio-state.edu/htbin/rfc/rfc855.html
IETF RFC 857 1983-05 http://www.cis.ohio-state.edu/htbin/rfc/rfc857.html

TELNET is the terminal emulation protocol of TCP/IP. Modern TELNET is a versatile terminal emulation due to the many options that have evolved over the past twenty years. Options give TELNET the ability to transfer binary data, support byte macros, emulate graphics terminals, and convey information to support centralized terminal management.

TELNET uses the TCP transport protocol to achieve a virtual connection between server and client. After connecting, TELNET server and client enter a phase of option negotiation that determines the options that each side can support for the connection. Each connected system can negotiate new options or renegotiate old options at any time. In general, each end of the TELNET connection attempts to implement all options that maximize performance for the systems involved.

In a typical implementation, the TELNET client sends single keystrokes, while the TELNET server can send one or more lines of characters in response. Where the Echo option is in use, the TELNET server echoes all keystrokes back to the TELNET client.

Dynamic Mode Negotiation

During the connection, enhanced characteristics other than those offered by the NVT may be negotiated either by the user or the application. This task is accomplished by embedded commands in the data stream. TELNET command codes are one or more octets in length and are preceded by an interpret as command (IAC) character, which is an octet with each bit set equal to one (FF hex). The following are the TELNET command codes:

Commands	*Code No.*		*Description*
	Dec	*Hex*	
data			All terminal input/output data.
End subNeg	240	FO	End of option subnegotiation command.
No Operation	241	F1	No operation command.
Data Mark	242	F2	End of urgent data stream.

Commands	*Code No. Dec*	*Hex*	*Description*
Break	243	F3	Operator pressed the Break key or the Attention key.
Int process	244	F4	Interrupt current process.
Abort output	245	F5	Cancel output from current process.
You there?	246	F6	Request acknowledgment.
Erase char	247	F7	Request that operator erase the previous character.
Erase line	248	F8	Request that operator erase the previous line.
Go ahead!	249	F9	End of input for half-duplex connections.
SubNegotiate	250	FA	Begin option subnegotiation.
Will Use	251	FB	Agreement to use the specified option.
Won't Use	252	FC	Reject the proposed option.
Start use	253	FD	Request to start using specified option.
Stop Use	254	FE	Demand to stop using specified option.
IAC	255	FF	Interpret as command.

Each negotiable option has an ID, which immediately follows the command for option negotiation, that is, IAC, command, option code. The following is a list of TELNET option codes:

Option ID Dec	*Hex*	*Option Codes*	*Description*
0	0	Binary Xmit	Allows transmission of binary data.
1	1	Echo Data	Causes server to echo back all keystrokes.
2	2	Reconnect	Reconnects to another TELNET host.
3	3	Suppress GA	Disables Go Ahead! command.
4	4	Message Sz	Conveys approximate message size.
5	5	Opt Status	Lists status of options.
6	6	Timing Mark	Marks a data stream position for reference.
7	7	R/C XmtEcho	Allows remote control of terminal printers.
8	8	Line Width	Sets output line width.
9	9	Page Length	Sets page length in lines.
10	A	CR Use	Determines handling of carriage returns.
11	B	Horiz Tabs	Sets horizontal tabs.

Option ID		*Option Codes*	*Description*
Dec	*Hex*		
12	C	Hor Tab Use	Determines handling of horizontal tabs.
13	D	FF Use	Determines handling of form feeds.
14	E	Vert Tabs	Sets vertical tabs.
15	F	Ver Tab Use	Determines handling of vertical tabs.
16	10	Lf Use	Determines handling of line feeds.
17	11	Ext ASCII	Defines extended ASCII characters.
18	12	Logout	Allows for forced log-off.
19	13	Byte Macro	Defines byte macros.
20	14	Data Term	Allows subcommands for Data Entry to be sent.
21	15	SUPDUP	Allows use of SUPDUP display protocol.
22	16	SUPDUP Outp	Allows sending of SUPDUP output.
23	17	Send Locate	Allows terminal location to be sent.
24	18	Term Type	Allows exchange of terminal type information.
25	19	End Record	Allows use of the End of record code (0xEF).
26	1A	TACACS ID	User ID exchange used to avoid more than 1 log-in.
27	1B	Output Mark	Allows banner markings to be sent on output.
28	1C	Term Loc#	A numeric ID used to identify terminals.
29	1D	3270 Regime	Allows emulation of 3270 family terminals.
30	1E	X.3 PAD	Allows use of X.3 protocol emulation.
31	1F	Window Size	Conveys window size for emulation screen.
32	20	Term Speed	Conveys baud rate information.
33	21	Remote Flow	Provides flow control (XON, XOFF).
34	22	Linemode	Provides linemode bulk character transactions.
255	FF	Extended options list	Extended options list.

WCCP

draft-ietf-wrec-web-pro-00.txt

The Web Cache Coordination Protocol (WCCP) has 2 main functions. The first is to allow a router enabled for transparent redirection to discover, verify and advertise connectivity to one or more web-caches.

Transparent redirection is a technique used to deploy web-caching without the need for reconfiguration of web-clients. It involves the interception and redirection of HTTP traffic to one or more web-caches by a router or switch, transparently to the web-client.

The second function of WCCP is to allow one of the web-caches, the designated web-cache, to dictate how the router distributes redirected traffic across the web-cache farm. The web-cache with the lowest IP address should be elected as designated web-cache for a farm.

Each WCCP protocol packet is carried in a UDP packet with a destination port of 2048.

Packets can be of the following types; HERE I AM, I SEE YOU, ASSIGN BUCKETS.

HERE I AM

The format of the Here I am message is:

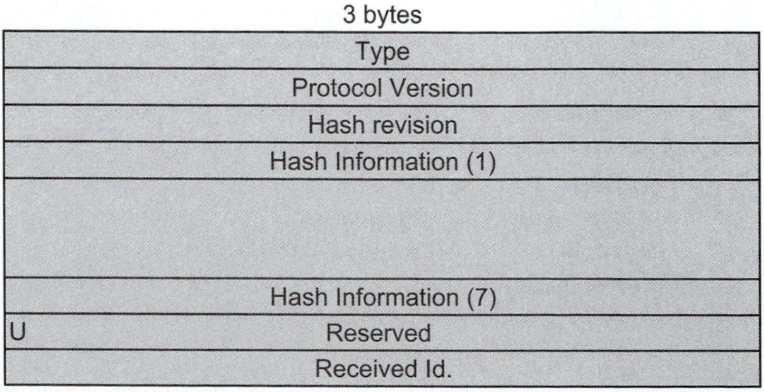

3 bytes
Type
Protocol Version
Hash revision
Hash Information (1)
Hash Information (7)
U Reserved
Received Id.

Here I am message format

Type
WCCP HERE I AM

Protocol version
This field has a value of 4.

Hash revision
The value of this field is 0.

Hash information
A 256-element bit-vector. A set bit indicates that the corresponding bucket in the Redirection Hash Table is assigned to this web-cache.

U
The value of the U flag present in the last WCCP I SEE YOU message received by this cache. Set in first WCCP HERE I AM to indicate that Hash Information is historical.

Received ID
The value of the Received ID present in the last WCCP I SEE YOU received by this web-cache.

I SEE YOU Message

The format of the I SEE YOU message is:

3 bytes
Type
Protocol version
Change number
Received Id.
Number of WCs
Web-Cache list entry(0)
Web-Cache list entry (n) v

I see you message format

Type
WCCP I SEE YOU

Protocol version
4

Change number
Incremented if a Web-Cache List Entry has been added, removed or its hash information has been modified since the last WCCP I SEE YOU sent by the router.

Received ID
Incremented each time the router generates a WCCP I SEE YOU.

Number of WCs
Number of Web-Cache List Entry elements in the packet.

Web cached list entry
The Web-Cache List Entry describes a Web-Cache by IP Address and lists the redirection hash table entries assigned to it.

WCCP ASSIGN BUCKET

The format of the WCCP ASSIGN BUCKET message is:

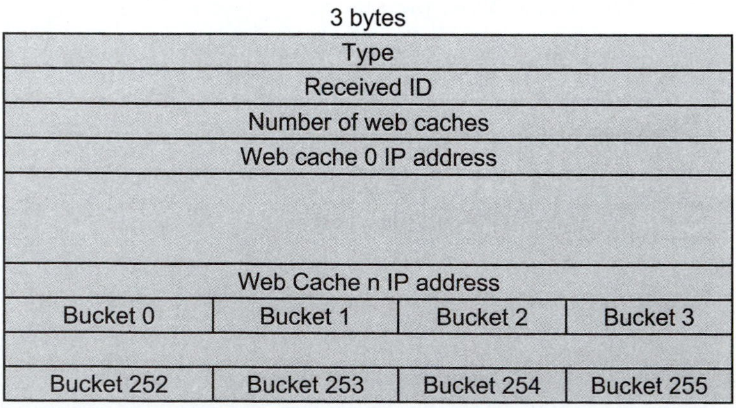

3 bytes

Type			
Received ID			
Number of web caches			
Web cache 0 IP address			
Web Cache n IP address			
Bucket 0	Bucket 1	Bucket 2	Bucket 3
Bucket 252	Bucket 253	Bucket 254	Bucket 255

WCCP ASSIGN BUCKET message format

Type
WCCP ASSIGN BUCKET

Received ID
Value of Received ID in last WCCP I SEE YOU received from router.

Number of Web Caches
Number of Web Caches to which redirect traffic can be sent.

Web Cache IP address 0-n
IP Addresses of Web-Caches to which redirect traffic can be sent. The position of a Web-Cache's IP Address in this list is the Web-Cache's index number. The first entry in the list has an index number of zero.

Buckets 0-255
These 256 buckets represent the redirection hash table. The value of each bucket may be 0xFF (Unassigned) or a Web-Cache index number (0-31).

X-Window

The X-Window protocol provides a remote windowing interface to distributed network applications. It is an application layer protocol which uses TCP/IP or DECnet protocols for transport.

The X-Window networking protocol is client-server based, where the server is the control program running on the user workstation and the client is an application running elsewhere on the network. An X-server control program running on a workstation can simultaneously handle display windows for multiple applications, with each application asynchronously updating its window with information carried by the X-Window networking protocol.

To provide user interaction with remote applications, the X-server program running on the workstation generates events in response to user input such as mouse movement or a keystroke. When multiple applications display, the system sends mouse movements or click events to the application currently highlighted by the mouse pointer. The current input focus selects which application receives keystroke events. In certain cases, applications can also generate events directed at the X-server control program.

Request and Reply Frames

Request and reply frames can use the following commands:

Command	*Description*
BackRGB	Background colors listed in red, green and blue components.
BackPM	Pixel map used for the window background.
BellPitch	Bell pitch.
BellVol	Bell volume in percent.
BM	Bit mask assigned to a drawable item.
BordPM	Border pixel map. Pixel map used for the window border.
b	Border width of the drawable item.
Click	Key click volume in percent.
Ord	Click order. Drawable clip order, as <Unsorted>, <Y-sorted>, <YX-sorted> or <YX-banded>.
CMap	Color map. Code representing the colors in use for a drawable.

Command	*Description*
CID	Context ID. Identifier for a particular graphics context.
Cur	Cursor. Reference code identifying a specific cursor.
d	Depth. Current window depth.
DD	Destination drawable. Target item in a bitmap copy.
D	Drawable. Reference code used to identify a specific window or pixel map.
Exp	Exposures. Drawable currently exposed.
Fam	Protocol family in use, as Internet, DECnet, or CHAOSnet.
Font	Reference code used to specify a font.
Font(a,d)	Font ascent/descent. The vertical bounds of a font.
ForeRGB	Foreground colors listed in red, green, and blue components.
Fmt	Format of the current window.
GC	Graphics context. Reference code used to identify a particular graphical definition.
h	Height of the drawable item.
Key	Key code. Specific key code value.
KeySym	Code used to identify the family of key codes in use.
MinOp	X-Windows minor operation code.
MajOp	X-Windows major operation code.
N	Number of drawable items in the list.
P	Parent window. Window that produced the current window.
PixMap	Pixel map. Reference code used to identify a bitmap region.
p	Plane. Bit plane in use.
PM	Plane max. Bit plane mask assigned to a drawable item.
Prop	Property. Specified window property.
SW	Sibling window. Window produced from this window.
SD	Source drawable. Source item in a bitmap copy.
T/O	Screen saver time out.
Typ	Type of current window.

Command	*Description*
w	Width of drawable item.
W	Window. Reference code used to identify a particular window.
X	X-coordinate for a drawable item.
Y	Y-coordinate for a drawable item.

Event Frames

Event frames can have the following commands:

Command	*Description*
Btn	Button number pressed.
C	Child window associated with the event.
F	Event flags. Set flags display in upper-case and inactive flags display in lower-case: f,F Input focus applies to the event. s,S Event is on the same screen.
E(x,y)	Event location. The X and Y coordinates of the event.
E	Event window. Window where the event occurred.
Key	Key number. Number associated with the pressed key.
O	Owner of the window associated with the event.
R	Root window associated with the event.
R(x,y)	Root location. X and Y coordinates of the root position.
SN	Sequence number used to serialize events.

31

UMTS

Third Generation Cellular Networks (commonly referred to as 3G) represent the next phase in the evolution of cellular technology, evolution from the analog systems (1st generation) and digital systems (2nd generation). 3G networks will represent a shift from voice-centric services to converged services, including voice, data, video, fax and so forth.

UMTS is the dominant 3G solution being developed, representing an evolution from the GSM network standards, interoperating with a GSM core network. The 3G will implement a new access network, utilizing both improved radio interfaces and different technologies for the interface between the access network and the radio network.

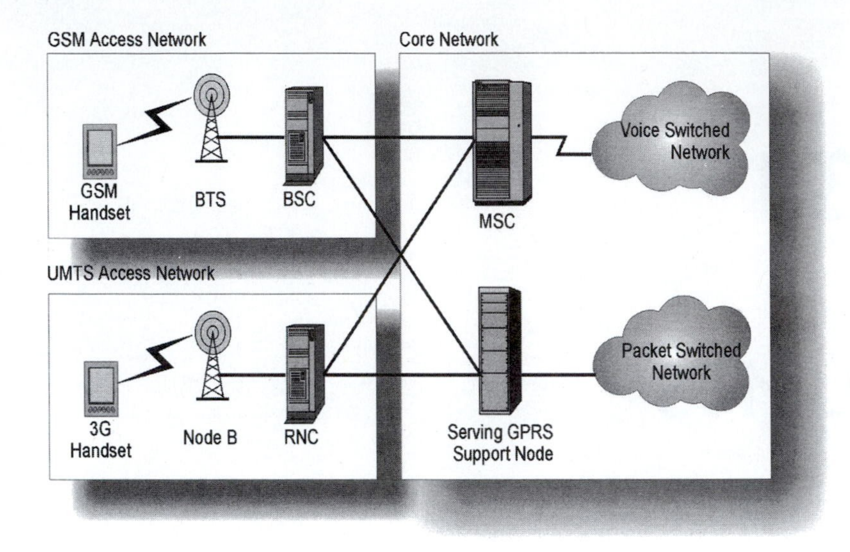

UMTS network structure

UMTS will use a wideband CDMA technology for transmission, and a more efficient modulation than GSM. This will allow UMTS to reach higher utilization, and offer higher bandwidth to the end-user. UMTS also implements an ATM infrastructure for the wireline interface, using both AAL2 and AAL5 adaptations; AAL2 for real-time traffic and AAL5 for data and signaling.

The following Protocols appear in this family
AAL2 see *ATM*
AAL5 see *ATM*
SSCOP (Q.2110)
SSCF-NNI (Q.2140)
BCC: Broadcast Call Control.
GCC: Group Call Control
GMM: GPRS Mobility Management.
MM: Mobility Management.
MTP-3B: Message Transfer Part Level 3B:
RANAP: Radio Access Network Application Protocol.
SCCP: Signalling Connection Control Part.
SCTP: Stream Control Transmission Protocol.
SNDCP: Sub-Network Dependant Convergence Protocol.
SM: Session Management.

BCC

3G TS 24.069 version 3.1.0 www.3gpp.org/ftp/specs

This protocol is a variant of the GPRS BCC protocol. The Broadcast Call Control (BCC) protocol is used by the Voice Group Call Service (VGCS) on the radio interface. It is one of the protocols of the Connection Management (CM) sublayer (see GSM 04.07).

Generally a number of mobile stations (MS) participate in a broadcast call. Consequently, there is in general more than one MS with a BCC entity engaged in the same broadcast call, and there is one BCC entity in the network engaged in that broadcast call.

The MS ignores BCC messages sent in unacknowledged mode and which specify as destination a mobile identity which is not a mobile identity of that MS. Higher layers and the MM sub-layer decide when to accept parallel BCC transactions and when/whether to accept BCC transactions in parallel to other CM transactions.

The broadcast call may be initiated by a mobile user or by a dispatcher. The originator of the BCC transaction chooses the Transaction Identifier (TI).

The call control entities are described as communicating finite state machines which exchange messages across the radio interface and communicate internally with other protocol (sub)layers. In particular, the BCC protocol uses the MM and RR sublayer specified in GSM 04.08. The network should apply supervisory functions to verify that the BCC procedures are progressing and if not, take appropriate means to resolve the problems.

The elementary procedures in the BCC include:

- Broadcast call establishment procedures,
- Broadcast call termination procedures
- Broadcast call information phase procedures
- Various miscellaneous procedures.

All messages have the following header:

8	7	6	5	4	3	2	1	Octet
Transaction identifier				Protocol discriminator				1
Message type								2
Information elements								3-n

BCC header structure

Protocol discriminator
The protocol discriminator specifies the message being transferred

Transaction identifier
Distinguishes multiple parallel activities (transactions) within one mobile station. The format of the transaction identifier is as follows:

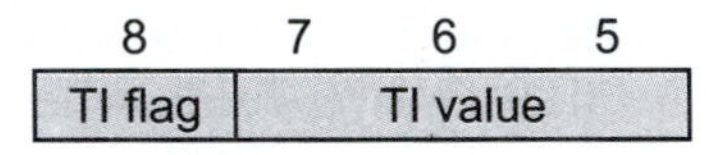

Transaction identifier

TI flag
Identifies who allocated the TI value for this transaction. The purpose of the TI flag is to resolve simultaneous attempts to allocate the same TI value.

TI value
The side of the interface initiating a transaction assigns TI values. At the beginning of a transaction, a free TI value is chosen and assigned to this transaction. It then remains fixed for the lifetime of the transaction. After a transaction ends, the associated TI value is free and may be reassigned to a later transaction. Two identical transaction identifier values may be used when each value pertains to a transaction originated at opposite ends of the interface.

Message type
The message type defines the function of each BCC message. The message type defines the function of each BCC message. The following message types exist:

0x110001	IMMEDIATE SETUP
0x110010	SETUP
0x110011	CONNECT
0x110100	TERMINATION
0x110101	TERMINATION REQUEST
0x110110	TERMINATION REJECT
0x111000	STATUS
0x111001	GET STATUS
0x111010	SET PARAMETER

Information elements

Each information element has a name which is coded as a single octet. The length of an information element may be fixed or variable and a length indicator for each one may be included.

GCC

3G TS 24.068 version 3.1.0 www.3gpp.org/ftp/specs

This protocol is a variant of the GPRS GCC protocol. The Group Call Control (GCC) protocol is used by the Voice Group Call Service (VGCS) on the radio interface within the 3GPP system. It is one of the protocols of the Connection Management (CM) sublayer (see GSM 04.07).

Generally a number of mobile stations (MS) participate in a group call. Consequently, there is in general more than one MS with a GCC entity engaged in the same group call, and there is one GCC entity in the network engaged in that group call.

The MS ignores GCC messages sent in unacknowledged mode and which specify as destination a mobile identity which is not a mobile identity of that MS. Higher layers and the MM sub-layer decide when to accept parallel GCC transactions and when/whether to accept GCC transactions in parallel to other CM transactions.

The group call may be initiated by a mobile user or by a dispatcher. In certain situations, a MS is assumed to be the originator of a group call without being the originator. The originator of the GCC transaction chooses the Transaction Identifier (TI).

The call control entities are described as communicating finite state machines which exchange messages across the radio interface and communicate internally with other protocol (sub) layers. In particular, the GCC protocol uses the MM and RR sublayer specified in GSM 04.08. The network should apply supervisory functions to verify that the GCC procedures are progressing and if not, take appropriate means to resolve the problems.

The elementary procedures in the GCC include:

- Group call establishment procedures
- Group call termination procedures
- Call information phase procedures
- Various miscellaneous procedures.

All messages have the following header:

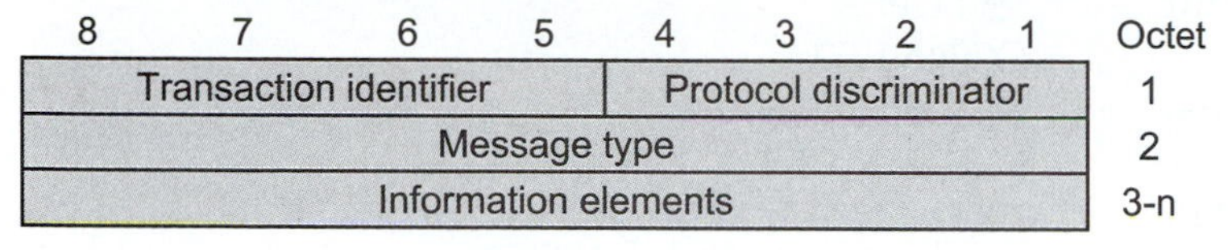

GCC header structure

Protocol discriminator

The protocol discriminator specifies the message being transferred

Transaction identifier

Distinguishes multiple parallel activities (transactions) within one mobile station. The format of the transaction identifier is as follows:

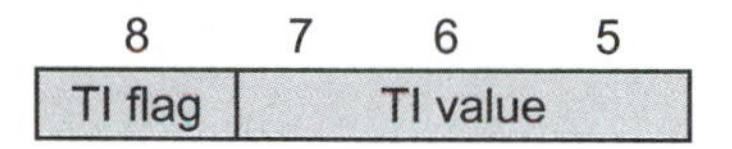

Transaction identifier

TI flag

Identifies who allocated the TI value for this transaction. The purpose of the TI flag is to resolve simultaneous attempts to allocate the same TI value.

TI value

The side of the interface initiating a transaction assigns TI values. At the beginning of a transaction, a free TI value is chosen and assigned to this transaction. It then remains fixed for the lifetime of the transaction. After a transaction ends, the associated TI value is free and may be reassigned to a later transaction. Two identical transaction identifier values may be used when each value pertains to a transaction originated at opposite ends of the interface.

Message type

The message type defines the function of each GCC message. The following message types exist:

0x110001	IMMEDIATE SETUP
0x110010	SETUP
0x110011	CONNECT
0x110100	TERMINATION
0x110101	TERMINATION REQUEST
0x110110	TERMINATION REJECT
0x111000	STATUS
0x111001	GET STATUS
0x111010	SET PARAMETER

Information elements

Each information element has a name which is coded as a single octet. The length of an information element may be fixed or variable and a length indicator for each one may be included.

GMM

3G.TS.24.008 v3.2.1:www.3gpp.org/ftp/specs

This protocol is a variant of the GPRS GMM protocol. UMTS and GPRS use the GSM MM (Mobility Management) protocol. Here it is known as the GPRS MM protocol (GMM). The main function of the MM sub-layer is to support the mobility of user terminals, such as informing the network of its present location and providing user identity confidentiality. A further function of the GMM sub-layer is to provide connection management services to the different entities of the upper Connection Management (CM) sub-layer.

The format of the header is shown in the following illustration:

8	7	6	5	4	3	2	1	Octet
Protocol discriminator				Skip indicator				1
Message type								2
Information elements								3-n

GMM header structure

Protocol discriminator

1000 identifies the GMM protocol.

Skip indicator

The value of this field is 0000.

Message type

Uniquely defines the function and format of each GMM message. The message type is mandatory for all messages. Bit 8 is reserved for possible future use as an extension bit. Bit 7 is reserved for the send sequence number in messages sent from the mobile station. GMM message types may be:

0 0 0 0 0 0 0 1 Attach request
0 0 0 0 0 0 1 0 Attach accept
0 0 0 0 0 0 1 1 Attach complete
0 0 0 0 0 1 0 0 Attach reject
0 0 0 0 0 1 0 1 Detach request
0 0 0 0 0 1 1 0 Detach accept

0 0 0 0 1 0 0 0	Routing area update request
0 0 0 0 1 0 0 1	Routing area update accept
0 0 0 0 1 0 1 0	Routing area update complete
0 0 0 0 1 0 1 1	Routing area update reject
0 0 0 1 0 0 0 0	P-TMSI reallocation command
0 0 0 1 0 0 0 1	P-TMSI reallocation complete
0 0 0 1 0 0 1 0	Authentication and ciphering req
0 0 0 1 0 0 1 1	Authentication and ciphering resp
0 0 0 1 0 1 0 0	Authentication and ciphering rej
0 0 0 1 0 1 0 1	Identity request
0 0 0 1 0 1 1 0	Identity response
0 0 1 0 0 0 0 0	GMM status
0 0 1 0 0 0 0 1	GMM information

Information elements

Various information elements.

MM

3G.TS.24.008 v.3.3.1www.3gpp.org/ftp/specs

The main function of the Mobility Management (MM) sub-layer is to support the mobility of user terminals, for instance; informing the network of its present location and providing user identity confidentiality. A further function of the MM sub-layer is to provide connection management services to the different entities of the upper Connection Management (CM) sub-layer.

The format of the header is shown in the following illustration:

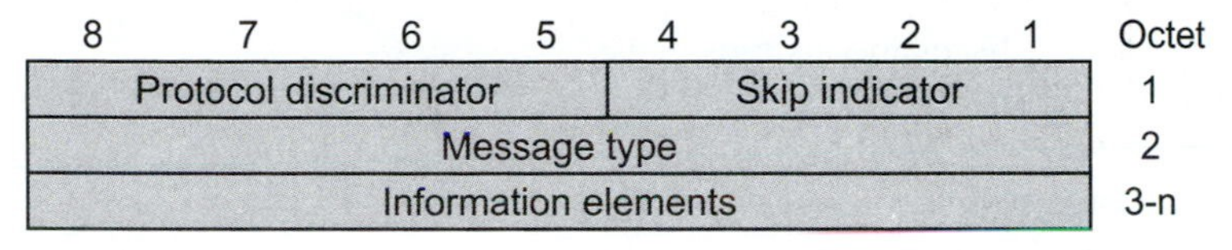

MM header structure

Protocol discriminator

0101 identifies the MM protocol.

Skip indicator

The value of this field is 0000.

Message type

Uniquely defines the function and format of each MM message. The message type is mandatory for all messages. Bit 8 is reserved for possible future use as an extension bit. Bit 7 is reserved for the send sequence number in messages sent from the mobile station. MM message types may be:

0x00	xxxx	Registration messages:
	0001	IMSI DETACH INDICATION
	0010	LOCATION UPDATING ACCEPT
	0100	LOCATION UPDATING REJECT
	1000	LOCATION UPDATING REQUEST
0x01	xxxx	Security messages:
	0001	AUTHENTICATION REJECT
	0010	AUTHENTICATION REQUEST

	0100	AUTHENTICATION RESPONSE
	1000	IDENTITY REQUEST
	1001	IDENTITY RESPONSE
	1010	TMSI REALLOCATION COMMAND
	1011	TMSI REALLOCATION COMPLETE
0x10	xxxx	Connection management messages:
	0001	CM SERVICE ACCEPT
	0010	CM SERVICE REJECT
	0011	CM SERVICE ABORT
	0100	CM SERVICE REQUEST
	1000	CM REESTABLISHMENT REQUEST
	1001	ABORT
0x11	xxxx	Miscellaneous messages:
	0001	MM STATUS

Information elements

Various information elements.

MTP-3B

Q.210 doc

Message Transfer Part - Level 3B (MTP-3B) connects Q.SAAL to the users. It is used to support MAP and BSSAP+ in circuit switched PLMNs. It transfers messages between the nodes of the signalling network. MTP-3B ensures reliable transfer of the signalling messages, even in the case of the failure of the signalling links and signalling transfer points. The protocol therefore includes the appropriate functions and procedures necessary both to inform the remote parts of the signalling network of the consequences of a fault, and to appropriately reconfigure the routing of messages through the signalling network.

The structure of the MTP-3B header is shown in the following illustration:

Service indicator	Subservice field
4 bits	4 bits

MTP-3B header structure

Service indicator

Used to perform message distribution and in some cases to perform message routing. The service indicator codes are used in international signalling networks for the following purposes:

- Signalling network management messages.
- Signalling network testing and maintenance messages.
- SCCP.
- Telephone user part.
- ISDN user part.
- Data user part.
- Reserved for MTP testing user part.

Sub-service field

The sub-service field contains the network indicator and two spare bits to discriminate between national and international messages.

RANAP

3G TS 25.413 V3.1.0 www.3gpp.org/ftp/specs

RANAP (Radio Access Network Application Protocol) encapsulates higher layer signalling. It manages the signalling and GTP connections between RNC and 3G-SGSN and signalling and circuit-switched connections between RNC and 3G MSC on the lu interface.

RANAP gives 3 sorts of services:

- General control services
- Notification services
- Dedicated control services

All messages are text messages in ASN.1 format.

SCCP

3G.TS.23.003 http://www.3gpp.org

The Signalling Connection Control Part (SCCP) offers enhancements to MTP level 3 to provide connectionless and connection-oriented network services, as well as to address translation capabilities. It is used to support MAP and BSSAP+ in circuit switched PLMNs. The SCCP enhancements to MTP provide a network service which is equivalent to the OSI Network layer 3.

The format of the header is shown in the following illustration:

Field	Octets
Routing label	3-4
Message type code	1
Mandatory fixed part	
Mandatory variable part	
Optional part	

SCCP header structure

Routing label

A standard routing label.

Message type code

A one octet code which is mandatory for all messages. The message type code uniquely defines the function and format of each SCCP message. Existing Message Type Codes are:

CR	Connection Request.
CC	Connection Confirm.
CREF	Connection Refused.
RLSD	Released.
RLC	Release Complete.
DT1	Data Form 1.
DT2	Data Form 2.
AK	Data Acknowledgment.
UDT	Unidata.
UDTS	Unidata Service.
ED	Expedited Data.
EA	Expedited Data Acknowledgment.

RSR Reset Request.
RSC Reset Confirm.
ERR Protocol Data Unit Error.
IT Inactivity Test.
XUDT Extended Unidata.
XUDTS Extended Unidata Service.

Mandatory fixed part

The parts that are mandatory and of fixed length for a particular message type will be contained in the mandatory fixed part.

Mandatory variable part

Mandatory parameters of variable length will be included in the mandatory variable part. The name of each parameter and the order in which the pointers are sent is implicit in the message type.

Optional part

The optional part consists of parameters that may or may not occur in any particular message type. Both fixed length and variable length parameters may be included. Optional parameters may be transmitted in any order. Each optional parameter will include the parameter name (one octet) and the length indicator (one octet) followed by the parameter contents.

SCTP

The Stream Control Transmission Protocol (SCTP) is designed to transport PSTN signalling messages over IP networks, but is capable of broader applications. SCTP is an application-level datagram transfer protocol operating on top of an unreliable datagram service such as UDP. It offers the following services to its users:

- Acknowledged error-free non-duplicated transfer of user data.
- Application-level segmentation to conform to discovered MTU size.
- Sequenced delivery of user datagrams within multiple streams, with an option for order-of-arrival delivery of individual datagrams.
- Optional multiplexing of user datagrams into SCTP datagrams, subject to MTU size restrictions.
- Enhanced reliability through support of multi-homing at either or both ends of the association.

The design of SCTP includes appropriate congestion avoidance behaviour and resistance to flooding and masquerade attacks. The SCTP datagram is comprised of a common header and chunks. The chunks contain either control information or user data.

The following is the format of the SCTP header.

2 bytes	2 bytes
Source Port Number	Destination Port Number
Verification Tag	
Adler 32 Checksum	

Source Port Number

This is the SCTP sender's port number. It can be used by the receiver, in combination with the source IP Address, to identify the association to which this datagram belongs.

Destination Port Number

This is the SCTP port number to which this datagram is destined. The receiving host will use this port number to de-multiplex the SCTP datagram to the correct receiving endpoint/application.

Verification Tag

The receiver of this 32 bit datagram uses the Verification tag to identify the association. On transmit, the value of this Verification tag must be set to the value of the Initiate tag received from the peer endpoint during the association initialization.

For datagrams carrying the INIT chunk, the transmitter sets the Verification tag to all 0's. If the receiver receives a datagram with an all-zeros Verification tag field, it checks the Chunk ID immediately following the common header. If the chunk type is not INIT or SHUTDOWN ACK, the receiver drops the datagram. For datagrams carrying the SHUTDOWN-ACK chunk, the transmitter sets the Verification tag to the Initiate tag received from the peer endpoint during the association initialization, if known. Otherwise the Verification tag is set to all 0's.

Adler 32 Checksum

This field contains an Adler-32 checksum on this SCTP datagram.

Chunk Field Descriptions

The following is the field format for the chunks transmitted in the SCTP datagram. Each chunk has a chunk ID field, a chunk specific Flag field, a Length field and a Value field.

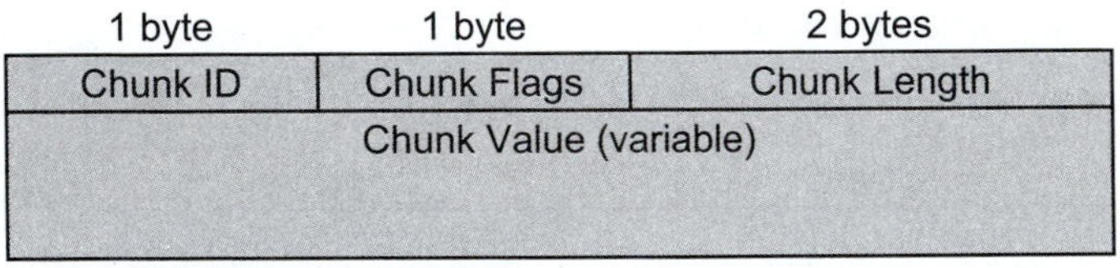

1 byte	1 byte	2 bytes
Chunk ID	Chunk Flags	Chunk Length
Chunk Value (variable)		

Chunk ID

The type of information contained in the chunk value field. The values of the chunk ID are defined as follows:

ID ValueChunk Type

00000000	Payload Data (DATA)
00000001	Initiation (INIT)
00000010	Initiation Acknowledgment (INIT ACK)
00000011	Selective Acknowledgment (SACK)
00000100	Heartbeat Request (HEARTBEAT)
00000101	Heartbeat Acknowledgment (HEARTBEAT ACK)
00000110	Abort (ABORT)

00000111	Shutdown (SHUTDOWN)
00001000	Shutdown Acknowledgment (SHUTDOWN ACK)
00001001	Operation Error (ERROR)
00001010	State Cookie (COOKIE)
00001011	Cookie Acknowledgment (COOKIE ACK)
00001100	Reserved for Explicit Congestion Notification Echo (ECNE)
00001101	Reserved for Congestion Window Reduced (CWR)
00001110 to 11111101 - reserved by IETF	
11111110	Vendor-specific Chunk Extensions
11111111 -	IETF-defined Chunk Extensions

Chunk Flags

The type of chunk flag as defined in the chunk ID defines whether these bits will be used. Their value is generally 0 unless otherwise specified.

Chunk Length

The size of the chunk in octets including the Chunk ID, Flags, Length and Value fields.

Chunk Value

This field contains the actual information to be transferred in the chunk. This is dependent on the chunk ID.

Chunk Types

Initiation (INIT)

This chunk is used to initiate a SCTP association between two endpoints. The INIT chunk contains the following parameters. Unless otherwise noted, each parameter is only be included once in the INIT chunk.

Fixed Parameters	**Status**
Initiate Tag	Mandatory
Receiver Window Credit	Mandatory
Number of Outbound Streams	Mandatory
Number of Inbound Streams	Mandatory
Initial TSN	Mandatory

Variable Parameters	**Status**
IPv4 Address/Port	Optional
IPv6 Address/Port	Optional

Cookie Preservative	Optional
Reserved For ECN Capable	Optional
Host Name Address	Optional
Supported Address Types	Optional

Initiate Acknowledgement (INIT ACK)

The INIT ACK chunk is used to acknowledge the initiation of a SCTP association. The parameter part of INIT ACK is formatted similarly to the INIT chunk. It uses two extra variable parameters: The Responder Cookie and the Unrecognized Parameter.

Selective Acknowledgement (SACK)

This chunk is sent to the remote endpoint to acknowledge received *Data* chunks and to inform the remote endpoint of gaps in the received subsequences of *Data* chunks as represented by their TSNs.

The selective acknowledgement chunk contains the highest consecutive TSN ACK and Rcv Window Credit (rwnd) parameters. By definition, the value of the highest consecutive TSN ACK parameter is the last TSN received at the time the Selective ACK is sent, before a break in the sequence of received TSNs occurs; the next TSN value following this one has not yet been received at the reporting end. This parameter therefore acknowledges receipt of all TSNs up to and including the value given.

The Selective ACK also contains zero or more fragment reports. Each fragment report acknowledges a sub-sequence of TSNs received following a break in the sequence of received TSNs. By definition, all TSNs acknowledged by fragment reports are higher than the value of the Highest Consecutive TSN ACK.

Heartbeat Request (HEARTBEAT)

An endpoint should send this chunk to its peer endpoint of the current association to probe the reachability of a particular destination transport address defined in the present association. The parameter fields contain the time values.

Heartbeat Acknowledgement (HEARTBEAT ACK)

An endpoint should send this chunk to its peer endpoint as a response to a Heartbeat Request. The parameter field contains the time values.

Abort Association (ABORT)

The Abort Association chunk is sent to the peer of an association to terminate the association. The Abort chunk may contain cause parameters to inform the receiver the reason for the abort. Data chunks are not bundled with the abort, control chunks may be bundled with an abort, but must be placed before the abort in the SCTP datagram or they will be ignored.

SHUTDOWN

An endpoint in an association uses this chunk to initiate a graceful termination of the association with its peer.

Shutdown Acknowledgement (SHUTDOWN ACK)

This chunk is used to acknowledge the receipt of the SHUTDOWN chunk at the completion of the shutdown process. The SHUTDOWN ACK chunk has no parameters.

Operation Error (ERROR)

This chunk is sent to the other endpoint in the association to notify certain error conditions. It contains one or more error causes.

State Cookie (COOKIE)

This chunk is used only during the initialization of an association. It is sent by the initiator of an association to its peer to complete the initialization process. This chunk precedes any Data chunk sent within the association, but may be bundled with one or more Data chunks in the same datagram.

Cookie Acknowledgement (COOKIE ACK)

This chunk is used only during the initialization of an association. It is used to acknowledge the receipt of a COOKIE chunk. This chunk precedes any Data chunk sent within the association, but may be bundled with one or more Data chunks in the same SCTP datagram.

Payload Data (DATA)

This contains the user data.

Vendor Specific Chunk Extensions

This chunk type is available to allow vendors to support their own extended data formats not defined by the IETF. It must not affect the operation of SCTP. Endpoints not equipped to interpret the vendor-specific chunk sent by a remote endpoint must ignore it. Endpoints that do not receive desired vendor specific information should make an attempt to operate without it, although they may do so (and report they are doing so) in a degraded mode.

SM

3G.TS.24.0008 v3.2.1:www.3gpp.org/ftp/specs

This protocol is a variant of the GPRS SM protocol. SM handles mobility issues such as roaming, authentication, selection of encryption algorithms and maintains PDP context. The main function of the session management (SM) is to support PDP context handling of the user terminal. The SM comprises procedures for: identified PDP context activation, deactivation and modification; and anonymous PDP context activation and deactivation. The format of the header is shown in the following illustration:

8	7	6	5	4	3	2	1	Octet
Protocol discriminator				Skip indicator				1
Message type								2
Information elements								3-n

SM header structure

Protocol discriminator

1010 identifies the SM protocol.

Skip indicator

The value of this field is 0000.

Message type

Uniquely defines the function and format of each SM message. The message type is mandatory for all messages. Bit 8 is reserved for possible future use as an extension bit. Bit 7 is reserved for the send sequence number in messages sent from the mobile station. SM message types may be:

0 1 x x x x x x	Session management messages
0 1 0 0 0 0 0 1	Activate PDP context request
0 1 0 0 0 0 1 0	Activate PDP context accept
0 1 0 0 0 0 1 1	Activate PDP context reject
0 1 0 0 0 1 0 0	Request PDP context activation
0 1 0 0 0 1 0 1	Request PDP context activation rej.
0 1 0 0 0 1 1 0	Deactivate PDP context request
0 1 0 0 0 1 1 1	Deactivate PDP context accept
0 1 0 0 1 0 0 0	Modify PDP context request

0 1 0 0 1 0 0 1	Modify PDP context accept
0 1 0 1 0 0 0 0	Activate AA PDP context request
0 1 0 1 0 0 0 1	Activate AA PDP context accept
0 1 0 1 0 0 1 0	Activate AA PDP context reject
0 1 0 1 0 0 1 1	Deactivate AA PDP context request
0 1 0 1 0 1 0 0	Deactivate AA PDP context accept
0 1 0 1 0 1 0 1	SM Status

Information elements

Various information elements.

SNDCP

GSM 04.65 version 6.1.0 www.3gpp.org/ftp/specs

Sub-Network Dependant Convergence Protocol (SNDCP) uses the services provided by the Logical Link Control (LLC) layer and the Session Management (SM) sub-layer. SNDCP splits into either IP or X.25 and maps them on to the LLC. It also provides fintions such as the compresssion, segmentation and multiplexing of network-layer messages to a single virtual connection.

The main functions of SNDCP are:

- Multiplexing of several PDPs (packet data protocol).
- Compression/decompression of user data.
- Compression/decompression of protocol control information.
- Segmentation of a network protocol data unit (N-PDU) into Logical Link Control Protocol Data Units (LL-PDUs) and re-assembly of LL-PDUs into a N-PDU.

The SN-DATA PDU is used for acknowledged data transfer. Its format is as follows:

8	7	6	5	4	3	2	1	Octet
X	C	T	M	NSAPI				1
DCOMP				PCOMP				2
Data								3-n

SN-DATA PDU structure

The SN-UNITDATA PDU is used for unacknowledged data transfer. Its format is as follows:

8	7	6	5	4	3	2	1	Octet
X	C	T	M	NSAPI				1
DCOMP				PCOMP				2
Segment offset				N-PDU number				3
E	N-PDU number (continued)							4
N-PDU number (extended)								5
Data								6-n

SN-UNITDATA PDU structure

NSAPI

Network service access point identifier. Values may be:

0 Escape mechanisms for future extensions.
1 Point-to-multipoint multicast (PTM-M) information.
2-4 Reserved for future use.
5-15 Dynamically allocated NSAPI value.

M

More bit. Values may be:

0 Last segment of N-PDU.
1 Not the last segment of N-PDU, more segments to follow.

T

SN-PDU type specifies whether the PDU is SN-DATA (0) or SN-UNITDATA (1).

C

Compression indicator. A value of 0 indicates that compression fields, DCOMP and PCOMP, are not included. A value of 1 indicates that these fields are included.

X

Spare bit is set to 0.

DCOMP

Data compression coding, included if C-bit set. Values are as follows:

0 No compression.
1-14 Points to the data compression identifier negotiated dynamically.
15 Reserved for future extensions.

PCOMP

Protocol control information compression coding, included if C-bit set. Values are as follows:

0 No compression.
1-14 Points to the protocol control information compression identifier negotiated dynamically.
15 Reserved for future extensions.

Segment offset

Segment offset from the beginning of the N-PDU in units of 128 octets.

N-PDU number

0-2047 when the extension bit is set to 0.
2048-524287 if the extension bit is set to 1.

E

Extension bit for N-PDU number.
0 Next octet is used for data.

32

V5 Protocols

The V5 protocol stack is used to connect an Access Network (AN) to a Local Exchange (LE). It is used for the following access methods:

- Analogue telephone access.
- ISDN basic rate access.
- ISDN primary rate access (V5.2).
- Other analog or digital accesses for semi-permanent connections without associated outband signalling information.

V5 uses 2048 kbps links; V5.2 may use up to 16 such interface links. For analog access, on the LE side signalling from the PSTN user port is converted into a functional part of the V5 protocol for signalling to the AN side. For ISDN users a control protocol is defined in the V5 for the exchange of the individual functions and messages required for coordination with the call control procedures in the Local Exchange.

In order to support more traffic and dynamic allocation of links, the V5.2 protocol has several additions:

- A bearer channel connection protocol establishes and de-establishes bearer connections required on demand, identified by the signalling information, under the control of the Local Exchange.
- A link control protocol is defined for the multi-link management to control link identification, link blocking and link failure conditions.
- A protection protocol, operated on two separate data links for security reasons, is defined to manage the protection switching of communication channels in case of link failures.

The following protocols are defined for the various protocol layers: LAPV5-EF, LAPV5, V5-Link Control, V5-BCC, V5-PSTN, V5-Control and V5-Protection.

```
WAN:Capture Buffer Display
Filter: All Frames
Protocol: V5-BCC

Frame: 1 Captured at:  +00:00.000
Length: 28    From: User    Status: Ok
V5-BCC: Protocol Discriminator: 0x48
V5-BCC: Layer 3 Header: 62719
V5-BCC:  1............... Source ID: AN
V5-BCC:  .1110100..111111 BCC Reference Number Value: 7487
V5-BCC:  ........11...... Reserved
V5-BCC: Message Type: 32 ALLOCATION
V5-BCC:  IE: 64 User Port Identification
V5-BCC:   Length: 4
V5-BCC:   User Port Identification Value: 2840
V5-BCC:  IE: 65 ISDN Port Channel Identification
V5-BCC:   Length: 3
V5-BCC:   Octet 3: 138
V5-BCC:   100..... Reserved
V5-BCC:   ...01010 ISDN User Port Time Slot Number: 10
V5-BCC:  IE: 66 V5-Time Slot Identification
V5-BCC:   Length: 4
V5-BCC:   V5 2048 Kbit/s Link Identifier: 87
V5-BCC:   Octet 4: 55
V5-BCC:   00...... Reserved
V5-BCC:   ..1..... Override: Override Request
V5-BCC:   ...10111 V5-Time Slot Number: 23
V5-BCC:  IE: 67 Multi-Slot Map
V5-BCC:   Length: 11

Options...  Search...  Restart  Setup...  Done
```

V5-BCC decode

The following diagram illustrates the V5 and other telephony protocols in relation to the OSI model:

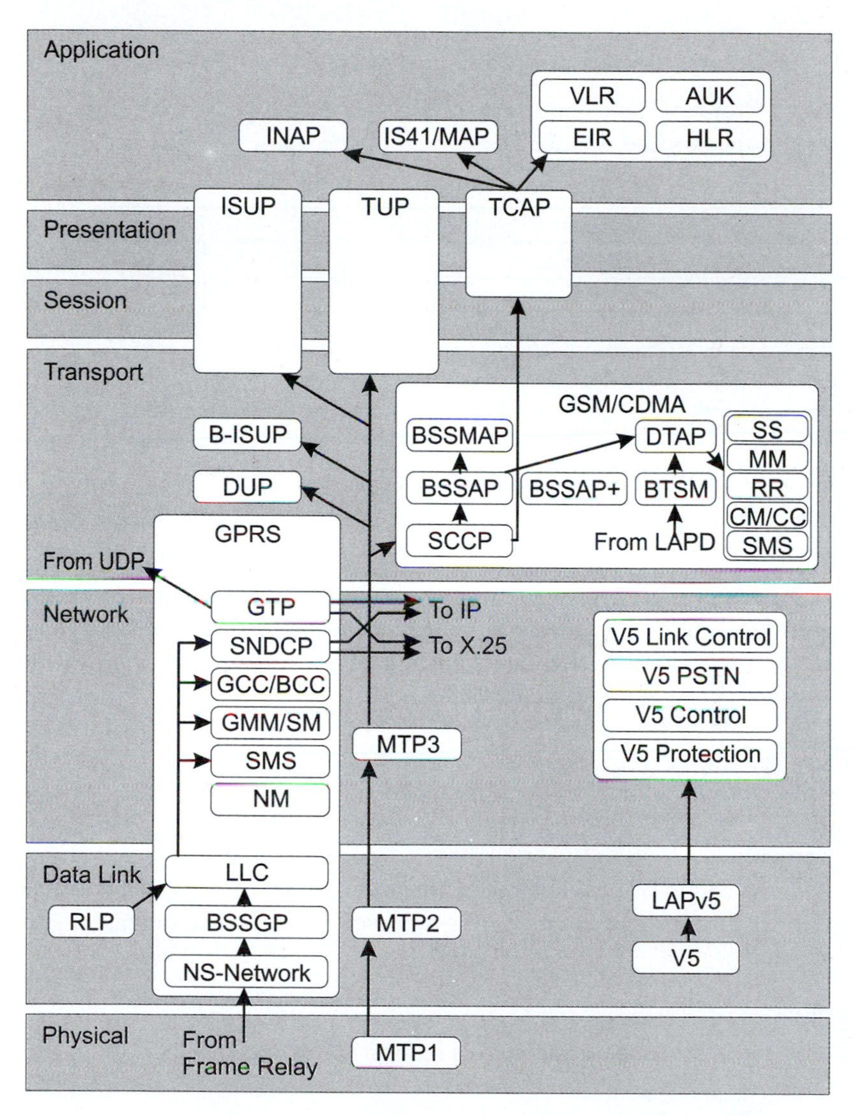

Telephony protocol suite in relation to the OSI model

LAPV5-EF

ITU G.964: http://www.itu.int/itudoc/itu-t/rec/g/g800up/g964.html

The V5 Protocol Envelope Function Sublayer exchanges information between the AN and the LE. The format of the frame is as follows:

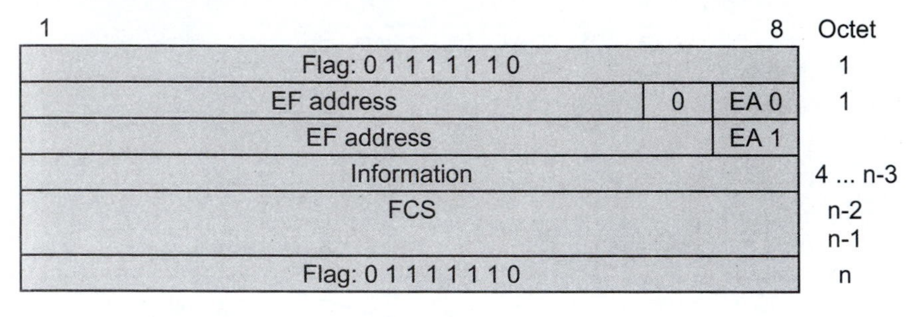

V5 envelope function sublayer structure

EF address

13-bit address field. The range from 0 to 8175 is used to identify an ISDN user port within the V5-interface. Values from 8176 to 8191 are reserved and are used to identify a point at which data link layer services are provided by the V5 layer 2 entity to a layer 3.

EA

Address extension bits.

Information

The information field consists of an integral number of octets. The default value for the maximum number of octets is 533 octets. The minimum size is 3 octets.

FCS

Frame check sequence as defined in section 2.1 of standard G.921.

LAPV5-DL

ITU G.964: http://www.itu.int/itudoc/itu-t/rec/g/g800up/g964.html

The LAPV5 Data Link Sublayer defines peer-to-peer exchanges of information between the AN and the LE .

LAPV5-DL frames can be of format A without an information field or format B with an information field. The format of these frames is shown in the following illustration:

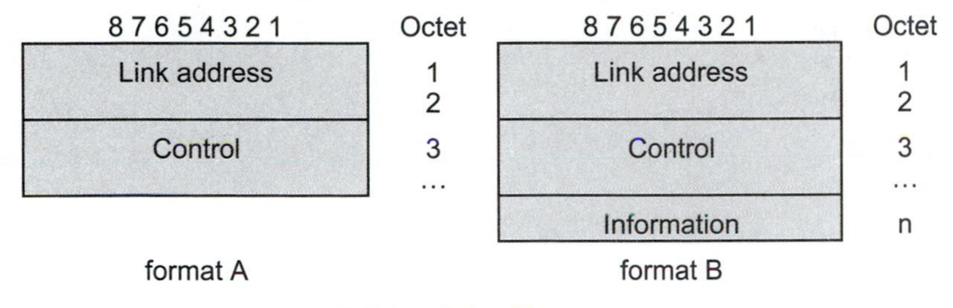

V5 data link sublayer structure

Link address

The link address field consists of two octets. The format of the link address is as follows:

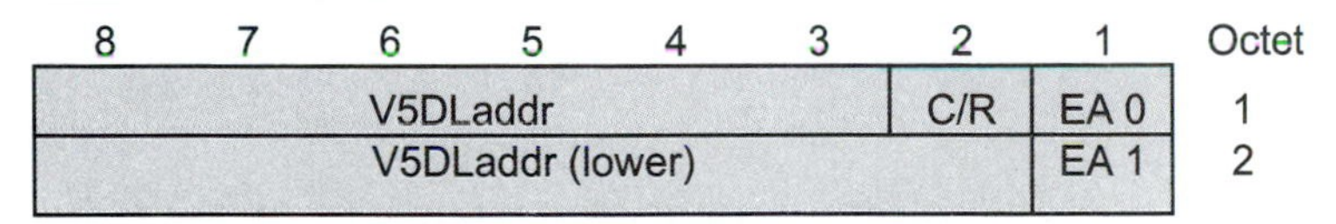

V5 Link address structure

V5DLaddr

13-bit address field. Values in the range of 0 to 8175 are used to identify ISDN user ports. Defined values are as follows:

8	7	6	5	4	3	2	1	
1	1	1	1	1	1	C/R	EA	Octet 1
								Octet 2
1	1	1	0	0	0	0	EA	PSTN signalling
1	1	1	0	0	0	1	EA	Control protocol

EA
Address extension bits.

Control
The control field identifies the type of frame which is either a command or response. The control field contains sequence numbers, where applicable.

Information
The information field of a frame, when present, consists of an integer number of octets. The maximum number of octets in the information field is 260.

V5-Link Control

ITU G.965: http://www.itu.int/itudoc/itu-t/rec/g/g800up/g965.html

The V5-Link Control Protocol is sent by the AN or LE to convey information required for the coordination of the link control functions for each individual 2048 kbps link.

The format of the link control protocol header is shown in the following illustration:

8	7	6	5	4	3	2	1	Octet
Protocol discriminator								1
Layer 3 address								2
Layer 3 address (lower)								3
0	Message type							4
Other IEs								etc.

V5 link control structure

Protocol discriminator
Distinguishes between messages corresponding to one of the V5 protocols.

Layer 3 address
Identifies the layer 3 entity, within the V5 interface to which the transmitted or received message applies.

Message type
Identifies the function of the message being sent or received. The message type may be LINK CONTROL or LINK CONTROL ACK.

Other IEs
Information elements. The only IE for the link control protocol is the Link Control Function. Its format is as follows:

8	7	6	5	4	3	2	1	Octet
0 0 1 0 0 0 0 1								1
Length of link control function								2
1	Link control function							4

Link Control Function IE

Link control function

The link control function conveyed by the message.

V5-BCC

ITU G.965: http://www.itu.int/itudoc/itu-t/rec/g/g800up/g965.html

The V5-BCC protocol provides the means for the LE to request the AN to establish and release connections, between specified AN user ports and specified V5-interface time slots. It enables V5 interface bearer channels to be allocated or de-allocated by independent processes (on a per call, preconnected or semi-permanent basis). There may be more than one process active at any one time for a given user port.

The format of the header is shown in the following illustration:

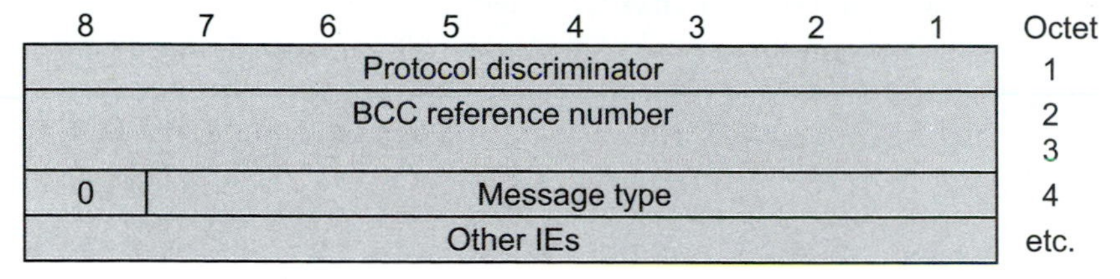

BCC header structure

Protocol discriminator

The Protocol discriminator distinguishes between messages corresponding to one of the V5 protocols.

BCC reference number

The BCC reference number information element is specific to the BCC protocol and uses the location of the Layer 3 address information element within the general V5 message structure.

The BCC reference number identifies the BCC protocol process, within the V5.2 interface, to which the transmitted or received message applies.

The BCC reference number value is a random value generated by the entity (AN or LE) creating the new BCC protocol process (this random value is implemented as a sequential generation of values). It is essential that values are not repeated in messages for which a different BCC process is required (in the same direction), until the old BCC process has been finished and the number deleted. In the case of any process generating error indications, the BCC reference number should not be re-used until sufficient time has elapsed for delayed arrival of messages containing the same BCC reference number. The length of the BCC reference number IE is 2 octets. Its format is shown here:

8	7	6	5	4	3	2	1	Octet
Src ID	BCC reference number value							4
0	0	BCC reference number value						

BCC reference number IE structure

Src ID

Source Identification is a field of one bit specifying the entity (LE or AN) that has created the BCC reference number (i.e. the entity that has created the BCC protocol process). The coding of this field is zero for an LE created process and one for an AN created process.

BCC reference number

13-bit field used for providing the binary coding that identifies the BCC process.

Message type

The Message type identifies the function of the message being sent or received. Message types may be as follows:

- ALLOCATION.
- ALLOCATION COMPLETE.
- ALLOCATION REJECT.
- DE-ALLOCATION.
- DE-ALLOCATION COMPLETE.
- DE-ALLOCATION REJECT.
- AUDIT.
- AUDIT COMPLETE.
- AN FAULT.
- AN FAULT ACKNOWLEDGE.
- PROTOCOL ERROR.

Other IEs

Various elements concerned with the V5-BCC messages. The following IEs are specific to BCC:

- User port identification.
- ISDN port channel identification.
- V5-time slot identification.
- Multi-slot map.
- Reject cause.
- Protocol error cause.
- Connection incomplete.

V5-PSTN

ITU G.964: http://www.itu.int/itudoc/itu-t/rec/g/g800up/g964.html

The signalling protocol specification and layer multiplexing PSTN protocol on the V5-interface is basically a stimulus protocol in that it does not control the call procedures in the AN, but rather transfers information about the analogue line state over the V5 interface. The V5-PSTN protocol is used in conjunction with the national protocol entity in the LE. The national protocol entity in the LE which is used for customer lines which are connected directly to the LE, is also used to control calls on customer lines which are connected via the V5-interface. For time critical sequences, it is necessary to extract certain signalling sequences (e.g., compelled sequences) from the national protocol entity into an 'AN part' of the national protocol entity. The V5-PSTN protocol has a relatively small functional part which is concerned with path setup, release of the path on the V5 interface, call collision resolution on the V5 interface and handling of new calls in case of overload conditions in the LE. The majority of line signals are not interpreted by the V5-PSTN protocol, but simply transferred transparently between the user port in the AN and national protocol entity in the LE.

The format of the header is shown in the following illustration:

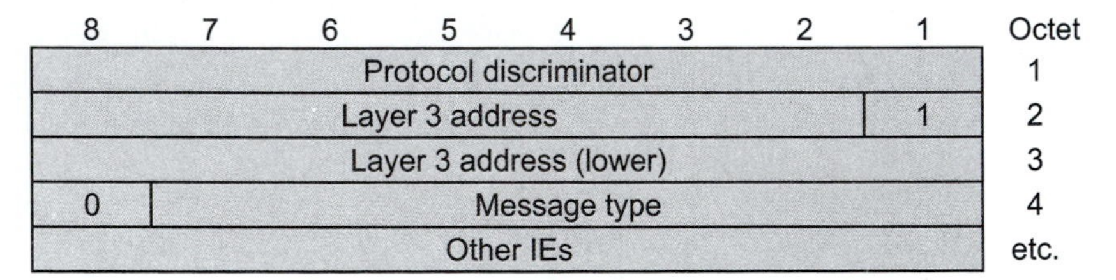

PSTN header structure

Protocol discriminator

Distinguishes between messages corresponding to one of the V5 protocols. The value of the protocol discriminator for PSTN is 01001000.

Layer 3 address

This identifies the layer 3 entity, within the V5 interface to which the transmitted or received message applies.

Message type

Identifies the function of the message being sent or received. Message types for PSTN may be as follows:

- ESTABLISH.
- ESTABLISH ACK.
- SIGNAL.
- SIGNAL ACK.
- STATUS.
- STATUS ENQUIRY.
- DISCONNECT.
- DISCONNECT COMPLETE.
- PROTOCOL PARAMETER.

Other IEs

For PSTN the following information elements may appear:
Single Octet IEs:

- Pulse-notification.
- Line-information.
- State.
- Autonomous-signalling-sequence.
- Sequence-response.

Variable Length IEs:

- Sequence-number.
- Cadenced-ringing.
- Pulsed-signal.
- Steady-signal.
- Digit-signal.
- Recognition-time.
- Enable-autonomous-acknowledge.
- Disable-autonomous-acknowledge.
- Cause.
- Resource-unavailable.

V5-Control

ITU G.964: http://www.itu.int/itudoc/itu-t/rec/g/g800up/g964.html

The V5-Control user port status indication is based on the defined split of responsibilities between AN and LE. The structure of the V5-Control protocol is shown in the following illustration:

8	7	6	5	4	3	2	1	Octet
Protocol discriminator								1
Layer 3 address							0	2
Layer 3 address (lower)							1	3
0	Message type							4
Other IEs								etc.

Control header structure

Protocol discriminator

Distinguishes between messages corresponding to one of the V5 protocols.

Layer 3 address

Identifies the layer 3 entity, within the V5 interface to which the transmitted or received message applies.

Message type

Identifies the function of the message being sent or received. Message types for the control protocol may be as follows:

- PORT CONTROL.
- PORT CONTROL ACK.
- COMMON CONTROL.
- COMMON CONTROL ACK.

Other IEs

For the control protocol the following information element may appear:

Single Octet IEs:

- Performance grading.
- Rejection cause.

Variable Length IEs:

- Control function element.
- Control function ID.
- Variant Interface ID.

V5-Protection

ITU G.965: http://www.itu.int/itudoc/itu-t/rec/g/g800up/g965.html

A single V5 interface may consist of up to sixteen 2048 kbps links. According to the protocol architecture and multiplexing structure a communication path may carry information associated with several 2048 kbps links (non-associated information transfer). The failure of a communication path could therefore impact the service of a large number of customers in an unacceptable way. This is particularly true for the BCC protocol, the control protocol and the link control protocol, where all user ports are affected in the event of a failure of the relevant communication path. In order to improve the reliability of the V5 interface, protection procedures for the switch-over of communication paths under failure are provided.

The protection mechanisms are used to protect all active C-channels. The protection mechanism also protects the protection protocol C-path (itself) which is used to control the protection switch-over procedures. In addition, flags are continuously monitored on all physical C-channels (active and standby C-channels) in order to protect against failures which are not already detected by Layer 1 detection mechanisms. If a failure is detected on a standby C-channel, the system management will be notified and as a result, will not switch a logical C-channel to that non-operational standby C-channel.

The format of the header is shown in the following illustration:

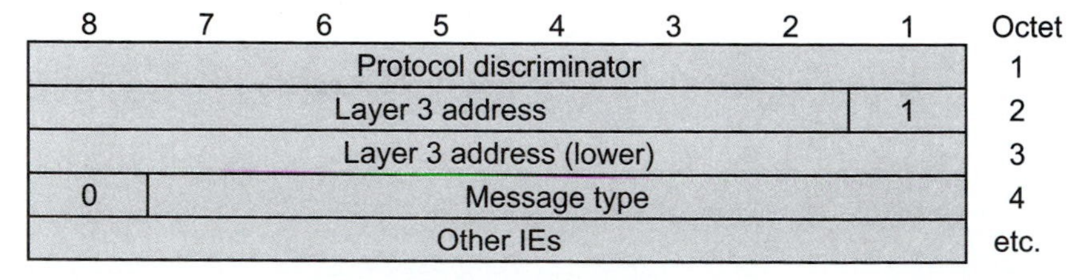

Protection protocol structure

Protocol discriminator

Distinguishes between messages corresponding to one of the V5 protocols.

Layer 3 address

Identifies the layer 3 entity, within the V5 interface to which the transmitted or received message applies.

Message type

Identifies the function of the message being sent or received. Message types for the protection protocol may be as follows:

- SWITCH-OVER REQ.
- SWITCH-OVER COM.
- OS-SWITCH-OVER COM.
- SWITCH-OVER ACK.
- SWITCH-OVER REJECT.
- PROTOCOL ERROR.
- RESET SN COM.
- RESET SN ACK.

Other IEs

For the protection protocol, the following information elements may appear:

Variable Length IEs:

- Sequence number.
- Physical C-channel identification.
- Rejection cause.
- Protocol error cause.

33

VB5

ETSI EN 301 005-1 V1.1.4 (1998-05)

The VB5 reference point concept, based on ITU-T Recommendation G.902, was split into two variants. The VB5.1 reference point, the first variant, is based on an ATM cross-connect with provisioned connectivity. The VB5.1 protocol is based on the Real Time Management Co-ordination (RTMC) protocol.

The VB5.1 message format is as shown in the following illustration:

8	7	6	5	4	3	2	1	Octet
Protocol discriminator								1
0	0	0	0	TID length				2
TAID								3
Transaction identifier value								4-5
Message type								6
Message compatibility instruction indicator								7
Message length								8-9
Information elements								10-n

VB5.1 message structure

Protocol discriminator

Distinguishes between VB5 specific protocols and other non-VB5 protocols. This field is coded as follows for RTMC:

8	7	6	5	4	3	2	1
0	1	0	0	1	0	0	1

Transaction identifier

4-byte field which identifies the transaction at the VB5.1 protocol virtual channel to which the particular message applies. This field includes the TAID flag and the TID length.

TID length

Transaction identifier length in octets.

TAID flag

When set, indicates that the message is sent to the side that originated the transaction identifier. Otherwise, the message is sent from the side that originated the transaction identifier.

Message type

Identifies the specific VB5 protocol the message belongs to and the function of the message being sent. Values may be as follows:

0 0 0 0 0 0 0 0 Reserved.
0 0 0 0 0 0 1 0 BLOCK_RSC.
0 0 0 0 0 0 1 1 BLOCK_RSC_ACK.
0 0 0 0 0 1 0 0 CONS_CHECK_REQ.
0 0 0 0 0 1 0 1 CONS_CHECK_REQ_ACK.
0 0 0 0 0 1 1 0 CONS_CHECK_END.
0 0 0 0 0 1 1 1 CONS_CHECK_END_ACK.
0 0 0 0 1 0 0 0 REQ_LSPID.
0 0 0 0 1 0 0 1 LSPID.
0 0 0 0 1 0 1 0 PROTOCOL_ERROR.
0 0 0 0 1 1 0 0 RESET_RSC.
0 0 0 0 1 1 0 1 RESET_RSC_ACK.
0 0 0 0 1 1 1 0 AWAIT_CLEAR.
0 0 0 0 1 1 1 1 AWAIT_CLEAR_ACK.
0 0 0 1 0 0 0 0 AWAIT_CLEAR_COMP.
0 0 0 1 0 0 0 1 AWAIT_CLEAR_COMP_ACK.
0 0 0 1 0 0 1 0 UNBLOCK_RSC.
0 0 0 1 0 0 1 1 UNBLOCK_RSC_ACK.
1 1 1 1 1 1 1 1 Reserved.

Message compatibility instruction indicator
Defines the behavior of the peer network element if the message is not understood.

Message length
Identifies the length of the contents of a message. It is the binary coding of the number of octets of the message contents.

Information element
Each information element contains the following elements:

8	7	6	5	4	3	2	1	Octet
Information element type								1
Information element compatibility instruction indicator								2
Information element length								3-4
Information element content								5-n

Information element type
Identifies the specific VB5 protocol the information element belongs to and the function of the information element being sent.

Information element compatibility instruction indicator
Defines the behavior of the peer network element if the information element is not understood.

Information element length
Length of the contents of the information element. It is the binary coding of the number of octets of the information element contents, i.e., the number of octets following the information element length octets.

34

VoIP

Voice-over-IP Overview

Voice-over-IP (VoIP) implementations enables users to carry voice traffic (for example, telephone calls and faxes) over an IP network.

There are 3 main causes for the evolution of the Voice over IP market:

- Low cost phone calls
- Add-on services and unified messaging
- Merging of data/voice infrastructures

A VoIP system consists of a number of different components: Gateway/Media Gateway, Gatekeeper, Call agent, Media Gateway Controller, Signaling Gateway and a Call manager

The Gateway converts media provided in one type of network to the format required for another type of network. For example, a Gateway could terminate bearer channels from a switched circuit network (i.e., DS0s) and media streams from a packet network (e.g., RTP streams in an IP network). This gateway may be capable of processing audio, video and T.120 alone or in any combination, and is capable of full duplex media translations. The

Gateway may also play audio/video messages and performs other IVR functions, or may perform media conferencing.

In VoIP, the digital signal processor (DSP) segments the voice signal into frames and stores them in voice packets. These voice packets are transported using IP in compliance with one of the specifications for transmitting multimedia (voice, video, fax and data) across a network: H.323 (ITU), MGCP (level 3,Bellcore, Cisco, Nortel), MEGACO/H.GCP (IETF), SIP (IETF), T.38 (ITU), SIGTRAN (IETF), Skinny (Cisco) etc.

Coders are used for efficient bandwidth utilization. Different coding techniques for telephony and voice packet are standardized by the ITU-T in its G-series recommendations: G.723.1, G.729, G.729A etc.

The coder-decoder compression schemes (CODECs) are enabled for both ends of the connection and the conversation proceeds using Real-Time Transport Protocol/User Datagram Protocol/Internet Protocol (RTP/UDP/IP) as the protocol stack.

Quality of Service

A number of advanced methods are used to overcome the hostile environment of the IP net and to provide an acceptable Quality of Service. Example of these methods are: delay, jitter, echo, congestion, packet loss, and missordered packets arrival. As VoIP is a delay-sensitive application, a well-engineered, end-to-end network is necessary to use VoIP successfully. The Mean Opinion Score is one of the most important parameters that determine the QoS.

There are several methods and sophisticated algorithms developed to evaluate the QoS: PSQM (ITU P.861), PAMS (BT) and PESQ.

Each CODEC provides a certain quality of service. The quality of transmitted speech is a subjective response of the listener (human or artificial means). A common benchmark used to determine the quality of sound produced by specific CODECs is the mean opinion score (MOS). With MOS, a wide range of listeners judge the quality of a voice sample (corresponding to a particular CODEC) on a scale of 1 (bad) to 5 (excellent).

Services

The following are examples of services provided by a Voice over IP network according to market requirements:

Phone to phone, PC to phone, phone to PC, fax to e-mail, e-mail to fax, fax to fax, voice to e-mail, IP Phone, transparent CCS (TCCS), toll free number (1-800), class services, call center applications, VPN, Unified Messaging, Wireless Connectivity, IN Applications using SS7, IP PABX and soft switch implementations. This the following VoIP protocols are described in this book:

- H.323: see *H.323 (Chapter 15)*
- IPDC.
- MGCP: Media Gateway Control Protocol.
- Megaco.
- RVP over IP.
- SAPv2: Session Announcement Protocol.
- SDP: Session Description Protocol.
- SGCP: Simple Gateway Control Protocol
- SIP: Session Initiation Protocol.
- Skinny.

Megaco (H.248)

Internet draft: draft-ietf-megaco-merged-00.txt

The Media Gateway Control Protocol, (Megaco) is a result of joint efforts of the IETF and the ITU-T Study Group 16. The protocol definition of this protocol is common text with ITU-T Recommendation H.248.

The Megaco protocol is used between elements of a physically decomposed multimedia gateway. There are no functional differences from a system view between a decomposed gateway, with distributed sub-components potentially on more than one physical device, and a monolithic gateway such as described in H.246. This protocol creates a general framework suitable for gateways, multipoint control units and interactive voice response units (IVRs).

Packet network interfaces may include IP, ATM or possibly others. The interfaces support a variety of SCN signalling systems, including tone signalling, ISDN, ISUP, QSIG and GSM. National variants of these signalling systems are supported where applicable.

All messages are in the format of ASN.1 text messages.

MGCP

RFC: 2705

Media Gateway Control Protocol (MGCP) is used for controlling telephony gateways from external call control elements called media gateway controllers or call agents. A telephony gateway is a network element that provides conversion between the audio signals carried on telephone circuits and data packets carried over the Internet or over other packet networks.

MGCP assumes a call control architecture where the call control intelligence is outside the gateways and handled by external call control elements. The MGCP assumes that these call control elements, or Call Agents, will synchronize with each other to send coherent commands to the gateways under their control. MGCP is, in essence, a master/slave protocol, where the gateways are expected to execute commands sent by the Call Agents.

The MGCP implements the media gateway control interface as a set of transactions. The transactions are composed of a command and a mandatory response. There are eight types of commands:

- CreateConnection.
- ModifyConnection.
- DeleteConnection.
- NotificationRequest.
- Notify.
- AuditEndpoint.
- AuditConnection.
- RestartInProgress.

The first four commands are sent by the Call Agent to a gateway. The Notify command is sent by the gateway to the Call Agent. The gateway may also send a DeleteConnection. The Call Agent may send either of the Audit commands to the gateway. The Gateway may send a RestartInProgress command to the Call Agent.

All commands are composed of a command header, optionally followed by a session description. All responses are composed of a response header, optionally followed by a session description. Headers and session descriptions are encoded as a set of text lines, separated by a carriage return and line feed character (or, optionally, a single line-feed character). The headers are separated from the session description by an empty line.

MGCP uses a transaction identifier to correlate commands and responses. Transaction identifiers have values between 1 and 999999999. An MGCP entity cannot reuse a transaction identifier sooner than 3 minutes after completion of the previous command in which the identifier was used.

The command header is composed of:

- A command line, identifying the requested action or verb, the transaction identifier, the endpoint towards which the action is requested, and the MGCP protocol version,
- A set of parameter lines, composed of a parameter name followed by a parameter value.

The command line is composed of:

- Name of the requested verb.
- Transaction identifier correlates commands and responses. Values may be between 1 and 999999999. An MGCP entity cannot reuse a transaction identifier sooner than 3 minutes after completion of the previous command in which the identifier was used.
- Name of the endpoint that should execute the command (in notifications, the name of the endpoint that is issuing the notification).
- Protocol version.

These four items are encoded as strings of printable ASCII characters, separated by white spaces, i.e., the ASCII space (0x20) or tabulation (0x09) characters. It is recommended to use exactly one ASCII space separator.

RVP over IP

RVP Over IP Specification, MCK Communications (Proprietary)

Remote Voice Protocol (RVP) is MCK Communications' protocol for transporting digital telephony sessions over packet or circuit based data networks. The protocol is used primarily in MCK's Extender product family, which extends PBX services over Wide Area Networks (WANs). RVP provides facilities for connection establishment and configuration between a client (or remote station set) device and a server (or phone switch) device.

RVP/IP uses TCP to transport signalling and control data, and UDP to transport voice data.

Signalling and Control Packets

Control and signalling packets carried over TCP are encapsulated using the following format, a header followed by signalling or control messages:

1 byte	1 byte	
Length	Protocol code	RVP/IP message . . .

RVP over IP packet structure

Length

A one byte field containing the length of the header (protocol code and the entire RVP/IP message). The length field allows recognition of message boundaries in a continuous TCP data stream.

Protocol code

Identifies the RVP/IP protocol:

35 RVP/IP control messages (see *RVP Control Protocol*).

36 RVP/IP signalling data (see *RVP Signalling Operations*).

RVP/IP messages

RVP/IP messages include *RVP Control Protocol* (RVPCP) and *RVP Signalling Operations* described below.

RVP Control Protocol (RVPCP)

RVP Control Protocol is for control messages that configure and maintain the data link between the client and the server. The control protocol was originally developed for point-to-point data applications; most of its functionality is unnecessary when using TCP/IP. During an RVP/IP session, only one class of RVP/IP control message are exchanged: RVPCP ADD VOICE (operation code 12) packet, used to send the UDP port used by the client (for subsequent voice data packets) to the server. This message always takes a single parameter of type RVPCP UDP PORT (type code 9), which always has a length of exactly two and a value that is the two-byte UDP port to which voice data packets should be addressed. The server responds with a packet containing the code RVPCP ADD VOICE ACK (operation code 13) which contains exactly one parameter, the server's voice UDP port. If RVP/IP is operating in "dynamic voice" mode, this exchange must be repeated whenever the voice channel needs to be reestablished, i.e., whenever the phone goes off-hook.

The structure of the control messages is described below:

2 bytes	2 bytes	
Operation code	Parameter count	Parameters . . .

RVP over IP control message structure

Operation code

The operation code defines the class of RVP/IP control messages Possible classes are:

12 RVPCP ADD VOICE
13 RVPCP ADD VOICE ACK

Parameter count

The parameter count equals exactly one parameter.

Parameters

Parameters of all control messages are passed as Type, Length and Value (TLV) structures as described below:

2 bytes	2 bytes	
Type	Length	Value . . .

RVP over IP control message parameters

Type

RVPCP UDP PORT (or type code 9).

Length

The number of bytes in the value field.

Value

The UDP port number.

RVP Signalling Operations

The structure of RVP signalling data (protocol type 36) is described below:

7	8	8	8
Packet Length	Protocol	Message Length	Data

RVP over IP signalling message structure

RVP signalling data packets always begin with a length byte immediately after the RVP/IP encapsulation header. The packets contain two classes of data, either raw digital telephone signalling packets or high-level RVP session commands. Session commands are differentiated from raw signalling data by adding an offset of 130 in the "Message Length" field. All raw signalling data has a true length field of less than or equal to 128. The true length of a session command message is calculated by subtracting 130 from the length field.

For all session commands, the Command Code (one-byte) follows the message length field. Bit seven of the command code is considered the "ACK" bit. All other bits in this field are part of the command code itself.

Voice Data Packets

The structure of voice data packets, carried over UDP datagrams, is described below:

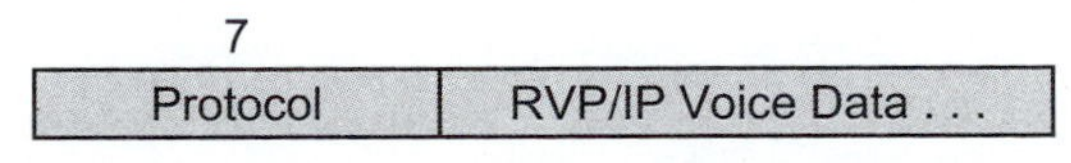

RVP over IP Voice packet structure

Protocol

The protocol code is always 37 for RVP/IP voice data packets.

RVP/IP voice data

A single voice packet is carried in each UDP datagram.

SAPv2

Internet draft: draft-ietf-mmusic-sap-v2-04.txt

SAP is an announcement protocol used by session directory clients. A SAP announcer periodically multicasts an announcement packet to a well-known multicast address and port. The announcement is multicast with the same scope as the session it is announcing, ensuring that the recipients of the announcement can also be potential recipients of the session the announcement describes (bandwidth and other such constraints permitting). This is also important for the scalability of the protocol, as it keeps local session announcements local.

The following is the format of the SAP data packet.

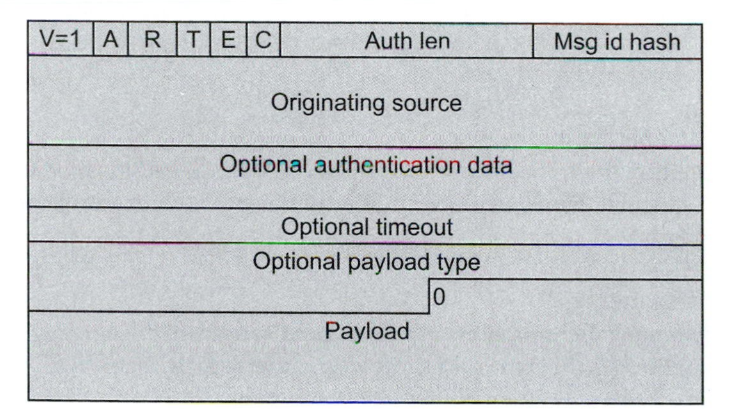

SAP data packet structure

V: version number

The version number field is three bits and must be set to 1.

A: address type

The Address type field is one bit. It can have a value of 0 or 1:

0 The originating source field contains a 32-bit IPv4 address.

1 The originating source contains a 128-bit IPv6 address.

R: reserved

SAP announcers set this to 0. SAP listeners ignore the contents of this field.

T: message type

The Message Type field is one bit. It can have a value of 0 or 1:

0 Session announcement packet

1 Session deletion packet.

E: encryption bit

The encryption bit may be 0 or 1.

1 The payload of the SAP packet is encrypted and the timeout field must be added to the packet header.

0 The packet is not encrypted and the timeout must not be present.

C: compressed bit

If the compressed bit is set to 1, the payload is compressed.

Authentication length

An 8 bit unsigned quantity giving the number of 32 bit words, following the main SAP header, that contain authentication data. If it is zero, no authentication header is present.

Message identifier hash

A 16-bit quantity that, used in combination with the originating source, provides a globally unique identifier indicating the precise version of this announcement.

Originating source

This field contains the IP address of the original source of the message. This is an IPv4 address if the A field is set to zero; otherwise, it is an IPv6 address. The address is stored in network byte order.

Timeout

When the session payload is encrypted, the detailed timing fields in the payload are not available to listeners not trusted with the decryption key. Under such circumstances, the header includes an additional 32-bit timestamp field stating when the session should be timed out. The value is an unsigned quantity giving the NTP time in seconds at which time the session is timed out. It is in network byte order.

Payload type

The payload type field is a MIME content type specifier, describing the format of the payload. This is a variable length ASCII text string, followed by a single zero byte (ASCII NUL).

Payload

The Payload field includes various sub fields:

Version number (V)

The version number of the authentication format is 1.

Padding bit (P)

If necessary, the authentication data is padded to be a multiple of 32 bits and the padding bit is set. In this case, the last byte of the authentication data contains the number of padding bytes (including the last byte) that must be discarded.

Authentication type (Auth)

The authentication type is a 4 bit encoded field that denotes the authentication infrastructure the sender expects the recipients to use to check the authenticity and integrity of the information. This defines the format of the authentication sub-header and can take the values: 0=PGP format, 1=CMS format. All other values are undefined.

SDP

RFC 2327

The Session Description Protocol (SDP) describes multimedia sessions for the purpose of session announcement, session invitation and other forms of multimedia session initiation.

On Internet Multicast backbone (Mbone) a session directory tool is used to advertise multimedia conferences and communicate the conference addresses and conference tool-specific information necessary for participation. The SDP does this. It communicates the existence of a session and conveys sufficient information to enable participation in the session. Many of the SDP messages are sent by periodically multicasting an announcement packet to a well-known multicast address and port using SAP (session announcement protocol). These messages are UDP packets with a SAP header and a text payload. The text payload is the SDP session description. Messages can also be sent using email or the WWW (World Wide Web).

The SDP text messages include:

- Session name and purpose
- Time the session is active
- Media comprising the session
- Information to receive the media (address etc.)

SDP messages are text messages using the ISO 10646 character set in UTF-8 encoding.

SIP

RFC 2543

Session Initiation Protocol (SIP) is a application layer control simple signalling protocol for VoIP implementations using the Redirect Mode.

SIP is a textual client-server base protocol and provides the necessary protocol mechanisms so that the end user systems and proxy servers can provide different services:

1 Call forwarding in several scenarios: no answer, busy , unconditional, address manipulations (as 700, 800 , 900- type calls).
2 Callee and calling number identification
3 Personal mobility
4 Caller and callee authentication
5 Invitations to multicast conference
6 Basic Automatic Call Distribution (ACD)

SIP addresses (URL) can be embedded in Web pages and therefore can be integrated as part of powerful implementations (*Click to talk,* for example).

SIP using simple protocol structure, provides the market with fast operation, flexibility, scalability and multiservice support.

SIP provides its own reliability mechanism. SIP creates, modifies and terminates sessions with one or more participants. These sessions include Internet multimedia conferences, Internet telephone calls and multimedia distribution. Members in a session can communicate using multicast or using a mesh of unicast relations, or a combination of these. SIP invitations used to create sessions carry session descriptions which allow participants to agree on a set of compatible media types. It supports user mobility by proxying and redirecting requests to the user's current location. Users can register their current location. SIP is not tied to any particular conference control protocol. It is designed to be independent of the lower-layer transport protocol and can be extended with additional capabilities.

SIP transparently supports name mapping and redirection services, allowing the implementation of ISDN and Intelligent Network telephony subscriber services. These facilities also enable personal mobility which is based on the use of a unique personal identity

SIP supports five facets of establishing and terminating multimedia communications:

User location
User capabilities
User availability
Call setup
Call handling

SIP can also initiate multi-party calls using a multipoint control unit (MCU) or fully-meshed interconnection instead of multicast. Internet telephony gateways that connect Public Switched Telephone Network (PSTN) parties can also use SIP to set up calls between them.

SIP is designed as part of the overall IETF multimedia data and control architecture currently incorporating protocols such as RSVP, RTP RTSP, SAP and SDP. However, the functionality and operation of SIP does not depend on any of these protocols.

SIP can also be used in conjunction with other call setup and signalling protocols. In that mode, an end system uses SIP exchanges to determine the appropriate end system address and protocol from a given address that is protocol-independent. For example, SIP could be used to determine that the party can be reached using H.323 to find the H.245 gateway and user address and then use H.225.0 to establish the call.

SIP Operation

Sip works as follows:

Callers and callees are identified by SIP addresses. When making a SIP call, a caller first locates the appropriate server and then sends a SIP request. The most common SIP operation is the invitation. Instead of directly reaching the intended callee, a SIP request may be redirected or may trigger a chain of new SIP requests by proxies. Users can register their location(s) with SIP servers.

SIP messages can be transmitted either over TCP or UDP

SIP messages are text based and use the ISO 10646 character set in UTF-8 encoding. Lines must be terminated with CRLF. Much of the message syntax and header field are similar to HTTP. Messages can be request messages or response messages.

Protocol header structure.

The protocol is composed of a start line, message header, an empty line and an optional message body.

Request Messages

The format of the Request packet header is shown in the following illustration:

Method	Request URI	SIP version

SIP request packet structure

Method
The method to be performed on the resource. Possible methods are Invite, Ack, Options, Bye, Cancel, Register

Request-URI
A SIP URL or a general Uniform Resource Identifier, this is the user or service to which this request is being addressed.

SIP version
The SIP version being used; this should be version 2.0

Response Message

The format of the Response message header is shown in the following illustration:

SIP version	Status code	Reason phrase

SIP response packet structure

SIP version
The SIP version being used.

Status-code
A 3-digit integer result code of the attempt to understand and satisfy the request.

Reason-phrase
A textual description of the status code.

SGCP

IETF draft: http://www.ietf.org/internet-drafts/draft-huitema-sgcp-v1-02.txt

Simple Gateway Control Protocol (SGCP) is used to control telephony gateways from external call control elements. A telephony gateway is a network element that provides conversion between the audio signals carried on telephone circuits and data packets carried over the Internet or over other packet networks.

The SGCP assumes a call control architecture where the call control intelligence is outside the gateways and is handled by external call control elements. The SGCP assumes that these call control elements, or Call Agents, will synchronize with each other to send coherent commands to the gateways under their control.

The SGCP implements the simple gateway control interface as a set of transactions. The transactions are composed of a command and a mandatory response. There are five types of commands:

- CreateConnection.
- ModifyConnection.
- DeleteConnection.
- NotificationRequest.
- Notify.

The first four commands are sent by the Call Agent to a gateway. The Notify command is sent by the gateway to the Call Agent. The gateway may also send a DeleteConnection.

All commands are composed of a Command header, optionally followed by a session description. All responses are composed of a Response header, optionally followed by a session description. Headers and session descriptions are encoded as a set of text lines, separated by a line feed character. The headers are separated from the session description by an empty line.

The command header is composed of:

- Command line.
- A set of parameter lines, composed of a parameter name followed by a parameter value.

The command line is composed of:

- Name of the requested verb.
- Transaction identifier, correlates commands and responses. Transaction identifiers may have values between 1 and 999999999 and transaction identifiers are not reused sooner than 3 minutes after completion of the previous command in which the identifier was used.
- Name of the endpoint that should execute the command (in notifications, the name of the endpoint that is issuing the notification).
- Protocol version.

These four items are encoded as strings of printable ASCII characters, separated by white spaces, i.e. the ASCII space (0x20) or tabulation (0x09) characters. It is recommended to use exactly one ASCII space separator.

Skinny

Cisco protocol

Skinny Client Control Protocol (SCCP). Telephony systems are moving to a common wiring plant. The end station of a LAN or IP- based PBX must be simple to use, familiar and relatively cheap. The H.323 recommendations are quite an expensive system. An H.323 proxy can be used to communicate with the Skinny Client using the SCCP. In such a case the telephone is a skinny client over IP, in the context of H.323. A proxy is used for the H.225 and H.245 signalling.

The skinny client (i.e. an Ethernet Phone) uses TCP/IP to transmit and receive calls and RTP/UDP/IP to/from a Skinny Client or H.323 terminal for audio. Skinny messages are carried above TCP and use port 2000.

The messages consist of Station message ID messages.

They can be of the following types:

Code	Station Message ID Message
0x0000	Keep Alive Message
0x0001	Station Register Message
0x0002	Station IP Port Message
0x0003	Station Key Pad Button Message
0x0004	Station Enbloc Call Message
0x0005	Station Stimulus Message
0x0006	Station Off Hook Message
0x0007	Station On Hook Message
0x0008	Station Hook Flash Message
0x0009	Station Forward Status Request Message
0x11	Station Media Port List Message
0x000A	Station Speed Dial Status Request Message
0x000B	Station Line Status Request Message
0x000C	Station Configuration Status Request Message
0x000D	Station Time Date Request Message
0x000E	Station Button Template Request Message
0x000F	Station Version Request Message
0x0010	Station Capabilities Response Message
0x0012	Station Server Request Message
0x0020	Station Alarm Message

Code	Station Message ID Message
0x0021	Station Multicast Media Reception Ack Message
0x0024	Station Off Hook With Calling Party Number Message
0x22	Station Open Receive Channel Ack Message
0x23	Station Connection Statistics Response Message
0x25	Station Soft Key Template Request Message
0x26	Station Soft Key Set Request Message
0x27	Station Soft Key Event Message
0x28	Station Unregister Message
0x0081	Station Keep Alive Message
0x0082	Station Start Tone Message
0x0083	Station Stop Tone Message
0x0085	Station Set Ringer Message
0x0086	Station Set Lamp Message
0x0087	Station Set Hook Flash Detect Message
0x0088	Station Set Speaker Mode Message
0x0089	Station Set Microphone Mode Message
0x008A	Station Start Media Transmission
0x008B	Station Stop Media Transmission
0x008F	Station Call Information Message
0x009D	Station Register Reject Message
0x009F	Station Reset Message
0x0090	Station Forward Status Message
0x0091	Station Speed Dial Status Message
0x0092	Station Line Status Message
0x0093	Station Configuration Status Message
0x0094	Station Define Time & Date Message
0x0095	Station Start Session Transmission Message
0x0096	Station Stop Session Transmission Message
0x0097	Station Button Template Message
0x0098	Station Version Message
0x0099	Station Display Text Message
0x009A	Station Clear Display Message
0x009B	Station Capabilities Request Message
0x009C	Station Enunciator Command Message
0x009E	Station Server Respond Message
0x0101	Station Start Multicast Media Reception Message
0x0102	Station Start Multicast Media Transmission Message
0x0103	Station Stop Multicast Media Reception Message
0x0104	Station Stop Multicast Media Transmission Message

Code	Station Message ID Message
0x105	Station Open Receive Channel Message
0x0106	Station Close Receive Channel Message
0x107	Station Connection Statistics Request Message
0x0108	Station Soft Key Template Respond Message
0x109	Station Soft Key Set Respond Message
0x0110	Station Select Soft Keys Message
0x0111	Station Call State Message
0x0112	Station Display Prompt Message
0x0113	Station Clear Prompt Message
0x0114	Station Display Notify Message
0x0115	Station Clear Notify Message
0x0116	Station Activate Call Plane Message
0x0117	Station Deactivate Call Plane Message
0x118	Station Unregister Ack Message

35

WAP

WAP Forum (www.wapforum.com)

WAP, Wireless Application Protocol aims to provide Internet content and advanced telephony services to digital mobile phones, pagers and other wireless terminals. The protocol family works across different wireless network environments and makes web pages visible on low-resolution and low-bandwidth devices. WAP phones are "smart phones" allowing their users to respond to e-mail, access computer databases and to empower the phone to interact with Internet-based content and e-mail.

WAP specifies a Wireless application Environment and Wireless Protocols. The Wireless application environment (WAE) is based on WSP (Wireless Session Protocol) and WTP (Wireless Transaction Protocol).

The OSI Model for Wireless Communication

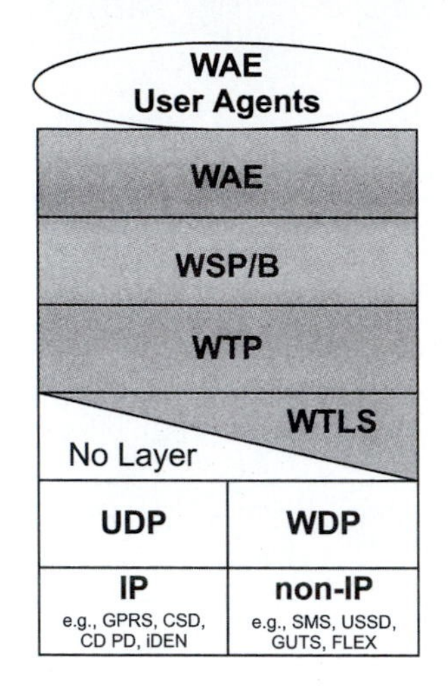

WAP Protocol stack

The basic construction of WAP architecture can be explained using the following model. The order of the independent levels – which are a hierarchy - has the advantage that the system is very flexible and can be scaled up or down. Because of the different levels – or stacks - this is called the "WAP Stack", which is divided into 5 different levels.

- Application Layer: Wireless Application Environment (WAE).
- Session Layer: Wireless Session Protocol (WSP).
- Transaction Layer: Wireless Transaction Protocol (WTP).
- Security Layer: Wireless Transport Layer Security (WTLS).
- Transport Layer: Wireless Datagram Protocol (WDP).

Each stack overlaps with the stack below. This stack architecture makes it possible for software manufacturers to develop applications and services for certain stacks. They may even develop services for stacks which are not specified yet.

The WAP stack is an entity of protocols which cover the wireless data transfer. The diagram above shows the order of the different stacks and their protocols. This includes the stacks responsible for the layout as well as the stacks resposible for the actual data transfer. The highest level or stack is the one which deals with the layout. A lower stack is responsible for the transfer and the security through WTLS (Wireless Transport Layer Security).

All stacks lower than this one are being called network stack. Due to this hierarchy of stacks any changes made in the network stacks will have no influence over the stacks above

Application Layer (WAE and WTA)

The environment for wireless applications (Wireless Application Environment WAE) and the application for wireless phones (Wireless Telephony Application WTA) are the highest layer in the hierarchy of WAP architechture. These two are the main interface to the client device, which gives and controls the description language, the script language of any application and the specifics of the telephony. WAE and WTA have only a few easy functions on the client device, like the maintenance of a history list, for example.

Session Layer (Wireless Session Protocol WSP)

The Wireless Session Protocol (WSP) has all the specifications for a session. It is the interface between the application layer and the transfer layer and delivers all functions that are needed for wireless connections. A session mainly consists of 3 phases: start of the session, transfering information back and forth and the end of the session. Additionally, a session can be interrupted and started again (from the point where it was interrupted.)

Transaction Layer (Wireless Transaction Protocol WTP)

The specifications for the transfer layer are in the Wireless Transaction Protocol (WTP). Like the User Datagramm Protocol (UDP), the WTP runs at the head of the datagramm service. Both the UDP and the WTP are a part of the standard application from the TCP/IP to make the simplified protocol compatible to mobile terminals. WTP supports chaining together protocol data and the delayed response to reduce the number of transmissions. The protocol tries to optimize user interaction in order that information can be received when needed.

Wireless Transport Layer Security WTLS

The Wireless Transport Layer Security (WTLS) is a optional layer or stack which consists of description devices. A secure transmission is crucial for certain applications such as e-commerce or WAP-banking and is a standard in these days. Furthermore WTLS contains a check for data integrity, user authentification and gateway security.

Transport Layer (Wireless Datagram Protocol WDP)

The Wireless Datagram Protocol (WDP) represents the transfer or transmission layer and is also the interface of the network layer to all the above stacks/layers. With the help of WDP the transmission layer can be assimilated to the specifications of a network operator. This means that WAP is completely independent from any network operator. The transmission of SMS, USSD, CSD, CDPD, IS-136 packet data and GPRS is supported. The Wireless Control Message Protocol (WCMP) is an optional addition to WAP, which will inform users about occurred errors.

WTLS

Wapforum version 11/99

Wireless Transport Layer Security is a protocol based on the TLS protocol. It is used with the WAP transport protocols and has been optimised for use over narrow-band communication channels. The WTLs layer is above the transport protocol layer. The required security layer of the protocol determines whether it is used or not. It provides a secure transport service interface that preserves the transport service interface below; additionally it provides an interface for managing secure connections. WTLS aims to provide privacy, data integrity and authentication between two communication applications. Among its features are datagram support, optimised handshaking and dynamic key refreshing. It is optimised for low-bandwidth bearer networks with relatively long latency.

The WTLS Record Protocol is a layered protocol. The Record Protocol takes messages to be transmitted, optionally compresses the data, applies a MAC, encrypts, and transmits the result. Received data is decrypted, verified, and decompressed, then delivered to higher-level clients. Four record protocol clients are described in the WTLS standard; the change cipher spec protocol, the handshake protocol, the alert protocol and the application data protocol. If a WTLS implementation receives a record type it does not understand, it ignores it. Several records can be concatenated into one transport SDU. For example, several handshake messages can be transmitted in one transport SDU. This is particularly useful with packet-oriented transports such as GSM short messages.

Handshake protocols	Alert Protocol	Application Protocol	Change Cipher Spec Protocol
Record protocol			

The handshake protocol is made up of 3 sub-protocols. All messages are encapsulated in a plaintext structure.

WTP

WAPforum WTP 11/6/99

The Wireless Transaction Protocol provides the services necessary for interactive browsing applications. During a browsing session the client requests information from a server and the server responds with the information. This is referred to as a transaction. WTP runs on a datagram service and possible a security service.

Advantages of WTP include:

- Improved reliability over datagram services
- Imported efficiency over connection oriented services
- As a message oriented protocol, it is designed for services oriented towards transactions.

Main features:

- 3 kinds of transaction services.
    - Class 0 Unreliable invoke messages with no result messages
    - Class 1: Reliable invoke messages with no result messages
    - Class 2: Reliable invoke messages with exactly one reliable result message.
- Reliability achieved by using unique transaction identifiers, acknowledgements, duplicate removal; and retransmissions.
- No explicit set up or tear down phases.
- Optional user-to-user reliability.
- Optionally the last acknowledgement of the transaction may contain out-of-band information.
- Concatenation may be used to convey multiple PDUs in one service data unit of the datagram transport.
- The basic unit of interchange is an entire message, not a stream of bytes.
- Mechanisms are provided to minimize the number of transactions replayed as a result of duplicate packets.
- Abort of outstanding transactions.
- For reliable invoke messages, both success and failure reported.
- Asynchronous transactions allowed.

The protocol data unit (PDU) consists of the header and data (if present). The header contains a fixed part and a variable part; The variable parts are carried in the Transport Information Item (TPI). Each PDU has its own fixed header (the fixed headers vary slightly in structure). As an example, the structure of the invoke PDU fixed header appears below:

<table>
<tr><td>1</td><td colspan="3">2-5</td><td colspan="2">6</td><td>7</td><td>8</td></tr>
<tr><td>Con</td><td colspan="3">PDU Type</td><td colspan="2">GTR</td><td>TTR</td><td>RID</td></tr>
<tr><td colspan="8">TID</td></tr>
<tr><td>Version</td><td>TIDnew</td><td>U/P</td><td>RES</td><td>RES</td><td colspan="3">TCL</td></tr>
</table>

CON continue flag (1 bit):

The continue flag indicates the presence of any TPIs in the variable part. If the flag is set, there are one or more TPIs in the variable portion of the header. If the flag is clear, the variable part of the header is empty. This flag is also used as the first bit of a TPI, and indicates whether the TPI is the last of the variable header. If the flag is set, another TPI follows this TPI. If the flag is clear, the octet after this TPI is the first octet of the user data.

PDU type

The PDU type determines the length and structure of the header and dictates what type of WTP PDU the PDU is (Invoke, Ack, etc). This provides information to the receiving WTP provider as to how the PDU data should be interpreted and what action is required.

The following PDU types are defined:

PDU Code	PDU Type
0x01	Invoke
0x02	Result
0x03	Ack
0x04	Abort
0x05	Segmented Invoke
0x06	Segmented Result
0x07	Negative Ack

Group trailer (GTR) and Transmission trailer (TTR) flag (2 bit):

When segmentation and re-assembly is implemented, the TTR flag is used to indicate the last packet of the segmented message. The GTR flag is used to indicate the last packet of a packet group.

GTR/TTR flag combinations:

GTR TTR	Description
00	Not last packet
01	Last packet of message
10	Last packet of packet group
11	Segmentation and Re-assembly NOT supported.

The default setting should be GTR=1 and TTR=1, that is, WTP segmentation and re-assembly not supported.

RID Re-transmission Indicator (1 bit):
Enables the receiver to differentiate between packets duplicated by the network and packets re-transmitted by the sender. In the original message the RID is clear. When the message gets re-transmitted the RID is set.

TID Transaction identifier (16 bit):
The TID is used to associate a packet with a particular transaction.

Version
The current version is 0X00

TIDnew flag
This bit is set when the Initiator has wrapped the TID value, i.e. set it to be lower than the previous TID value.

U/P
When this flag is set it indicates that the Initiator requires a User acknowledgement from the server WTP user. The WTP user confirms every received message.

RES
This is a reserved bit and its value should be set to 0.

TCL
The transaction class shows the desired transaction class in the invoke message.

Packet sequence number (8 bit):
This is used by the PDUs belonging to the segmentation and re-assembly function. This number indicates the position of the packet in the segmented message. It appears in the TPI.

WSP

WAP WSP 5/11/99

The Session layer protocol family in the WAP architecture is called the Wireless Session Protocol, WSP. WSP provides the upper-level application layer of WAP with a consistent interface for two session services. The first is a connection-mode service that operates above a transaction layer protocol WTP, and the second is a connectionless service that operates above a secure or non-secure datagram transport service.

The Wireless Session Protocols currently offer services most suited for browsing applications. WSP provides HTTP 1.1 functionality (it is a binary form of HTTP) and incorporates new features such as long-lived sessions, a common facility for data push, capability negotiation and session suspend/resume. The protocols in the WSP family are optimized for low-bandwidth bearer networks with relatively long latency. Requests and responses can include both headers and data. WSP provides push and pull data transfer WSP functions on the transaction and datagram services.

Messages can be in connection mode or connectionless. Connection mode messages are carried over WTP. In this case the protocol consists of WTP protocol messages with WSP PDUs as their data. Connectionless messages consist only of the WSP PDUs.

The general structure of the WSP PDU is as follows:

1 byte	1 byte	
TID/PID	PDU Type	Type Specific Contents

TID/PID

Transaction ID or Push ID. The TID field is used to associate requests with replies in the connectionless session service. The presence of the TID is conditional. It is included in the connectionless WSP PDUs, and is not included in the connection-mode PDUs. In connectionless WSP, the TID is passed to and from the session user as the "Transaction Id" or "Push Id" parameters of the session primitive

PDU type

The Type field specifies the type and function of the PDU. The type numbers for the various PDUs are defined below. The rest of the PDU is type-specific information, referred to as the contents.

Number	Name Assigned
0x00	Reserved
0x01	Connect
0x02	ConnectReply
0x03	Redirect
0x04	Reply
0x05	Disconnect
0x06	Push
0x07	ConfirmedPush
0x08	Suspend
0x09	Resume
0x10–0x3F	Unassigned
0x40	Get
0x41	Options (Get PDU)
0x42	Head (Get PDU)
0x43	Delete (Get PDU)
0x44	Trace (Get PDU)
0x45-0x4F	Unassigned (Get PDU)
0x50-0x5F	Extended Method (Get PDU)
0x60	Post
0x61	Put (Post PDU)
0x62–0x6F	Unassigned (Post PDU)
0x70-0x7F	Extended Method (Post PDU)
0x80-0xFF	Reserved

36

X.25

X.25 is the CCITT's recommendation for the interface between a DTE and DCE over a Public Switched Telephone Network (PSTN). Generally, X.25 covers layers 1 to 3 of the ISO communication model, but the term is used here to refer specifically to packet layer 3. X.25 is carried within the Information Field of LAPB frames.

The following protocols are described in this chapter:

- LAPB.
- X.25.
- X.75.
- MLP.
- HDLC.

The following diagram shows X.25 in relation to the OSI model:

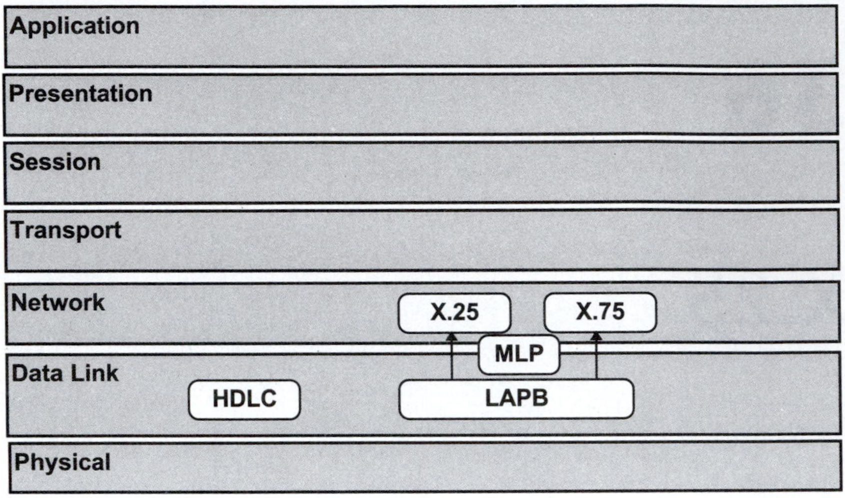

X.25 in relation to the OSI model

LAPB

LAPB is the layer 2 protocol used to carry X.25 packets. The format of a standard LAPB frame is as follows:

Flag	Address field	Control field	Information	FCS	Flag

LAPB frame structure

Flag

The value of the flag is always (0x7E). In order to ensure that the bit pattern of the frame delimiter flag does not appear in the data field of the frame (and therefore cause frame misalignment), a technique known as Bit Stuffing is used by both the transmitter and the receiver.

Address field

The first byte of the frame after the header flag is known as the Address Field. In LAPB, however, this field has no meaning since the protocol works in a point to point mode and the DTE network address is represented in the layer 3 packets. This byte is therefore put to a different use; it separates the link commands from the responses and can have only two values: 0x01 and 0x03. 01 identifies frames containing commands from DTE to DCE and responses to these commands from DCE to DTE. 03 is used for frames containing commands from DCE to DTE and for responses from DTE to DCE.

Control field

The field following the Address Field is called the Control Field and serves to identify the type of the frame. In addition, it includes sequence numbers, control features and error tracking according to the frame type.

In LAPB, since there is no master/slave relationship, the sender uses the Poll bit to insist on an immediate response. In the response frame this same bit becomes the receivers Final bit. The receiver always turns on the Final bit in its response to a command from the sender with the Poll bit set. The P/F bit is generally used when either end becomes unsure about proper frame sequencing because of a possible missing acknowledgement, and it is necessary to re-establish a point of reference.

Modes of operation

LAPB works in the Asynchronous Balanced Mode (ABM). This mode is totally balanced (i.e., no master/slave relationship) and is signified by the SABM(E) frame. Each station may initialize, supervise, recover from errors, and send frames at any time. The DTE and DCE are treated as equals.

FCS

The Frame Check Sequence (FCS) enables a high level of physical error control by allowing the integrity of the transmitted frame data to be checked. The sequence is first calculated by the transmitter using an algorithm based on the values of all the bits in the frame. The receiver then performs the same calculation on the received frame and compares its value to the CRC.

Window size

LAPB supports an extended window size (modulo 128) where the number of possible outstanding frames for acknowledgement is raised from 8 to 128. This extension is generally used for satellite transmissions where the acknowledgement delay is significantly greater than the frame transmission times. The type of the link initialization frame determines the modulo of the session and an "E" is added to the basic frame type name (e.g., SABM becomes SABME).

Frame types

The following are the Supervisory Frame Types in LAPB:

RR	Information frame acknowledgement and indication to receive more.
REJ	Request for retransmission of all frames after a given sequence number.
RNR	Indicates a state of temporary occupation of station (e.g., window full).

The following are the Unnumbered Frame Types in LAPB:

DISC	Request disconnection.
UA	Acknowledgement frame.
DM	Response to DISC indicating disconnected mode.
FRMR	Frame reject.
SABM	Initiator for asynchronous balanced mode. No master/slave.
SABME	SABM in extended mode.

There is one Information Frame Type in LAPB:

Info	Information frame.

X.25

The structure of the X.25 packet is as follows for Modulo 8:

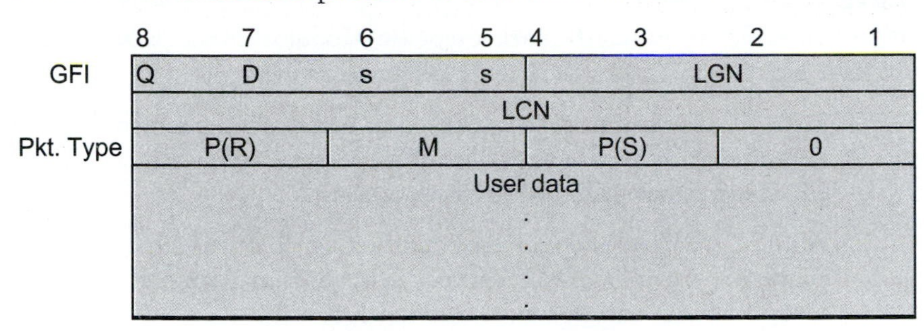

X.25 packet structure for Modulo 8

The structure of the X.25 packet is as follows for Modulo 128:

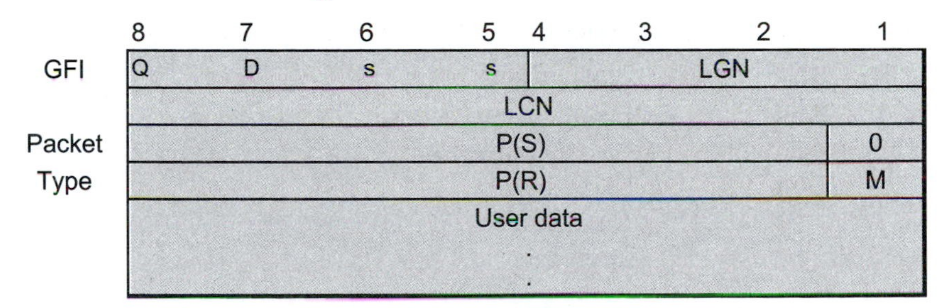

X.25 packet structure for Modulo 128

GFI

General format identifier which indicates the layout of the remainder of the packet header.

Q Q bit.

D D bit.

s Sequence scheme - specifies whether the frame is modulo 8 (ss=1) or module 128 (ss=2).

LGN

Logical channel group number is a 4-bit field which together with the LCN identifies the actual logical channel number.

LCN

Logical channel number is an 8-bit field which identifies the actual logical channel number of the DTE-DCE link.

Packet type

8-bit field which identifies the packet type (in Modulo 128 the packet type is 16 bits). See below for a list of all available packet types.

> P(R) Packet receive sequence number which appears in data and flow control packets or the called DTE address which may appear in call setup, clearing and registration packets.
>
> P(S) Packet send sequence number which appears in data packets or the calling DTE address field which may appear in call setup, clearing, and registration packets.
>
> M More data bit which appears only in data packets. The field is set to 1 to indicate that the packet is part of a sequence of packets that should be treated as a logical whole.

Packet types may be as follows:

CALL ACC	Call Accept
CALL REQ	Call Request
CLR CNF	Clear Confirmation
CLR REQ	Clear Request
DATA	Data Packet
DIAG	Diagnostic
INT CNF	Interrupt Confirmation
INT REQ	Interrupt Request
REJ	Reject
RES CNF	Reset Confirmation
RES REQ	Reset Request
RNR	Receive Not Ready
RR	Receive Ready
RSTR CNF	Restart Confirmation
RSTR REQ	Restart Request
REG REQ	Registration Request
REG CNF	Registration Confirmation

X.75

X.75 is a signalling system which is used to connect packet switched networks (such as X.25) on international circuits. It permits the transfer of call control and network control information and user traffic.

On layer 2, X.75 uses LAPB in the same way as X.25. On layer 3, X.75 is identical to X.25 with one exception. X.25 has a field (of variable length) for facilities, while X.75 has a field (of variable length) for network utilities followed by facilities.

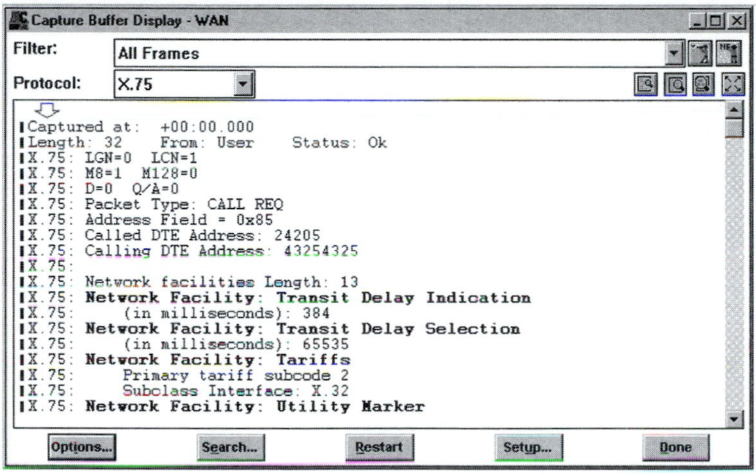

```
Capture Buffer Display - WAN
Filter:    All Frames
Protocol:  X.75

Captured at:  +00:00.000
Length: 32    From: User    Status: Ok
X.75: LGN=0  LCN=1
X.75: M8=1  M128=0
X.75: D=0  Q/A=0
X.75: Packet Type: CALL REQ
X.75: Address Field = 0x85
X.75: Called DTE Address: 24205
X.75: Calling DTE Address: 43254325
X.75:
X.75: Network facilities Length: 13
X.75: Network Facility: Transit Delay Indication
X.75:      (in milliseconds): 384
X.75: Network Facility: Transit Delay Selection
X.75:      (in milliseconds): 65535
X.75: Network Facility: Tariffs
X.75:      Primary tariff subcode 2
X.75:      Subclass Interface: X.32
X.75: Network Facility: Utility Marker

Options...   Search...   Restart   Setup...   Done
```

X.75 decode

MLP

Multilink procedure (MLP) exists as an added upper sublayer of the data link layer (LAPB), operating between the packet layer and a multiplicity of single data link protocol functions (SLPs) in the data link layer.

When specified in the LAPB frame, the MLP frame is inserted before the LAPB information field. The contents of the MLC (multilink control field) are as follows:

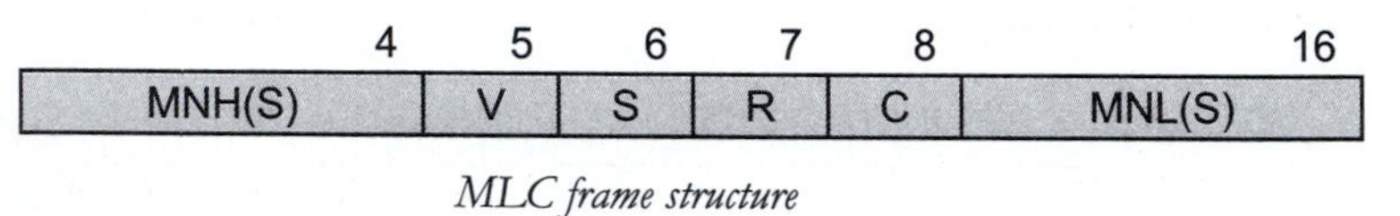

MLC frame structure

MNH(S)

Bits 9-12 of 12-bit multilink send sequence number MN(S).

MNL(S)

Bits 1-8 of 12-bit multilink send sequence number MN(S).

V

Void sequencing bit.

S

Sequence check option bit.

R

MLP reset request bit.

C

MLP reset confirmation bit.

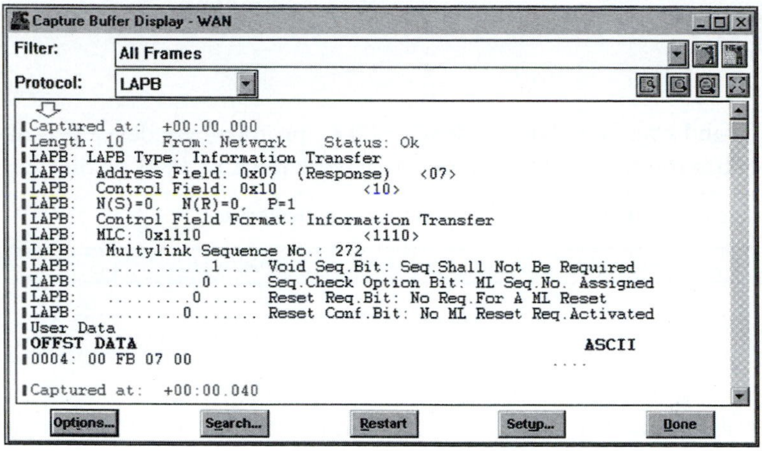

Capture Buffer Display - WAN

Filter: All Frames

Protocol: LAPB

```
Captured at:  +00:00.000
Length: 10     From: Network     Status: Ok
LAPB: LAPB Type: Information Transfer
LAPB:   Address Field: 0x07  (Response)    <07>
LAPB:   Control Field: 0x10           <10>
LAPB:   N(S)=0,  N(R)=0,  P=1
LAPB:   Control Field Format: Information Transfer
LAPB:   MLC: 0x1110                   <1110>
LAPB:    Multylink Sequence No.: 272
LAPB:    ...........1.... Void Seq.Bit: Seq.Shall Not Be Required
LAPB:    ..........0..... Seq.Check Option Bit: ML Seq.No. Assigned
LAPB:    .........0...... Reset Req.Bit: No Req.For A ML Reset
LAPB:    ........0....... Reset Conf.Bit: No ML Reset Req.Activated
User Data
OFFST DATA                                               ASCII
0004: 00 FB 07 00                                        ....

Captured at:  +00:00.040
```

Options... Search... Restart Setup... Done

Decode of MLP in LAPB

HDLC

The High Level Data Link Control (HDLC) protocol was developed by ISO and is based primarily on the pioneering work done by IBM on SDLC.

The format of a standard HDLC frame is as follows:

Flag	Address field	Control field	Information	FCS	Flag

HDLC frame structure

Flag

The value of the flag is always (0x7E). In order to ensure that the bit pattern of the frame delimiter flag does not appear in the data field of the frame (and therefore cause frame misalignment), a technique known as Bit Stuffing is used by both the transmitter and the receiver.

Address field

The first byte of the frame after the header flag is known as the Address Field. HDLC is used on multipoint lines and it can support as many as 256 terminal control units or secondary stations per line. The address field defines the address of the secondary station which is sending the frame or the destination of the frame sent by the primary station.

Control field

The field following the Address Field is called the Control Field and serves to identify the type of the frame. In addition, it includes sequence numbers, control features and error tracking according to the frame type.

Every frame holds a one bit field called the Poll/Final bit. In the NRM mode of HDLC this bit signals which side is 'talking', and provides control over who will speak next and when. When a primary station has finished transmitting a series of frames, it sets the Poll bit, thus giving control to the secondary station. At this time the secondary station may reply to the primary station. When the secondary station finishes transmitting its frames, it sets the Final bit and control returns to the primary station.

Modes of operation

HDLC has 3 modes of operation according to the strength of the master/slave relationship. This is determined by a unique frame type specifier:

- Normal Response Mode (NRM): This mode is totally master/slave and is signified by the SNRM(E) frame. The primary station initiates the session and full polling is used for all frame transmissions.
- Asynchronous Response Mode (ARM): This mode is similar to NRM and is signified by the SARM(E) frame. The difference, however, is that secondary stations can transmit freely without waiting for a poll.
- Asynchronous Balanced Mode (ABM): This mode is totally balanced (i.e., no master/slave relationship) and is signified by the SABM(E) frame. Each station can initialize, supervise, recover from errors and send frames at any time.

FCS

The Frame Check Sequence (FCS) enables a high level of physical error control by allowing the integrity of the transmitted frame data to be checked. The sequence is first calculated by the transmitter using an algorithm based on the values of all the bits in the frame. The receiver then performs the same calculation on the received frame and compares its value to the CRC.

Window size

HDLC supports an extended window size (modulo 128) where the number of possible outstanding frames for acknowledgement is raised from 8 to 128. This extension is generally used for satellite transmissions where the acknowledgement delay is significantly greater than the frame transmission times. The type of the link initialization frame determines the modulo of the session and an "E" is added to the basic frame type name (e.g., SABM becomes SABME).

Extended address

HDLC provides another type of extension to the basic format. The address field may be extended to more than one byte by agreement between the involved parties. When an address extension is used, the presence of a 1 bit in the first bit of an address byte indicates that the following byte is also an address byte. The last byte of the string of address bytes is signalled by a 0 bit in the first position of the byte.

Frame types

The following are the Supervisory Frame Types in HDLC:

RR Information frame acknowledgement and indication to receive more.

REJ Request for retransmission of all frames after a given sequence number.

RNR Indicates a state of temporary occupation of station (e.g., window full).

SREJ Request for retransmission of one given frame sequence number.

The following are the Unnumbered Frame Types in HDLC:

DISC	Request disconnection.
UA	Acknowledgement frame.
DM	Response to DISC indicating disconnected mode.
FRMR	Frame reject.
SABM	Initiator for asynchronous balanced mode. No master/slave relationship.
SABME	SABM in extended mode.
SARM	Initiator for asynchronous response mode. Semi master/slave relationship.
SARME	SAMR in extended mode.
REST	Reset sequence numbers.
CMDR	Command reject.
SNRM	Initiator for normal response mode. Full master/slave relationship.
SNRME	SNRM in extended mode.
RD	Request disconnect.
RIM	Secondary station request for initialization after disconnection.
SIM	Set initialization mode.
UP	Unnumbered poll.
UI	Unnumbered information. Sends state information/data.
XID	Identification exchange command.

There is one Information Frame Type in HDLC:

Info Information frame.

X.25 Terminology

Command

LAPB frame types are defined as commands or responses depending on the address field and the direction of the frame. A command frame from the DTE has an address of 1; a command frame from the DCE has an address of 3.

Control frames

All LAPB frame types except for information frames.

Control packets

All X.25 packet types except for DATA packets.

D bit

Delivery confirmation bit of the X.25 packet header. Used to indicate whether packet layer acknowledgements have local (0) or global (1) significance between the DTE and DCE. This is not a universally accepted bit and is generally set to 0.

Information frame

A specific LAPB frame type that is used to transport data across the link, inside the information field of the frame. The information field contains the X.25 packet header and data according to the standards in the X.25 layer 3 protocol.

LAPB

Link Access Procedure-Balanced protocol. The CCITT's adaptation of the layer 2 HDLC protocol for the X.25 interface. The LAPB address field, in contrast to HDLC, only has two permissible values.

LCN

Logical Channel Number. Together with the LGN (in the X.25 packet header), identifies the actual logical channel number of the DTE-DCE link. The LCN is an 8-bit field and consequently represents a number between 0 and 255.

LGN

Logical Channel Group Number. Together with the LCN (in the X.25 packet header), identifies the actual logical channel number of the DTE-DCE link. The LGN is a 4-bit field and consequently represents a number between 0 and 15, inclusive.

M bit

More data bit. A bit contained within the X.25 data packet which is set by layers higher than layer 3 to inform the destination DTE that more data will follow in the next packet. Using the M bit, packets can be logically grouped together to convey a large block of related information.

Modulo

Window Size. Represents the maximum number of frames (layer 2) or packets (layer 3) that can be left outstanding and unacknowledged for a DTE after transmission. In Modulo 8, sequence numbers can be represented by the digits 0 to 7 and are encoded in 3 bits of information. In Modulo 128, sequence numbers can be represented by the digits 0 to 127 and are encoded in 7 bits of information.

N(R)

A frame sequence number used in both LAPB information and supervisory frames in order to indicate to the transmitter the status of sent information frames. The meaning of N(R) is dependent on the type of the frame and can include acknowledgements, missed frames or a busy state. A 3-bit N(R) field is used to identify sequence numbers for a modulo 8 transmission; a 7-bit field is used for modulo 128.

N(S)

A frame sequence number used only within an LAPB information frame to identify each frame sent to the receiver. A 3-bit N(S) field is used to identify sequence numbers for a modulo 8 transmission; a 7-bit field is used for modulo 128.

P(R)

A packet sequence number used in the X.25 data packets in order to indicate to the transmitter the status of sent data packets. The meaning of P(R) is dependent on the type of the packet and can include acknowledgements, missed frames or a busy state. A 3-bit P(R) field is used to identify the sequence number of modulo 8 transmission; a 7-bit field is used for modulo 128.

P(S)

A packet sequence number used only within an X.25 data packet in order to identify each packet sent to the receiver. A 3-bit P(R) field is used to identify the sequence number of modulo 8 transmission; a 7-bit field is used for modulo 128.

P/F bit

The Poll/Final bit is set within the LAPB information frame by the transmitter to insist on an immediate response from the receiver. The receiver always turns on the bit in its response to a command from the sender with a poll bit set.

Q bit

A bit within the X.25 packet header used to signify X.25 control packets used by asynchronous PADs. Mainly used as a "qualified" proprietary protocol within the packet.

Response

LAPB frame types are defined as commands or responses, depending on the address field and the direction of the frame. A response frame is specified as a frame with either an address of 1, (from the DCE), or an address of 3 (from the DTE).

Sequence number

A unique number given to specific frames and packets to ensure that they are received and interpreted in the correct order. The bit size of the sequence number field is dependent on the respective modulo (maximum window size) of the layer and determines the maximum number of frames or packets that can be outstanding by the transmitter at any given time.

Supervisory frames

A collection of LAPB frames used to control information flow, request retransmissions and acknowledge information frames. Receive Ready (RR) and Reject (REJ) are two examples.

Unnumbered frame

A collection of LAPB frames used to provide additional data-link control functions such as link initialization and disconnection, link reset after unrecoverable errors, and rejection of invalid frames. They are called unnumbered because they do not contain frame sequence numbers.

Disconnect (DISC) and Unnumbered Acknowledgement (UA) are two examples.

X.25

The CCITT's recommendation for the interface between a DTE and DEC over a Public Switched Telephone Network (PSTN). Generally, X.25 covers layers 1 to 3 of the ISO model, but the term is used in this document to refer specifically to packet layer 3.

37

XNS Protocols

The Xerox Network Systems (XNS) protocols provide routing capability and support for both sequenced and connectionless packet delivery. Novell and 3Com3Plus protocols use the lower layers of XNS for packet delivery. XNS includes the following protocols:

- IDP: Internet Datagram Protocol.
- RIP: Routing Information Protocol.
- PEP: Packet Exchange Protocol.
- SPP: Sequenced Packet Protocol.

The following diagram illustrates the XNS protocol suite in relation to the OSI model:

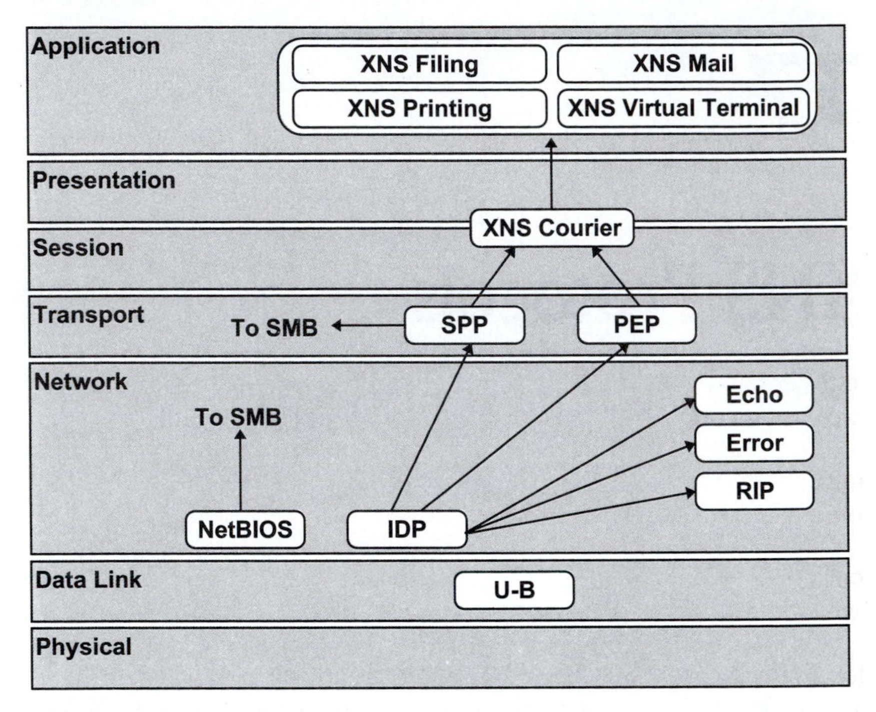

XNS in relation to the OSI model

IDP

The Internet Datagram Protocol (IDP) delivers a single frame as an independent entity to an Internet address, irrespective of other packets or addressee responses. XNS generally limits the IDP packets to a maximum size of 576 bytes, excluding the data link header.

The following parameters are available for IDP:

Destination network
Four-byte address of the destination network.

Destination socket
Two-byte socket number of the destination port.

Source network
Four-byte address of the source network.

Source socket
Two-byte socket number of the source port.

Hop count
Indicates the number of routers encountered during transport of the packet. Each router handling a packet increments the hop count by one. When the hop count reaches 16, this protocol discards the packet.

Packet type
Number indicating the higher level protocol in use. XNS defines the following packet types:

0 Unknown.
1 Routing Information Protocol.
2 Echo Protocol.
3 Error Protocol.
4 Packet Exchange Protocol.
5 Sequenced Packet Protocol.

RIP

XNS uses the Routing Information Protocol (RIP) to maintain a database of network hosts and exchange information about the topology of the network. Each router maintains a list of all networks known to that router along with the routing cost in hops required to reach each network. XNS distributes routing information on the network by routers broadcasting their routing tables every 30 seconds. This protocol sends routing tables as a result of changes in service or topology or in response to a request for routing information.

XNS generally uses the Echo protocol to demonstrate the existence and accessibility of another host on the network, while using the Error protocol to signal routing errors.

Frames

RIP frames may be one of the following commands:

[routing reqst]	Request for routing information.
[routing reply]	Routing information response.
[echo request]	Request to echo the data given.
[echo reply]	Echo of the data requested.
[error: unknown error]	Error of unknown nature.
[error: corrupt at dest]	Data corrupt at destination.
[error: unknown socket]	Socket number unknown.
[error: out of resources]	Router out of resources.
[error: routing error]	Unspecified routing error.
[error: corrupt en route]	Data corrupted in transit.
[error: dest unreachable]	Destination network unreachable.
[error: TTL expired]	Packet discarded after 15 hops.
[error: packet too large]	Packet larger than permitted.

Request and Reply Parameters

RIP routing request and routing reply parameters consist of the listing of networks and hop counts. [routing reqst] frames include the network numbers of the networks for which routing information is requested; [routing reply] frames list the networks known to the router. RIP routing parameters are in the following format: NNNNNNNN (CC), where NNNNNNNN is a 4-byte hexadecimal network number and CC is the cost in decimal hops. XNS interprets the network number FFFFFFFF as all networks. A cost of 16 or more hops implies that the network is unreachable.

The parameter for [echo request] and [echo reply] frames is a dump of the echo data.

Error Frames

Each [error:...] frame is followed by up to the first 42 bytes of the frame responsible for the error message. The message [error: packet too large] is followed by the maximum acceptable size parameter (Max=xxx).

PEP

The Packet Exchange Protocol (PEP) provides a semi-reliable packet delivery service that orients toward single-packet exchanges.

The following parameters are available for PEP:

Packet ID

A unique number used to identify responses as belonging to a particular request. The sending host sets the packet ID field to a fixed value, then looks for PEP responses containing the same packet ID value.

Client type

A registered code used to identify the particular application in use.

SPP

The Sequenced Packet Protocol (SPP) provides reliable transport delivery with flow control.

The following parameters are available for SPP:

Source connection ID

Reference number used to identify the source end of a transport connection. This protocol establishes Connection IDs at connect time to distinguish between multiple transport connections.

Destination connection ID

Reference number used to identify the target end of a transport connection.

Sequence number

Sequence number of the packet. Each successive packet transmitted and acknowledged on the transport connection must have a sequence number one higher than the previous sequence number.

Acknowledge number

Sequence number of the last packet that the protocol received properly. Each side of the transport connection uses its own sequence of numbers for transmitted packets, resulting in sequence and acknowledge numbers in the same packet generally being out of phase with each other.

Credit

Number of unacknowledged packets that the other side of the transport connection can send.

Transport control flag

When set (value of 1), the packet is used for transport control.

Acknowledge required flag

When set (value of 1), an immediate acknowledgement is requested.

Attention flag

When set (value of 1), the packet is sent regardless of the credit advertised by the destination.

EOM

End of message flag. When set (value of 1), a logical end of message stream is denoted.

Datastream type

Reserved field ignored by the SPP transport layer. SPP provides the datastream type for use by higher level protocols as control information.

PHYSICAL INTERFACES

38

WAN Physical Interfaces

This chapter describes the pin assignments for the following WAN interfaces:

- E1.
- RS-232 (V.24).
- RS-530.
- RS-449/RS-422.
- V.35.
- X.21.
- T1.

E1

E1 DA15 (D-type)

E1 DA15 cables have D-type 15 pin connectors. The pin assignments for the cables are as follows:

Pin 1 Transmit positive (TTIP).
Pin 3 Receive positive (RTIP).
Pin 9 Transmit negative (TRING).
Pin 11 Receive negative (RRING).
Pins 2, 4, 5-8, 10, and 12-15 are not used.

E1 RJ48

Pin assignments for RJ48 cables are as follows:

Pin 1 Transmit negative (TRING).
Pin 2 Transmit positive (TTIP).
Pin 4 Receive negative (RRING).
Pin 5 Receive positive (RTIP).
Pins 3, 6, 7, and 8 are not used.

E1 BANTAM

Pin assignments for BANTAM cables are as follows:

Tip - positive
Ring - negative
Sleeve - ground.

RS-232 (V.24)

RS-232 cables have D-type 25 pin connectors. Pin assignments for RS-232 cables are as follows:

Pin 1 Protective Ground (Shield).
Pin 2 Transmit Data.
Pin 3 Receive Data.
Pin 4 Request To Send.
Pin 5 Clear To Send.
Pin 6 Data Set Ready.
Pin 7 Signal Ground.
Pin 8 Data Carrier Detect.
Pin 15 Transmit Clock (from DCE).
Pin 17 Receive Clock.
Pin 18 Local Analog Loopback.
Pin 20 Data Terminal Ready.
Pin 21 Remote Digital Loopback.
Pin 22 Ring Indicator.
Pin 24 Transmit Clock (from DTE).
Pin 25 Test Mode.
Pins 9-14, 16, 19 and 23 are not used.

RS-530

RS-530 cables have D-type 25 pin connectors. Pin assignments for RS-530 cables are as follows:

Pin 1	Protective Ground (Shield).
Pins 2, 14	Transmit Data.
Pins 3, 16	Receive Data.
Pins 4, 19	Request To Send.
Pins 5, 13	Clear To Send.
Pins 6, 22	Data Set Ready.
Pin 7	Signal Ground.
Pins 8, 10	Data Carrier Detect.
Pins 9, 17	Receive Clock.
Pins 11, 24	Transmit Clock (from DTE).
Pins 12, 15	Transmit Clock (from DCE).
Pin 18	Local Analog Loopback.
Pins 20, 23	Data Terminal Ready.
Pin 21	Remote Digital Loopback.
Pin 25	Test Mode.

RS-449/RS-422

RS-449 cables have D-type 37 pin connectors. Pin assignments for RS-449 cables are as follows:

Pin 1	Protective Ground (Shield).
Pins 4, 22	Transmit Data .
Pins 5, 23	Transmit Clock (from DCE).
Pins 6, 24	Receive Data.
Pins 7, 25	Request To Send.
Pins 8, 26	Receive Clock.
Pins 9, 27	Clear To Send.
Pin 10	Local Analog Loopback.
Pins 11, 29	Data Set Ready.
Pins 12, 30	Data Terminal Ready.
Pins 13, 31	Data Carrier Detect.
Pin 14	Remote Digital Loopback.
Pin 15	Ring Indicator.
Pins 17, 35	Transmit Clock (from DTE).
Pin 18	Test Mode.
Pin 19, 37	Signal Ground.

Pins 2, 3, 16, 20, 21, 28, 32, 33, 34 and 36 are not used.

V.35

Pin assignments for V.35 cables are as follows:

Pin A	Protective Ground (Shield).
Pin B	Signal Ground.
Pin C	Request To Send.
Pin D	Clear To Send.
Pin E	Data Set Ready.
Pin F	Data Carrier Detect.
Pin H	Data Terminal Ready.
Pin J	Ring Indicator.
Pins P, S	Transmit Data.
Pins R, T	Receive Data.
Pins U, W	Transmit Clock (from DTE).
Pins V, X	Receive Clock.
Pins Y, AA	Transmit Clock (from DCE).
Pin N	Remote Digital Loopback.
Pin L	Local Analog Loopback.
Pin NN	Test Mode.

X.21

X.21 cables have D-type 15 pin connectors. Pin assignments for X.21 cables are as follows:

Pin 1	Protective Ground (Shield).
Pins 2, 9	Transmit.
Pins 3, 10	Control.
Pins 4, 11	Receive.
Pins 5, 12	Indication.
Pins 6, 13	Signal Timing.
Pins 7, 14	Terminal Transmit Clock.
Pin 8	Signal Ground.

Pin 15 is not used.

T1

T1 DA15

T1 DA15 cables have D-type 15 pin connectors. Pin assignments for T1 DA15 cables are as follows:
Pin 1 Transmit positive (TTIP).
Pin 3 Receive positive (RTIP).
Pin 9 Transmit negative (TRING).
Pin 1 Receive negative (RRING).
Pins 2, 4, 5-8, 10, and 12-15 are not used.

T1 RJ48

Pin assignments for RJ48 cables are as follows:
Pin 1 Transmit negative (TRING).
Pin 2 Transmit positive (TTIP).
Pin 4 Receive negative (RRING).
Pin 5 Receive positive (RTIP).
Pins 3, 6, 7, and 8 are not used.

T1 BANTAM

Pin assignments for BANTAM cables are as follows:
Tip - positive
Ring - negative
Sleeve - ground

39

LAN Physical Interfaces

This chapter describes the pin assignments for the following LAN interfaces:

- Ethernet AUI.
- Token Ring UTP.
- Token Ring STP.

Ethernet AUI

AUI cables have a D-type 15 pin connector. Pin assignments are as follows:

Pin 1	Control In circuit Shield.
Pin 2	Control In circuit A.
Pin 3	Data Out circuit A.
Pin 4	Data In circuit Shield.
Pin 5	Data In circuit A.
Pin 6	Voltage Common.
Pin 9	Control In circuit B.
Pin 10	Data Out circuit B.
Pin 11	Data Out circuit Shield.
Pin 12	Data In circuit B.
Pin 13	Voltage Plus.
Pin 14	Voltage Shield.

Pins 7, 8 and 15 are not used.

Token Ring

UTP

UTP Token Ring cables have an RJ45 connector with pin assignments as follows:

Pin 3 Transmit 1.
Pin 4 Receive 2.
Pin 5 Receive 1.
Pin 6 Transmit 2.

STP

STP Token Ring cables have a 9-pin D-type connector with pin assignments as follows:

Pin 1 Receive 1.
Pin 5 Transmit 1.
Pin 6 Receive 2.
Pin 9 Transmit 2.
Pins 2, 3, 4, 7 and 8 are not used.

40

DS-1 ATM Physical Interface

The DS-1 interface operates at 1.544 Mbps over UTP-3 cables, compliant with ATM Forum UNI specifications. It supports both PLCP and direct cell mapping and complies with the following standards: ANSI T1.403-1989, AT&T TR62411, CCITT G.704 and G.706. The interface has RJ45 connectors.

The pin assignments of the standard DS-1 UTP cable are as follows:

Pin	*Straight Cable*	*Cross Cable*
1	Tx- (ring)	Rx- (ring)
2	Tx+ (tip)	Rx+ (tip)
4	Rx- (ring)	Tx- (ring)
5	Rx+ (tip)	Tx+ (tip)

Pins 3, 6, 7 and 8 are not used.

The DS-1 frame is 193 bits long. The first bit (F-bit) is used for overhead. The remaining 192 bits comprise 8 bits of payload from each of 24 users.

Twelve frames are transmitted together as a Superframe (SF); 24 frames may be transmitted together as an Extended Superframe (ESF).

The structure of a DS-1 frame is shown in the following illustration:

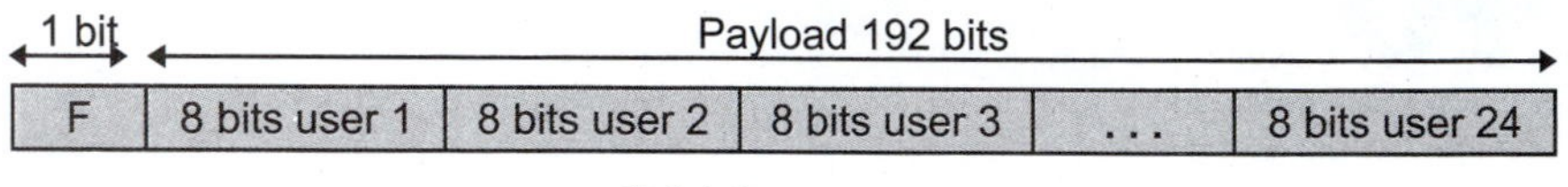

DS-1 frame structure

One frame may be transmitted every 125 microseconds. Thus, a superframe (12 frames) requires 1.5 milliseconds. An extended superframe (24 frames) requires 3.0 milliseconds for transmission. The net transmission rate is 1.391 Mbits/sec.

Direct Mapping

When direct mapping framing mode is used, cell delineation is used to locate the cell boundaries. Cell delineation is the process of framing to ATM cell boundaries using the header error checksum (HEC) field found in the ATM cell header. The HEC is a CRC-8 calculation over the first 4 bytes of the ATM cell header. When performing delineation, correct HEC calculations are assumed to indicate cell boundaries.

An initial bit by bit search is made for a correct HEC sequence (HUNT state). Once located, the particular cell boundary is noted (PRESYNC state) and the search continues to determine whether the following pattern is correct. Once no incorrect HEC is received within a set number of cells, the SYNC state is declared. In this state, synchronization is not relinquished until a set number of consecutive incorrect HEC patterns are received.

PLCP

The PLCP frame is octet aligned to the framing bit in the DS-1 frame. There is no relationship between the start of the PLCP frame and the start of the DS-1 frame. A 6-byte trailer is inserted at the end of each PLCP frame.

The DS-1 PLCP frame provides the transmission of ten ATM cells every 3 msec. The net transmission rate is thus 160 Kbytes/sec.

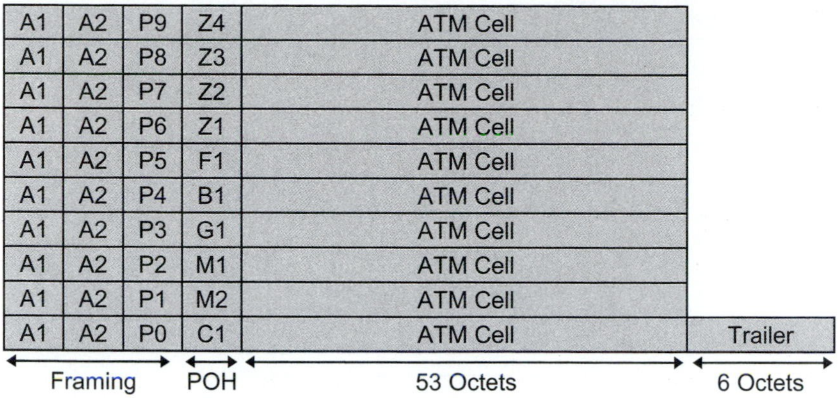

Framing		POH		53 Octets	6 Octets
A1	A2	P9	Z4	ATM Cell	
A1	A2	P8	Z3	ATM Cell	
A1	A2	P7	Z2	ATM Cell	
A1	A2	P6	Z1	ATM Cell	
A1	A2	P5	F1	ATM Cell	
A1	A2	P4	B1	ATM Cell	
A1	A2	P3	G1	ATM Cell	
A1	A2	P2	M1	ATM Cell	
A1	A2	P1	M2	ATM Cell	
A1	A2	P0	C1	ATM Cell	Trailer

DS-1 frame structure - PLCP cell mapping

A-bits
Framing pattern octet.

P-bits
Path overhead identifier.

C-bit
Pad bit counter.

M-bits
SIP layer 1 management information.

G-bit
PLCP path status.

B-bit
Bit-interleaved parity 8 (BIP-8).

F-bit
PLCP path user channel.

Z-bits
For future use.

41

DS-3 ATM Physical Interface

The DS-3 interface operates at 44.736 Mbps over coax cables, compliant with ATM Forum UNI specifications. It supports both PLCP and direct cell mapping and complies with the C-bit/M23 standards. The interface has BNC connectors.

There are three standards for DS-3 framing: M23, C-bit parity and SYNTRAN. C-bit parity is one block of data not muxed which uses C-bits for purposes other than dibble stuffing. M23 multiplex scheme provides for transmission of seven DS-2 channels. Since each DS-2 channel can contain 4 DS-1 signals, a total of 28 DS-1 signals (670 DS-0 signals) are transported in a DS-3 facility. The existing DS-3 signal format is a result of a multi-step, partially synchronous, partially asynchronous multiplexing sequence.

The DS-3 signal is partitioned into M-frames of 4,760 bits each. The M-frames are divided into 7 M-subframes each containing 680 bits. Each subframe is further divided into eight blocks of 85 bits each, with the first bit used for control and the rest for payload. There are 56 frame overhead bits which handle such functions as M-frame alignment, M-subframe alignment, performance monitoring, alarm and source application channels.

The following is an illustration of the DS-3 M-frame. The time length of the frame is 106.402 microseconds, thus making the net transmission rate 5.720 Mbits/sec.

← 680 bits (8 blocks of 84 + 1 bits) →

X	84 bits payload	F (1)	84 bits payload	CB	84 bits payload	F (0)	84 bits payload	CB	84 bits payload	F (0)	84 bits payload	CB	84 bits payload	F (1)	84 bits payload
X	84 bits payload	F (1)	84 bits payload	CB	84 bits payload	F (0)	84 bits payload	CB	84 bits payload	F (0)	84 bits payload	CB	84 bits payload	F (1)	84 bits payload
P	84 bits payload	F (1)	84 bits payload	CB	84 bits payload	F (0)	84 bits payload	CB	84 bits payload	F (0)	84 bits payload	CB	84 bits payload	F (1)	84 bits payload
P	84 bits payload	F (1)	84 bits payload	CB	84 bits payload	F (0)	84 bits payload	CB	84 bits payload	F (0)	84 bits payload	CB	84 bits payload	F (1)	84 bits payload
M (0)	84 bits payload	F (1)	84 bits payload	CB	84 bits payload	F (0)	84 bits payload	CB	84 bits payload	F (0)	84 bits payload	CB	84 bits payload	F (1)	84 bits payload
M (1)	84 bits payload	F (1)	84 bits payload	CB	84 bits payload	F (0)	84 bits payload	CB	84 bits payload	F (0)	84 bits payload	CB	84 bits payload	F (1)	84 bits payload
M (0)	84 bits payload	F (1)	84 bits payload	CB	84 bits payload	F (0)	84 bits payload	CB	84 bits payload	F (0)	84 bits payload	CB	84 bits payload	F (1)	84 bits payload

DS-3 M-frame

C-bit

The first C-bit in the M-frame is used to identify the type of framing used where 100=SYNTRAN, 111=C-bit and any other value represents M23. Remaining C-bits are used for dibble stuffing or C-bit frame applications.

M-bits

Multi-frame alignment bits are located in the fifth, sixth, and seventh subframes. M1=0, M2=1 and M3=0.

X-bits

Occupy the bit positions at the beginning of the first and second subframes. They are used for alarm functions and must be identical (00 or 11) within any M-frame. They can be used for low frequency signalling but cannot be changed more than once every second. When not used, X-bits should be set to 11; inserting an 01 or 10 can cause framing errors.

P-bits

The bit position at the beginning of the third and fourth subframes contains parity information. These 2 bits give the even parity. Valid values are 11 or 00.

F-bits

Frame alignment signal identifies all control bit positions within one M-subframe. F1=1, F2=0, F3=0 and F4=1.

C-Bit Framing

DS-3 C-bit parity format was advanced by AT&T to increase far-end performance monitoring. In this format, because stuff bits are used at every opportunity, C-bits can be used for purposes other than denoting the presence of stuff bits, including:

- Far End Alarm and Control signal (FEAC). In this case, C-bits are used to send alarm or status information from far end terminals to near end terminals and to initiate DS-3 and DS-1 remote loops. This is a repeating 16-bit word consisting of 0xxxxxx0 11111111. When no code is being transmitted, all 1's are transmitted. The code is transmitted for 10 times or the alarm state length, which ever is longer.
- Not used.
- DS-3 path parity information.
- Far end block errors.
- Terminal to terminal path maintenance data link (LAPD, subset of Q.921).

Direct Mapping

When direct mapping framing mode is used, cell delineation is used to locate the cell boundaries. Cell delineation is the process of framing to ATM cell boundaries using the header error checksum (HEC) field found in the ATM cell header. The HEC is a CRC-8 calculation over the first 4 bytes of the ATM cell header. When performing delineation, correct HEC calculations are assumed to indicate cell boundaries.

An initial bit by bit search is made for a correct HEC sequence (HUNT state). Once located, the particular cell boundary is noted (PRESYNC state) and the search continues to determine whether the following pattern is correct. Once no incorrect HEC is received within a set number of cells, the SYNC state is declared. In this state, synchronization is not relinquished until a set number of consecutive incorrect HEC patterns are received.

PLCP

The DS-3 PLCP frame provides the transmission of 12 ATM cells every 125 μsec; thus, the net transmission rate is 4.608 Mbytes/sec. The PLCP frame is nibble aligned to the overhead bits in the DS-3 frame. A trailer is inserted at the end of each PLCP frame. The number of nibbles, 13 or 14, is varied continuously so that the resulting PLCP frame rate can be locked to an 8 KHz reference.

A1	A2	P11	Z6	ATM Cell	
A1	A2	P10	Z5	ATM Cell	
A1	A2	P9	Z4	ATM Cell	
A1	A2	P8	Z3	ATM Cell	
A1	A2	P7	Z2	ATM Cell	
A1	A2	P6	Z1	ATM Cell	
A1	A2	P5	F1	ATM Cell	
A1	A2	P4	B1	ATM Cell	
A1	A2	P3	G1	ATM Cell	
A1	A2	P2	M1	ATM Cell	
A1	A2	P1	M2	ATM Cell	
A1	A2	P0	C1	ATM Cell	Trailer
1	1	1	1	53 bytes	13 or 14 nibbles

vDS-3 frame structure - PLCP cell mapping

A-bits

Framing pattern octets; A1=F6, A2=28 hex.

P-bits

Path overhead identifier octets.

C1

Pad bit counter.

M-bits

SIP layer 1 management information. If M2 contains a Type=1 field then M1 contains a Type=0 field and vice versa.

G1

PLCP path status.
Bits 1-4: FEBE code (up to 8 possible errors).
Bit 5: Yellow alarm. If there is an active high PLCP path layer failure condition (PLCP loss of frame) for 2.5 seconds, a yellow alarm results. When failure ends for 15 seconds, the yellow alarm status is removed.
Bits 6-8: Link status signal as follows:

LLS Code	*LLS Name*	*Link Status*
000	Connected	Received link connected
011	Rx_llink_down	Received link down, no input or forced down
110	Rx_link_up	Received link up

B1

Bip-8 error. Computed over a 12x54 octet structure consisting of the Path Overhead and ATM cell. The G1 byte provides the error count of the previous frames last frame.

F1

PLCP path user channel.

Z-bits

For future use.

42

E1 ATM Physical Interface

The E1 interface operates at 2 Mbps over coax cables, compliant with ATM Forum UNI specifications. It supports both PLCP and direct cell mapping and complies with the following standards: G.704, G.706, G.732. The interface has BNC connectors.

The E1 transmission link consists of 32 transmission channels (0-31), each of which is 64 Kbits/sec. The overall transmission rate is 2.048 Mbits/sec. Channels 0 and 16 are reserved for transmission management, while all other channels are used for payload. The payload bandwidth is thus 1.920 Mbits/sec. Since ATM uses 48 out of the possible 53 bytes for payload transmission, the net transmission rate becomes 1.738 Mbits/sec.

Channel 0 carries F3-OAM information, signals loss of frame or synchronization, and is responsible for transferring FERF and LOC messages. Channel 16 is reserved for signalling.

Direct Mapping

The direct mapping of ATM cells onto E1 transmission frames is specified in CCITT recommendation G.804. This specifies that ATM cells are to be carried in bits 9-28 and 137-256 (corresponding to channels 1-15 and 17-31).

The following is an illustration of the E1 frame format when direct mapping of ATM cells is used. The 53 byte ATM cell begins with a header and wraps around consecutive E1 frames.

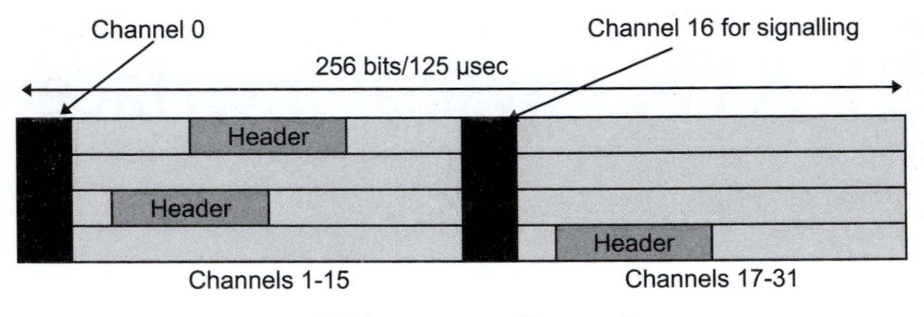

E1 frame structure - direct mapping

PLCP Cell Mapping

The PLCP format for E1 is described in ETSI document ETS 300 213, where an E1 PLCP frame is specified as consisting of ten rows of 57 bytes each. Four bytes are added to the cell length of 53 bytes to provide the various overhead functions.

The E1 frame structure with PLCP cell mapping is illustrated in the following diagram:

1	1	1	1	53 bytes
A1	A2	P9	Z4	First ATM Cell
A1	A2	P8	Z3	ATM Cell
A1	A2	P7	Z2	ATM Cell
A1	A2	P6	Z1	ATM Cell
A1	A2	P5	F1	ATM Cell
A1	A2	P4	B1	ATM Cell
A1	A2	P3	G1	ATM Cell
A1	A2	P2	M2	ATM Cell
A1	A2	P1	M1	ATM Cell
A1	A2	P0	C1	Last ATM Cell

E1 frame structure -PLCP cell mapping

A-bits
Separator bytes.

P-bits
Path overhead identifier.

C1
Pad bit counter.

M-bits
SIP layer 1 management information.

G1
PLCP path status.

B1
Bit-interleaved parity 8 (BIP-8).

F1
PLCP path user channel.

Z-bits
For future use.

Thirty of the available 32 E1 channels are used for transporting the PLCP frame. The remaining two channels are reserved for E1 framing and signalling functions. The PLCP frame is octet aligned to the channel boundaries in the E1 frame; thus the A1 octet of the first row of the PLCP frame is inserted into time slot 1 of the E1 frame.

43

E3 ATM Physical Interface

The E3 interface operates at 34.368 Mbps over coax cables, compliant with ATM Forum UNI specifications. It supports both PLCP and direct cell mapping and complies with the following standards: G.751, G.832. The interface has BNC connectors.

Direct Mapping

The E3 frame structure when direct mapping is employed is illustrated in the following diagram. The 530 octet payload consists of 59 columns and 9 rows. The 53 byte ATM cell wraps around the 59 bytes of the E3 frame. There is no relationship between the start of a direct mapping frame and the start of the ATM cell.

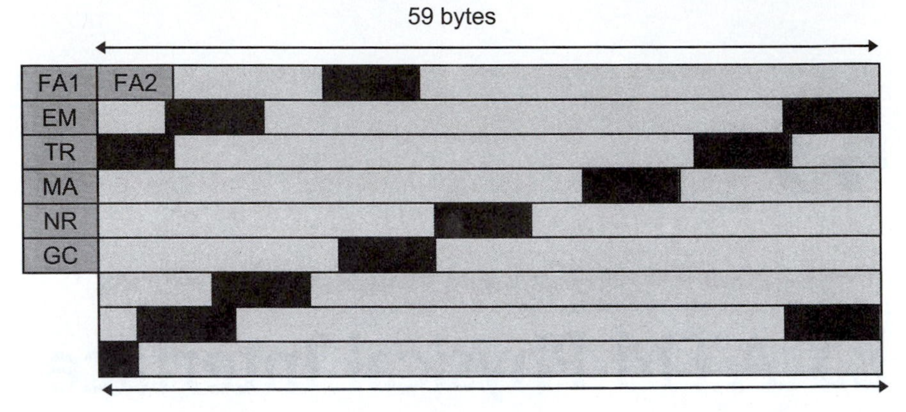

The 53 byte ATM cell is represented as follows:

53 bytes

E3 frame structure - direct mapping

FA1
Frame Alignment 1.

FA2
Frame Alignment 2.

EM
Error Monitor, BIP-8.

TR
Trail Trace.

MA
Maintenance and Adaptation.

NR
Network Operator.

GC

General Purpose Communications Channel.

PLCP Cell Mapping

The E3 PLCP frame provides the transmission of 9 ATM cells every 125 μsec. The net transmission rate is thus 3.456 Mbytes/sec. The PLCP frame is octet aligned to the 16 overhead bits in the ITU-T Recommendation G.751 E1 frame. There is no relationship between the start of the PLCP frame and the start of the E3 frame. A trailer is inserted at the end of each PLCP frame.

A1	A2	P8	Z3	ATM cell	
A1	A2	P7	Z2	ATM cell	
A1	A2	P6	Z1	ATM cell	
A1	A2	P5	F1	ATM cell	
A1	A2	P4	B1	ATM cell	
A1	A2	P3	G1	ATM cell	
A1	A2	P2	M1	ATM cell	
A1	A2	P1	M2	ATM cell	
A1	A2	P0	C1	ATM cell	Trailer
1	1	1	1	53 bytes	17-21 bytes

E3 frame structure - PLCP cell mapping

A-bits

Framing pattern octet.

P-bits

Path overhead identifier.

C1

Pad bit counter.

M-bits

SIP layer 1 management information.

G1

PLCP path status.

B1
Bit-interleaved parity 8 (BIP-8).

F1
PLCP path user channel.

Z-bits
For future use.

44

SONET OC-3c / SDH STM-1 ATM Physical Interface

Transmission of information at 155 Mbps over SONET or SDH interfaces is the transfer most associated with ATM. SONET/SDH, part of the ATM Forum UNI 3.0 specifications, is the fastest and one of the most popular ATM interfaces.

Connections to SONET/SDH lines may be via multi-mode, single-mode and UTP. The multi-mode and single-mode involve ATM links with SC-type optical connectors to multi-mode/single-mode fibers at 1300 nm. UTP links are with UTP-5 connectors.

The pin assignments of the standard UTP 155 cable are as follows:

Pin	*With Straight Cable*	*With Cross Cable*
1	Tx+ (tip)	Rx+ (tip)
2	Tx- (ring)	Rx- (ring)
7	Rx+ (tip)	Tx+ (or tip)
8	Rx- (ring)	Tx- (ring)

Pin 3, 4, 5 and 6 are not used.

Both SONET (Synchronous Optical NETwork) and SDH (Synchronous Digital Hierarchy) are based on transmission at speeds of multiples of 51.840 Mbps, or STS-1. The STS-1 frame is composed of octets which are nine rows high and 90 columns wide. The first three columns are used by the Transport Overhead (TOH) and contain framing, error monitoring, management and payload pointer information. The data (Payload) uses the remaining 87 columns, of which the first column is used for Path Overhead (POH). A pointer in the TOH identifies the start of the payload which is referred to as the Synchronous Payload Envelope or SPE.

OC-3c and STM-1 rates are an extension of the basic STS-1 speed and operate at 155.520 Mbps, carrying three interleaving STS-1 frames. Thus, the OC-3c frame has nine rows and 270 columns. Nine of these columns are TOH. In ATM, the three STS-1 frames are not independent, but rather concatenated to form one larger payload. One column of the 261 columns in the payload is used for the POH.

The payload may float inside the OC-3c frame in case the clock used to generate the payload is not synchronized with the clock used to generate the overhead. Pointers in the overhead always point to the start of the payload.

Of the 270 columns, ten are used for overhead information. Thus the actual useful information rate that can be carried inside the OC-3c payload is 149.76 Mbps. Since 5 bytes out of every 53-byte cell are the header, only 135.63 Mbps carry actual ATM payload.

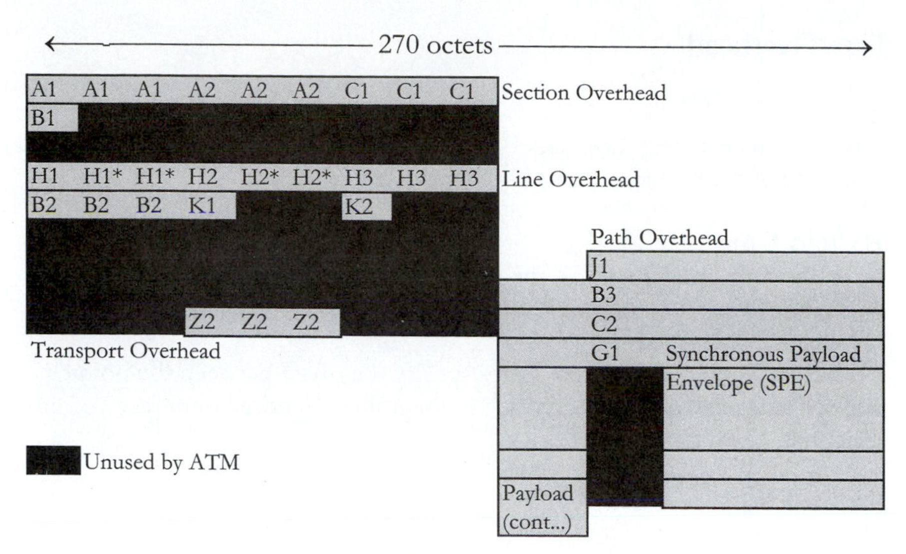

OC-3c frame structure

Section Overhead

A1, A2

Frame alignment. These octets contain the value of 0xF628. The receiver searches for these values in the incoming bit stream. These bytes are not scrambled.

C1

STS-1 identification. Since OC-3c and STM-1 contain three STS-1 streams, the three C1 bytes contain 0x01, 0x02 and 0x03, respectively.

B1

Section error monitoring. Contains BIP-8 of all bits in the previous frame using even parity, before scrambling.

Line Overhead

B2

Line error monitoring. Contains BIP-24 calculated over all bits of the line overhead of the previous frame with even parity.

H1 (bits 1-4)

New data flag (specifies when the pointer has changed), path AIS.

H1 and H2 (bits 7-16)

Pointer value, path AIS. These bytes specify the offset between the pointer and the first payload byte. A change in this value is ignored until received at least three consecutive times.

H1* and H2*

Concatenation indication, path AIS.

H3

Pointer action (used for frequency justification), path AIS.

K2 (bits 6-8)

Line AIS, line FERF, removal of line FERF.

Z2

Line FEBE. This contains the number of B2 (BIP-24) errors detected in the previous interval.

Path Overhead

J1

STS path trace. This byte is used repetitively to transmit a 64-byte fixed string so that the receiving terminal in a path can verify its continued connection to the transmitter. Its contents are unspecified.

B3

Path error monitoring. Path BIP-8 over all bits of the payload of the previous frame, using even parity before scrambling.

C2

Path signal level indicator. Contains one of two codes:
Code 0: indicates STS payload unequipped: no path originating equipment.
Code 1: indicates STS payload equipped: nonspecific payload for payloads that need no further differentiation.

G1 (bits 1-4)

Path FEBE. Allows monitoring of complete full-duplex path at any point along a complex path.

G1 (bit 5)

Path yellow alarm, path RDI (Remote Defect Indicator).

For further reference, consult:

1. The ATM Forum, ATM User-Network Interface Specifications 3.0 and 3.1, Prentice-Hall, 1993 and 1994.
2. Bell Communications Research Inc. (Bellcore), "Synchronous Optical Network (SONET) Transport Systems: Common Generic Criteria", TR-NWT-000253, December 1991.
3. American National Standards Institute (ANSI), "Digital Hierarchy - Optical Interface Rates and Formats Specifications (SONET)", T1.105, 1991.

45

SONET OC-12c / SDH STM-4 ATM Physical Interface

Transmission of information at 622 Mbps over SONET or SDH interfaces is the transfer most associated with ATM. SONET/SDH, part of the ATM Forum UNI 3.0 specifications, is the fastest and one of the most popular ATM interfaces.

Connections to SONET/SDH lines may be via multi-mode or single-mode. The multi-mode and single-mode involve ATM links with SC-type optical connectors to multi-mode/single-mode fibers at 1300 nm.

SONET/SDH Frame Structure

A SONET/SDH based TC is specified for the 622.08 Mbps interface. The SONET STS-12c (T1.646) and SDH STM-4 (G.708, G.709) frame formats are used to transport ATM cells (see the figure below). These two frame formats are largely identical except for the usage of certain overhead bytes.

Mapping of ATM cells into the SONET/SDH frame is accomplished by scrambling the ATM cell payload and mapping the resulting cell stream into a synchronous payload envelope (SPE) or equivalently a VC-4-4c, mapping the SPE into the SONET/SDH frame using the H1-H2 pointer, and finally applying the SONET/SDH frame synchronous scrambler to the resulting frame.

ATM cell extraction operates in the analogous reverse procedure, i.e. by descrambling the SONET/SDH frame, reading the H1-H2 pointer to locate the SPE, performing cell delineation and finally descrambling the ATM cell payload.

ATM cell mapping is performed by aligning by row, the byte structure of every cell with the byte structure of the SPE. The bit rate available for user information cells, signalling cells and OAM cells (excluding unassigned bytes and physical layer related maintenance information transported in SONET/SDH overhead bytes) is nominally 599.04 Mbps (refer to “Payload” depicted in the figure below).

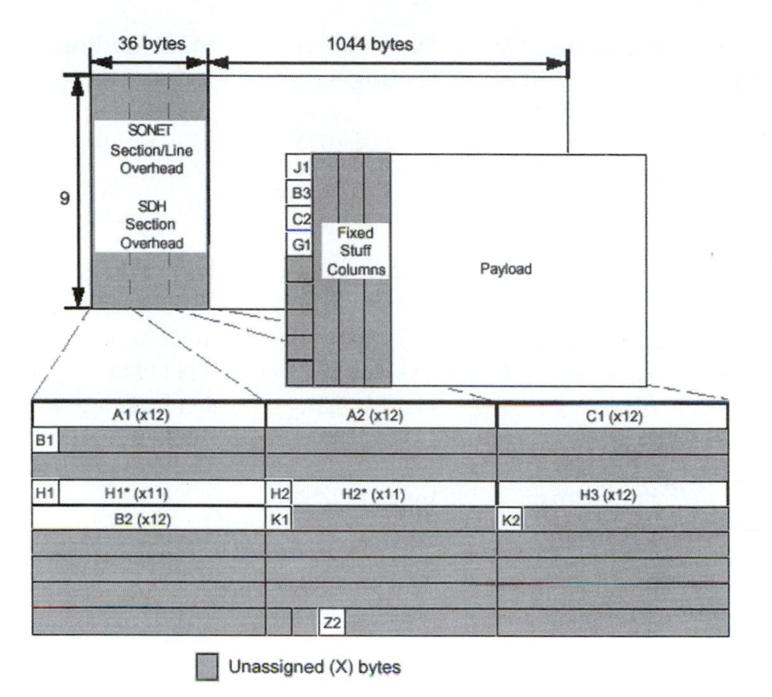

SONET STS-12c/STM-4 frame structure

Section Overhead

The defined overhead bytes, along with the coding differences between SONET and SDH, are summarized below:

Overhead Byte	Function	SONET Coding	SDH Coding
A1	Frame Alignment	11110110	11110110
A2	Frame Alignment	00101000	00101000
C1	Identification (Note 1)	00000001-00001100 (Bytes 1 – 12)	00000001-00000100 (Bytes 1 – 4)
B1	Section Error Monitoring	BIP-8	BIP-8
B2	Line Error	BIP-96	BIP-96

Overhead Byte	Function	SONET Coding	SDH Coding
	Monitoring		
H1 (bits 1-4)	NDF, Path AIS (Notes 2,3)	0110 or 1001	0110 or 1001
H1 (bits 5-6)	SS bits, Path AIS (Notes 2,3)	00	10
H1 (bits 7-8), H2	Pointer, Path AIS (Notes 2,3)	0000000000 – 1100001110	0000000000 – 1100001110
H1*	Concatenation	10010011	10010011
H2*	Concatenation	11111111	11111111
K1, K2	APS	Per T1.105	Per G.783
K2 (bits 6-8)	Line AIS, Line RDI	111, 110	111, 110
Z2	Line FEBE	B2 Error Count	B2 Error Count
J1	Path Trace	Note 4	Note 5
B3	Path Error Monitoring	BIP-8	BIP-8
C2	Path Signal Label	00010011	00010011
G1 (bits 1-4)	Path FEBE	B3 Error Count	B3 Error Count
G1 (bit 5)	Path RDI	1 (Note 6)	1 (Note 6)

Please note the following:

1. Receivers should not use this pattern for frame alignment identification because new functions may be defined for these bytes in the future. For SDH interfaces, bytes 5 to 12 are unused. These bytes are not scrambled and should be set to a balanced value (i.e. 11001100).

2. H1 and H2 are the first of twelve H1 and H2 bytes. H1* and H2* are the 2nd through 12th H1 and H2 bytes. The asterix indicates concatenation.

3. Path AIS is indicated by an all 1s condition in H1, H2, H1* and H2*.

4. A 64 ASCII COMMON LANGUAGE® Location Identifier (CLLI) (8-bit) code, padded with ASCII NULL characters and terminated with CR/LF, is a suitable trace message for SONET interfaces. If no message has been loaded, then 64 NULL characters shall be transmitted. COMMON LANGUAGE is a registered trademark and CLLI is a trademark of Bellcore.

5. SDH interfaces may use a 64 byte free format string (see Note 4) or the 16 byte E.164 format as described in G.709.

6. The assertion of path RDI in response to loss of pointer (LOP), path AIS and loss of cell delineation (LCD) is specified in in I.432.

For further reference, consult:

The ATM Forum, *ATM User-Network Interface Specifications 3.0 and 3.1*, Prentice-Hall, 1993 and 1994.

The ATM Forum, *622.08 Mbps Physical Layer Specification*, af-phy-0046.000.

Bell Communications Research Inc. (Bellcore), *Synchronous Optical Network (SONET) Transport Systems: Common Generic Criteria*, TR-NWT-000253, December 1991.

American National Standards Institute (ANSI), *Digital Hierarchy - Optical Interface Rates and Formats Specifications (SONET)*, T1.105, 1991.

46

25 Mbps ATM Physical Interface

The 25 Mbps interface operates at 25.6 Mbps over twisted pair cables, compliant with ATM Forum UNI specifications. The interface has 4B/5B NRZI coding with RJ48 electrical connectors.

The pin assignments of the standard 25 Mbps UTP cable are as follows:

Pin	*Straight Cable*	*Cross Cable*
1	Rx+ (tip)	Tx+ (tip)
2	Rx- (ring)	Tx- (ring)
7	Tx+ (tip)	Rx+ (tip)
8	Tx- (ring)	Rx- (ring)

Pins 3, 4, 5 and 6 are not used.

The pin assignments of older IBM cables are as follows:

Pin	*Straight Cable*	*Cross Cable*
3	Rx+ (tip)	Tx+ (tip)
4	Tx+ (tip)	Rx+ (tip)
5	Tx- (ring)	Rx- (ring)
6	Rx- (ring)	Tx- (ring)

Pin 1, 2, 7 and 8 are not used.

For further information about 25 Mbps, refer to the ATM Forum Physical Interface Specification for 25.6 Mbps over Twisted Pair Cable, June 1995 (af-phy-0040.00).

HEC generation

ATM HEC generation is used with the addition of COSET.

Cell scrambling

All 53 bytes of the ATM cell are scrambled prior to transmission; however, scrambling may be disabled. The scrambling polynomial is $x^{10} + x^7 + 1$ and is XORed with the incoming data every 4 clock periods.

Cell delineation

Cell delineation is performed using the 4B5B codes. X_X marks the start of cells with the scrambler reset; X_4 marks the start of cells with no scrambler reset.

8 KHz timing signal

A sync event timing marker is sent at the next octet boundary after an incoming sync signal is detected. The sync event has priority over all line activity.

Line coding

Line coding has the capability to be disabled. All 53 bytes of the cell are encoded prior to transmission. Commands are passed through the 4B5B codes as follows:

X_X Start of cell with scrambler reset.
X_4 Start of cell with no scrambler reset.
X_8 Sync event.

Reception of any other command is an error and can cause cell discard.

Physical medium connectors

The 25 Mbps interface supports category 3, 4, 5 UTP cables. A 100 Ohm resistive load is required.

47

TAXI ATM Physical Interface

AMD's TAXI is the physical layer for 100 Mbps multi-mode fiber transmission. It is specified in the ATM Forum, ATM User-Network Interface Specifications 3.0. The private UNI does not require the operation and maintenance complexity or link distance provided by telecommunications lines. TAXI takes advantage of existing chips for the FDDI LAN system of cell transport. It uses the same physical media as FDDI and the same lasers, cabling and AMD TAXI chips.

TAXI has the same data rate (100 Mbps) as FDDI, but it does not use the ring architecture. Links are full-duplex point-to-point, carrying 53-byte ATM cells with no physical framing structure.

48

Acronyms

A

AAL	ATM Adaptation Layer.
AARP	AppleTalk Address Resolution Protocol.
ABR	Available Bit Rate.
ADSP	AppleTalk Data Stream Protocol.
AEP	AppleTalk Echo Protocol.
AFP	AppleTalk Filing Protocol.
AH	Authentication Header.
AIS	Alarm Indication Signal.
AN	Access Network.
ANSI	American National Standards Institute.
APPN	Advanced Peer to Peer Network.

ARINC	Aeronautical Radio, Inc.
ARP/RARP	Address Resolution Protocol/Reverse Address Resolution Protocol.
ASN.1	Abstract Syntax Notation One.
ASP	AppleTalk Session Protocol.
ATCP	AppleTalk Control Protocol.
ATP	AppleTalk Transaction Protocol.
AVP	Attribute-Value Pair.

B

BACP	Bandwidth Allocation Control Protocol.
BAP	Bandwidth Allocation Protocol.
BCAST	Broadcast.
BCC	Broadcast Call Control.
BCP	Bridging Control Protocol.
BSC	Base Station Controller.
BECN	Backward Explicit Congestion Notification.
BERT	Bit Error Rate Tester.
BGP-4	Border Gateway Protocol.
B-HLI	Broadband High Layer Information.
B-ICI	BISDN Inter Carrier Interface.
B-LLI	Broadband Low Layer Information.
BMP	Burst Mode Protocol.
BOFL	Breath of Life.
BOM	Beginning of message.
BPDU	Bridge Protocol Data Unit.
BRE	Bridge Relay Encapsulation.
BRI	Basic Rate Interface.
BSMAP	Base Station Management Application Part.

BSS	Base Station System.
BSSAP	BSS Application Part.
BSSGP	Base Station System GPRS Protocol
BSSGP	Base Station System GPRS Protocol.
BSSMAP	BSS Management Application Part.
BTS	Base Transceiver Station.
BTSM	Base Station Controller to Base Transceiver Station.
BUS	Broadcast and Unknown Server.
BVCP	PPP Banyan Vines Control Protocol.

C

CAS	Channel Associated Signalling.
CBR	Constant Bit Rate.
CC	Call Control.
CCITT	International Telegraph and Telephone Consultative Committee
CCP	Compression Control Protocol.
CDMA	Code Division Multiple Access.
CDPD	Cellular Digital Packet Data.
CES	Circuit Emulation Service.
CGI	Common Gateway Interface.
CHAP	Challenge Handshake Authentication Protocol.
CIF	Cells In Frames.
CIR	Committed Information Rate.
CLLM	Consolidated Link Layer Management.
CLNP	Connectionless Network Protocol.
CLP	Cell Loss Priority.
CLR	Cell Loss Ratio.
COM	Continuation Of Message.

COPS	Common Open Policy Service.
CPCS	Common Part Convergence Sublayer.
CPE	Customer Premises Equipment.
CPN	Calling Party Number.
CPS	Common Part Sublayer.
CRC	Cyclic Redundancy Check.
CSU	Channel Service Unit.
CTERM	Command Terminal.

D

DAP	Data Access Protocol.
DAP	Directory Access Protocol.
DARPA	Defense Advance Research Projects Agency.
DCAP	Data Link Switching Client Access Protocol.
DCE	Data Communication Equipment.
DCP	Data Compression Protocol.
DCPCP	DCP Control Protocol.
DDP	Datagram Delivery Protocol.
DEC	Digital Equipment Corporation.
DES	Data Encryption Standard.
DHCP	Dynamic Host Configuration Protocol.
DIAG	Diagnostic Responder.
DLCI	Data Link Connection Identifier.
DLSw	Data Link Switching.
DNCP	PPP DECnet Phase IV Control Protocol.
DNS	Domain Name Service.
DPNSS1	Digital Private Network Signalling System No. 1.
DQDB	Distributed Queue Dual Bus.
DRiP	Cisco Duplicate Ring Protocol.

DSM-CC	Digital Storage Media Command and Control.
DTAP	Direct Transfer Application sub-Part.
DTE	Data Terminal Equipment.
DUP	Data User Part.
DVB	Digital Video Broadcasting.
DVMRP	Distance Vector Multicast Routing Protocol.
DXI	Data Exchange Interface.

E

ECN	Explicit Congestion Notification.
ECP	Encryption Control Protocol.
EF	Envelope Function.
EGP	Exterior Gateway Protocol.
EOM	End Of Message.
ES-IS	End-System to Intermediate System.
ESP	Encapsulating Security Payload.
ETSI	European Telecommunication Standards Institute

F

FCS	Frame Check Sequence.
FDDI	Fiber Distributed Data Interface.
FECN	Forward Explicit Congestion Notification.
FR-SSCS	Frame Relaying Service Specific Convergence Sublayer.
FTP	File Transfer Protocol.
FUNI	Frame-based User to Network Interface.

G

GARP	Generic Attribute Registration Protocol.
GCC	Group Call Control.

GFC	Generic Flow Control.
GID	GARP Information Declaration.
GMM	GSM MM protocol.
GMRP	GARP Multicast Registration Protocol.
GOB	Group Of Block.
GPRS	General Packet Radio Service.
GSM	Global System for Mobile telecommunications.
GSM	GPRS Session Management.
GSMP	General Switch Management Protocol.
GTP	GPRS Tunnelling Protocol.
GVRP	GARP VLAN Registration Protocol.

H

HDLC	High Level Data Link Control.
HEC	Header Error Control.
HSRP	Cisco Hot Standby Router Protocol.
HTTP	HyperText Transfer Protocol.

I

ICMP	Internet Control Message Protocol.
ICP	Internet Control Protocol.
IDLC	Integrated Digital Loop Carrier.
IDP	Internet Datagram Protocol.
IDU	Interface Data Unit.
IE	Information Element.
IEEE	Institute of Electrical and Electronic Engineers.
IETF	Internet Engineering Task Force.
IFMP	Ipsilon Flow Management Protocol.
IGMP	Internet Group Management Protocol.

IGRP	Interior Gateway Routing.
IISP	Interim Interswitch Signalling Protocol.
ILMI	Interim Local Management Interface.
IMAP4	Internet Message Access Protocol rev 4.
IP	Internet Protocol
IPC	InterProcess Communications Protocol.
IPCP	IP Control Protocol.
IPDC	IP Device Control.
IPHC	IP Header Compression.
IPv6CP	IPv6 Control Protocol.
IPX	Internetwork Packet Exchange.
IPXCP	IPX Control Protocol.
ISAKMP	Internet Security Association and Key Management Protocol.
ISDN	Integrated Services Digital Network.
IS-IS	Intermediate System to Intermediate System.
ISL	Inter-Switch Link.
ISO	International Standards Organization.
ISO-IP	ISO Internetworking Protocol.
ISUP	ISDN User Part.
ITU	International Telecommunications Union.

K

Kbps	Kilobits per second

L

L2F	Layer 2 Forwarding protocol.
L2TP	Layer 2 Tunneling Protocol.
LAN	Local Area Network.

LANE	LAN Emulation.
LAPB	Link Access Procedure-Balanced protocol.
LAPD	Link Access Protocol - Channel D.
LAT	Local Area Transport.
LAVC	Local Area VAX Cluster.
LCP	Link Control Protocol.
LDAP	Lightweight Directory Access Protocol.
LDP	Label Distribution Protocol.
LE	Local Exchange.
LE	LAN Emulation.
LEC	Local Exchange Carrier.
LEC	LAN Emulation Client.
LECS	LAN Emulation Configuration Server.
LES	LAN Emulation Server.
LEX	LAN EXtension interface protocol.
LEXCP	LAN EXtension interface Control Protocol.
LL-PDU	Logical Link control Protocol Data Unit.
LLC	Logical Link Control layer protocol.
LQR	Link Quality Report.
LU	Logical Unit.

M

MAC	Media Access Control.
MAP	Mobile Application Part.
MAPOS	Multiple Access Protocol over SONET/SDH.
MARS	Multicast Address Resolution Server.
MB	MacroBlock.
Mbps	Megabits per second.
MD-IS	Mobile Data Intermediate System.

MDLP	Mobile Data Link Protocol.
M-ES	Mobile End System.
Megaco	Media Gateway Control
MGCP	Media Gateway Control Protocol.
MIB	Management Information Base.
MIME	Multipurpose Internet Mail Extensions.
MLP	Multilink PPP.
MM	Mobility Management.
MO	Mobile Originating.
MOP	Maintenance Operation Protocol.
MPOA	Multiprotocol over ATM.
MPLS	Multi-Protocol Label Switching.
MPPC	Microsoft Point-to-Point Compression Protocol.
MS	Mobile Station.
MSC	Mobile Services switching Center.
MT	Mobile Terminating.
MTP	Message Transfer Part

N

NARP	NBMA Address Resolution Protocol.
NAU	Network Addressable Unit.
NBFCP	PPP NetBios Frames Control Protocol.
NBP	Name Binding Protocol.
NCP	NetWare Core Protocol.
NDS	NetWare Directory Services.
NetBIOS	Network Basic I/O System.
NetRPC	NetRemote Procedure Call.
NFS	Network File System.

NFSP	NetWare File Sharing Protocol.
NHDR	Network Layer Header.
NHRP	Next Hop Resolution Protocol.
NIS	Network Information Services.
NLPID	Network Level Protocol ID.
NLSP	NetWare Link Service Protocol.
NNI	Network Node Interface.
N-PDU	Network Protocol Data Unit.
NS	Network Service.
NSAP	Network Service Access Point.
NSP	Network Service Protocol.
NTP	Network Time Protocol.

O

OAM	Operations Administration and Maintenance.
OSI	Open System Interconnection.
OSINLCP	OSI Network Layer Control Protocol.
OSPF	Open Shortest Path First.
OUI	Organizationally Unique Identifier.

P

PAD	Packet Assembler and Disassembler.
PAP	Printer Access Protocol.
PAP	Password Authentication Protocol.
PBX	Private Branch eXchange.
PDU	Protocol Data Unit.
PEP	Packet Exchange Protocol.
PICS	Protocol Implementation Conformance Statement.

PID	Protocol ID.
PIM	Protocol Independent Multicast.
PLCP	Physical Layer Convergence Protocol.
PMAP	Port Mapper.
PNNI	Private Network-to-Network Interface.
POH	Path Overhead.
POP3	Post Office Protocol version 3.
PP	Presentation Protocol.
PPP	Point-to-Point Protocol.
PPP-BPDU	PPP Bridge Protocol Data Unit.
PPPoE	PPP over Ethernet.
PPTP	Point-to-Point Tunneling Protocol.
PRI	Primary Rate Interface.
PSTN	Public Switched Telephone Network.
PTI	Payload Type Indication.
PTSE	PNNI Topology State Element.
PTSP	PNNI Topology State Packet.
PU	Physical Unit.
PVC	Permanent Virtual Circuits.
PVCC	Permanent Virtual Channel Connection.

Q

QoS	Quality of Service.

R

RAS	Registration, Admission and Status.
RBOCs	Regional Bell Operating Companies.
RD	Routing Domain.
RFC	Request For Comment.

RIP	Request In Progress.
RIP	Routing Information Protocol.
RISC	Reduced Instruction Set Computing.
RLOGIN	Remote Login.
RLP	Radio Link Protocol.
RM	Resource Management.
RP	Routing Protocol.
RPC	Remote Procedure Call.
RR	Radio Resource.
RSVP	Resource reSerVation setup Protocol.
RTCP	Real-Time Transport Control Protocol.
RTMP	Routing Table Maintenance Protocol.
RTP	Real-Time Transport Protocol.
RTSP	Real-time Streaming Protocol.
RUDP	Reliable UDP.
RVP	Remote Voice Protocol.

S

S-HTTP	Secure Hypertext Transfer Protocol.
SAAL	Signalling ATM Adaptation Layer.
SAP	Service Advertising Protocol.
SAP	Session Announcement Protocol.
SAPI	Service Access Point Identifier.
SAR	Segmentation and Reassembly Sublayer.
SCCP	Signalling Connection Control Part.
SCP	Session Control Protocol.
SCTP	Stream Control Transmission Protocol.
SDCP	Serial Data Control Protocol.
SDH	Synchronous Digital Hierarchy.

SDP	Session Description Protocol
SDU	Service Data Unit.
SER	Serialization packet.
SGCP	Simple Gateway Control Protocol.
SGSN	Serving GPRS Support Node.
SIP	Session Initiation Protocol.
SLP	Service Location Protocol.
SM	Session Management.
SMB	Server Message Block.
SMDS	Switched Multimegabit Data Service.
SMI	Structure of Management Information.
SMS	Short Message Service.
SMTP	Simple Mail Transfer Protocol.
SNA.	Systems Network Architecture.
SNACP	SNA PPP Control Protocol.
SNAP	Sub-Network Access Protocol.
SNDCP	Subnetwork Dependent Convergence Protocol.
SNMP	Simple Network Management Protocol.
SP	Session Protocol.
SPANS	Simple Protocol for ATM Network Signalling.
SPP	Sequenced Packet Protocol.
SPX	Sequenced Packet Exchange.
SRB	Source Routing Bridging.
SS7	Signalling System No. 7.
SSCF	Service Specific Coordination Function.
SSCOP	Service Specific Connection Oriented Protocol.
SSCP	System Services Control Point.
SSCS	Service Specific Convergence Sublayer.
SSM	Single Segment Message.

STP Spanning Tree Protocol.

SVC Switched Virtual Circuits.

T

TACACS+. Terminal Access Controller Access Control System.

TALI Transport Adapter Layer Interface.

TCAP Transaction Capabilities Application Part.

TCP Transmission Control Protocol.

TDP Tag Distribution Protocol.

TE Terminal Equipment.

TEI Temporary Equipment Identifier.

TEI Terminal Endpoint Identifier.

TFTP Trivial File Transfer Protocol.

TH Transmission Header.

THDR RTP Transport header.

TIA Telecommunications Industry Association.

TM Traffic Management.

TP Transport Protocol.

TrCRF Token Ring Concentrator Relay Function.

TRPB Truncated Reverse Path Broadcasting.

TSDU Transport Service Data Unit.

TSR Tag Switching Routers.

U

UBR Unspecified Bit Rate.

UDP User Datagram Protocol.

UME UNI Management Entity.

UNI	User-Network Interface.
UTP	Unshielded Twisted Pair.

V

VARP	VINES Address Resolution Protocol.
VBR	Variable Bit Rate.
VCI	Virtual Channel Identifier.
VGCS	Voice Group Call Service.
VIP	VINES Internet Protocol.
VLAN	Virtual Local Area Network.
VOTA	Voice and Telephony Over ATM.
VPI	Virtual Path Identifier.
VRRP	Virtual Router Redundancy Protocol.

W

WAE	Wireless Application Enviroment.
WAN	Wide Area Network.
WAP	Wireless Application Protocol.
WCCP	Web Cache Coordination Protocol.
WDOG	Watchdog.
WML	
WSP	Wireless Session Protocol.
WTLS	Wireless Transport Layer Security.
WTP	Wireless Transaction Protocol

X Y

XNS	Xerox Network Systems.
XOT	X.25 Over TCP.
YP	Yellow Pages.

ZIP	Zone Information Protocol.

Index

K

L

M

T

U

V

W

X

Y

Z

About RADCOM

Company Overview

RADCOM Ltd., a member of the RAD Group and publicly traded on NASDAQ (RDCM), is a leading performance measurement and quality management solutions provider. The company designs, manufactures, markets and supports analysis and simulation solutions for Voice over IP (VoIP) and cellular convergence technologies.

PrismLite™ WAN/LAN/ATM protocol analyzer

Established in 1991, the company gained international recognition as a leading protocol analyzer manufacturer through its signature line of high quality, integrated protocol analyzers for WAN, LAN and ATM networks. The products of RADCOM's founding philosophy, these lightweight, truly portable multitechnology analyzers continue to successfully address the market's need for a single-solution method of testing existing and emerging network technologies.

RADCOM's test and measurement solutions are used in the development and manufacture of network devices, and in the installation and ongoing maintenance of operational networks to facilitate real-time isolation, diagnosis, and resolution of network problems.

RADCOM's success is based on a strong and dedicated sales channel comprised of over 60 distributors in 50 countries worldwide and 10 manufacturer's representatives across North America. Supported by the experience of hundreds of networking professionals, the company's worldwide sales channel offers top expertise and support for all customers, including international vendors developing communications equipment, service providers and large network operators.

Customers include AT&T, Cisco, Ericsson, Lucent, Nokia, Nortel, Motorola, Sprint, Worldcom, British Telecom, Deutshe Telecom, Telstra and many others.

For further information, visit RADCOM's Website at www.radcom-inc.com.

Omni-Q™ Voice Quality Management Solution

The migration of circuit switching traffic onto IP public networks presents service providers with great opportunities and equally great challenges. Opportunities include the ability to provide their customers with competitive pricing and advanced services. Challenges include managing PSTN and IP-related problems to maintain consistent and high levels of voice and signaling quality in order to fulfill the guaranteed service quality required to support their customers' mission-critical applications.

Omni-Q™ rack mount

RADCOM's Omni-Q voice quality management solution enables service providers to deliver the reliable, high-quality packet telephony services their customers want and to optimize their network resources. Using an innovative mix of proactive and non-intrusive network-wide measurements, including vProbes (passive monitor and partner verification), cProbes (end-to-end call quality) and iProbes (call quality on a packet network), this exclusive solution gives service providers the ability to pinpoint any sources of voice quality degradation. The cProbes and iProbes generate end-to-end circuit calls and edge-to-edge packet calls respectively, using standards-based algorithms. In addition, the passive vProbes monitor live traffic going through VoIP lines and conduct a set of call quality measurements. Together, all these probes are configured and controlled by the QManager centralized voice management system.

This innovative mix of proactive and non-intrusive network-wide measurements provides service providers the ability to pinpoint any sources of voice quality degradation and allows them to proactively manage quality throughout their networks efficiently and cost-effectively.

For further information, visit RADCOM's website at http://radcom-inc.com/radcom/managmnt/q_pro.htm.

The VoIP Performer™ Performance Analysis System

The VoIP Performer™ hardware and software suite is a next-generation network performance testing solution. Designed to support the development and predeployment testing of current and emerging convergence technology applications and services, the Performer's components generate realistic network environment stress levels on new VoIP devices and applications, then test the quality and grade of service delivered.

VoIP Performer™ rack mount

Highly accurate measurements and accelerated data output integrated with automation tools and specific applications and services testing programs, significantly shorten time-to-market, reduce R&D costs and simplify the evaluation process critical to successful deployment.

The complete Performer VoIP suite includes:

- **323Sim:** Voice over IP generator generating more than 2000 calls simultaneously, at a rate of over 80,000 calls per hour, emulating the functionality of an H.323 terminal. Several simulators can be simultaneously deployed to increase call volume.
- **SIPSim:** Session initiation protocol simulator with dual-mode functionality, capable of emulating several SIP phones and implementing sessions between two MGC or two IP/PBX servers. Also able to stress and check ability of registration servers to handle heavy registration loads.
- **MediaPro:** Real-Time Voice over Data monitor that analyzes media and signaling data generated from H.323/MGCP/SIP protocols and provides voice quality measurements.
- **QPro:** Voice quality evaluation tool that features MOS (mean opinion score) voice quality measurement on circuit-switched, PDH and many other interfaces.
- **InterSim:** Simulates various WAN impairments such as latency, jitter and packet loss to assess their effect on networks and components.

All components have a common user interface and are controlled from a single console, with the possibility of automated testing through scripting.

For further information, visit RADCOM's Web site at http://radcom-inc.com/radcom/test/ t_pro.htm.

www.protocols.com

Sponsored by RADCOM, www.protocols.com is a comprehensive website that offers network managers, developers and field service engineers an abundance of the latest information on data and telecommunications protocols and technologies.

Constructed in concise and clear format, www.protocols.com contains a protocol dictionary that lists the most common data and the telecommunications protocols in use today and indicates their function in respect to the OSI model. In particular, it provides information concerning the structure of the protocol (header, PDU, etc.), various errors and parameters. The protocol dictionary also describes numerous physical layer technologies which are used today in the communications industry.

A powerful search engine allows simplified access to the site's wealth of information, that also includes a technical acronym dictionary, important links to pertinent sites and an extensive library of technical papers.

RADCOM's www.protocols.com is an essential tool for all networking professionals contending with multiprotocol, multitechnology test solutions for WANs, LANs, ATM and VoD.

For further information, contact RADCOM:

International Headquarters:
RADCOM Ltd.
12 Hanechoshet Street
Tel Aviv 69710, Israel
Tel.: +972-3-6455055
Fax: +972-3-6474681
E-mail: info@radcom.co.il

US Office:
RADCOM Equipment, Inc.
6 Forest Avenue
Paramus, NJ 07652, USA